P9-BJL-272

The Hamlyn Encyclopedia of

PREHISTORIC ANIMALS

Published 1984 by
The Hamlyn Publishing Group Limited
London · New York · Sydney · Toronto
Astronaut House, Feltham, Middlesex, England
© Copyright Gruppo Editoriale Fabbri S.p.A., Milan 1982
for the Italian language.
© Copyright Librairie Ernest Flammarion, Paris 1981
for the French language.
© Copyright Gruppo Editoriale Fabbri S.p.A. – Librairie
Ernest Flammarion 1981 for all other languages.
© Copyright The Hamlyn Publishing Group Limited 1984
for the present English Language edition for United
Kingdom and Commonwealth.

All rights reserved. No part of this publication may
be reproduced, stored in a retrieval system, or transmitted,
in any form or by any means, electronic, mechanical,
photocopying, recording or otherwise, without the
permission of the Hamlyn Publishing Group Limited and
the Copyright Holders.

ISBN 0 600 38871 9
Printed in Czechoslovakia
52078

The Hamlyn Encyclopedia of PREHISTORIC ANIMALS

JEAN-JACQUES HUBLIN

Hamlyn
London · New York · Sydney · Toronto

Contents

AUGUSTANA UNIVERSITY COLLEGE
LIBRARY

Introduction

The subject of evolution – the process by which all living creatures on Earth slowly came about – has been dealt with in many different ways. There are books which discuss the evolution of just one group of plants or animals, and some which try to explain the evolution of all living things, both plant and animal. In the Encyclopedia of Prehistoric Animals *Jean-Jacques Hublin tells us how animal life evolved.*

However, even this thorough and detailed book cannot answer all the questions, for there is still much to be learned about the creatures that once lived on Earth, and scientists still disagree about some of the details. Sometimes we can only guess at the way in which these now extinct animals lived their lives, and there are probably many other animals once living on Earth whose remains were never preserved, and so whose appearance we will never know. But new discoveries are occurring all the time. As recently as 1983, in a claypit in Surrey, England, fossil hunters found first a claw and then

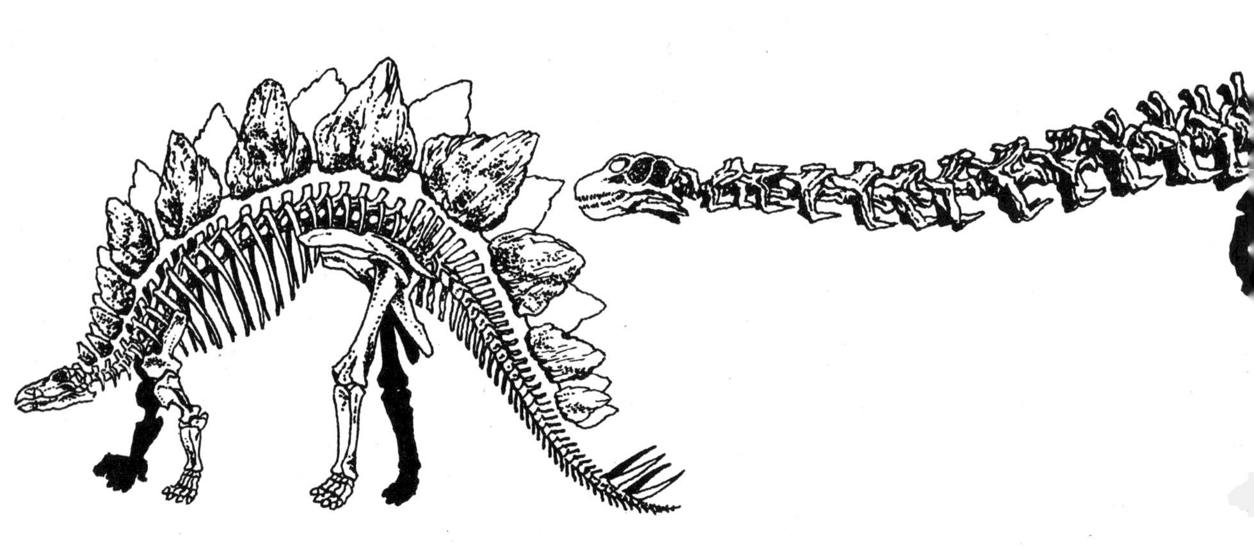

many other bones, of a hitherto unknown species of meat-eating dinosaur. Another small piece of the jigsaw puzzle of evolution!

The remains, or fossils, of the earliest known creatures are found in rocks dating back nearly 600 million years. From these first, simple forms of life came, in time, larger more complex creatures, until eventually the seas had seen the coming, and sometimes the passing, of many different species – sponges, jellyfishes, worms, molluscs, trilobites, fishes and others. Between 350 and 400 million years ago the conquest of the land took place, resulting in, among others, the amphibians, reptiles, birds and mammals.

Not all the illustrations in this book are of long-dead animals, however. By a skilful blend of evidence from both the fossil world and the living world, the author compares the appearances and lifestyles of some of the extinct animals with those living today.

The science of palaeontology, like many other sciences, has its share of long and complicated words and names. Sometimes these are impossible to avoid, since to do so would result in a long explanation of the meaning every time the word appeared in the text. For this reason, a glossary is provided at the back of the book, which may be consulted whenever such a recurring word is encountered.

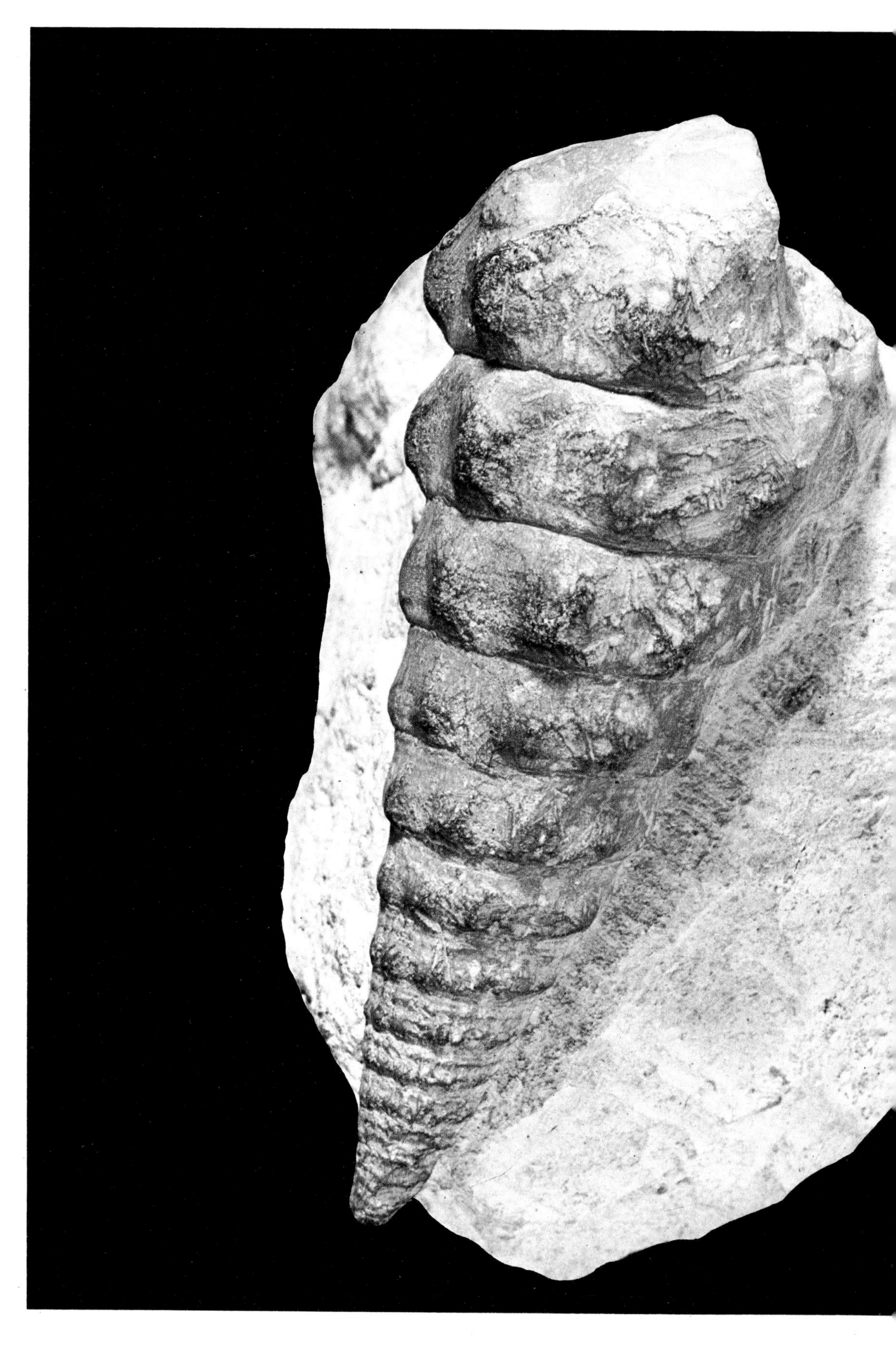

Chapter One

Fossils Tell The Story

Fossils tell the story

Legends of creation

Through the ages, probably nothing has given Man so much 'food for thought' as his own origins – and how the world itself came into existence. Questions about the birth of the Earth and the appearance of the human race have been answered in many different ways. Many of these explanations are based on scientific evidence, often the result of many years of study. However, even the most primitive cultures which lack any form of scientific measurements always have a myth or a legend that satisfies them about the creation of the Earth and of the beings which inhabit it. These myths or legends which try to explain the mystery of our origins are often very beautiful, and are usually closely linked to each particular culture's religious beliefs.

The ancient Norse people believed that the first being was a giant called Ymir, born out of nothing after the union between fire and ice. His 'foster-mother', the cow Adumbla, then gave birth to a god whose race quickly challenged the giants to battle. From Ymir's dead body the Earth was created, the oceans were created out of his blood, the mountains out of his bones, the forests out of his hair, and the rocks out of his teeth. The very first man, called Ask, was made by the gods from an ash tree; the first woman, called Embla, from an elm. Their offspring lived in a world governed by the gods, and they all waited for the vengeance of the giants which one day would destroy them.

In traditional Hindu tales, God created all things, starting with water into which he

People have always been interested in the fossilized bones and shells found in the ground, but it was not until the 18th century that they were recognized as the remains of animals that had lived on the Earth and in the oceans long before Man. Looking at fossils helps us to reconstruct the evolution of the living world over the hundreds of millions of years of its history.

placed a seed. This seed became a golden egg from which God himself was 'hatched', in the form of Brahma millions of years later. He then separated the egg into two, forming the Heavens and the Earth. Finally he gave life to all living things and created the elements – earth, fire, air and water. According to the legend, every time Brahma falls asleep, our world dissolves, and every time he awakes it is reborn.

The Ancient Greeks and Romans believed that the gods came from Heaven and Earth. The Greek and Latin mythologies describe a sort of paradise, now lost forever, in a story similar to the Bible's Garden of Eden. The story of Genesis in the Bible is in fact almost entirely copied from accounts made by the Chaldeans who were the ancient peoples of Iraq. According to Genesis, God created the World out of chaos in seven days. He made the sky, the continents and the seas, the fishes, the birds, the land animals and finally the first human couple: Adam was made from dust on the sixth day and Eve from a rib taken from Adam's side while he slept. But soon they were driven out of their earthly paradise and gave birth to a race of people which God decided to punish for their sins. He made them perish in the Great Flood, and only Noah was saved to become the father of the modern human race.

What is a fossil?

It was not until the end of the 18th century and during the 19th century that a revolutionary idea emerged which, as we shall see, overturned the traditional thoughts about creation. New evidence came to light suggesting that the different species of plants and animals were not fixed at the point of their creation, but altered during the course of time. New species arose and their appearance changed, as did the places they lived in and their way of life. Strange plants and animals, very different from those we see today, inhabited the Earth long before the eyes of men could have seen them. Much of this new understanding came about when the true nature of fossils was recognized, for fossils are the preserved remains of plants and animals which have long since died.

Fossils are found in geological formations. Fossil shells, for instance, are quite common in certain sedimentary rocks deposited by the seas or lakes. The organic origin of these by now familiar finds is no longer disputed – but it was not always that way.

In the Middle Ages the wise men of the time explained these odd-looking forms by means of a curious theory. They thought they arose as the result of a 'plastic force'. This 'plastic force' was dreamed up by an Arab naturalist philosopher called Ibn Sina who lived in the Persian city of Hamadan, around 1000 A.D. The work of Ibn Sina was translated into Latin, and soon his *vis plastica* was accepted by all men of science as the explanation for these mysterious 'picture-stones' which people picked up in quarries or at the roadside. His idea was that Nature was capable of producing only a limited number of shapes, and that she also imitated the appearance of living things: fishes, shells, bones and plants, as well as the Moon, Sun and stars. Although some early Greek philosophers had indeed concluded from the presence of the remains of marine life on dry land that the oceans had once covered the continents, their works were quickly forgotten.

The Bible and other historical texts were also used to try and help in the understanding of certain finds. Massive bones discovered from time to time were frequently attributed to giants or other huge legendary creatures. One of the most revealing, and at the same time most amusing, cases was that of the giant *Teutobochus*. In 1613 the bones of an enormous *Deinotherium* (see page 264) were found in a quarry in the Dauphiné region of France. A surgeon called Mazurier studied these relics and announced that they were the mortal remains of the king of the Cimbrians, *Teutobochus*. He embellished the tale and, with *Teutobochus* safely loaded on to a cart, began to exhibit him from town to town, obviously in return for money. He toured France and Germany with his prize, and made a large sum of money. But all good things come to an end, and his deceit was eventually exposed for what it was.

On other occasions many fossil skeletons were thought to be those of legendary beasts such as griffons or unicorns. Right up to modern times, bones of animals living in the Quaternary Period collected around Peking were used as 'dragons' teeth and bones by Chinese chemists. At the beginning of this

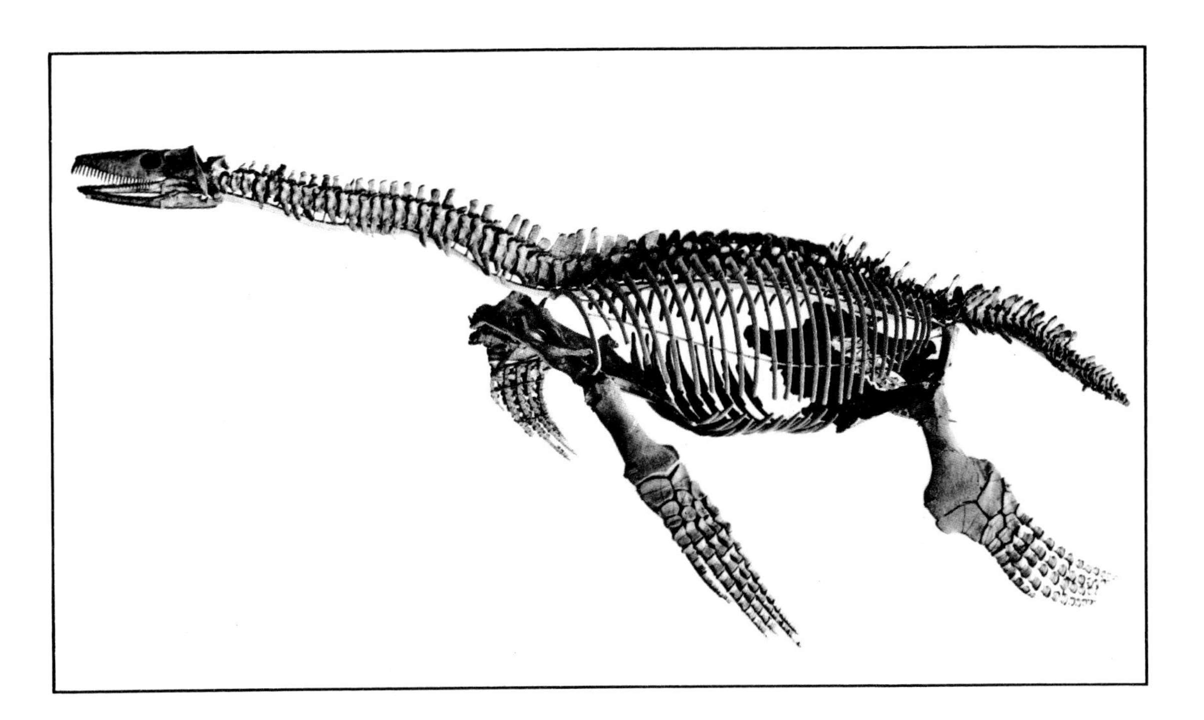

For a long time people thought that the enormous skeletons discovered by quarrymen in the rocks deposited during the Mesozoic Era belonged to the dragons of popular myths. Today, however, these remains are correctly identified as belonging to huge reptiles like this plesiosaur.

century in fact the very first teeth of *Sinanthropus*, a Chinese fossil man (see page 288) were found in Peking pharmacies.

The Great Flood mentioned in the Bible, together with the many deaths it would have caused, is often used to explain the presence of petrified bones in the ground. This theory, whilst totally conflicting with the theory of the *vis plastica*, did have the great advantage of being in perfect harmony with religious beliefs of the times. Early thinkers who suspected the true meaning of fossils often attracted unwanted attention to themselves. For instance one such man, Palissy, died a captive in prison for his beliefs.

One of the most famous victims of the Great Flood is a curious skeleton found on a slab of rock mined from a quarry near Oeningen in Baden during the 1720s. A Zurich physician, Canon Jean-Jacques Scheuchzer, took it to be the skeleton of a man drowned by the waters of the Great Flood exactly 4032 years before, according to his calculations. It was only many years later that the fossilized remains on this very same slab were found to be those of a giant salamander, but at the time hardly anyone doubted Scheuchzer's explanation of this most famous discovery. For a long time fossils were thought to be closely linked to the Great Flood, and even at the end of the last century, when the existence of Prehistoric Man was finally admitted, people still referred to periods of time 'before the Flood'.

Sometimes even Biblical times seemed too distant to account for the age of some fossils, and the remains of known animals were thought to belong to relatively recent history. On one such occasion, around 1715, an Englishman called Conyers, one of the very first to recognize a chipped axe-head for what it was, noticed that the ivory found with the tool came from an elephant's tusk and not from a unicorn or any other mythical beast. But he finally decided that what he had found was evidence of a fight between an elephant in the armies of Caesar and an ancient Briton who had killed it with the aid of his stone weapon!

It is perhaps worth telling the story of one of Scheuchzer's fiercest opponents, Jean-Barthélémy Beringer, a professor at the University of Würzburg, Germany. Beringer was a keen fossil collector who made his students accompany him on his hunting expeditions. Beringer's discoveries were extraordinary. Not only did he collect marvel-

lous examples of petrified birds, insects and reptiles, he also found mating frogs and even a spider catching a fly! Here was proof, he said, that these animals could not have been overtaken by the Great Flood. He thought the stone itself gave birth to living creatures of which these petrified remnants were the almost complete skeleton. Beringer's downfall came the day he found a similar stone bearing his own name. For a long time the students he had dragged around the quarries, having lost all interest in fossils, had made fun of him by littering the place with small terracotta statues – which explains the never before seen fossils of mating frogs, and spiders catching flies! His reputation ruined, the unfortunate professor spent the rest of his life, and all of his money, trying to buy back all the copies of the book he had published to explain his ideas.

In the so-called 'Age of Enlightenment' the question of fossils remained unanswered. Even the great French writer Voltaire sneered at the scientists and their disputes, claiming that the shellfish and remains of other marine life found on the tops of mountains had never been deposited in the oceans, but merely represented the leftovers of the meals of passing travellers and pilgrims.

In addition to the great thinkers of the age,

The most spectacular plant fossils are without doubt the petrified tree-trunks found in the cliffs of Tiguidit at In Gall, in the Republic of Niger (above). They have kept so well because silica has replaced the timber. As can be seen in the illustration on the left of a polished section of the trunk of an Eocene Period *Ginkgo biloba*, silicification was one form of fossilization that perfectly preserved plant structures. In fact it is hard to tell them apart from living timber – even under a microscope.

the ordinary people developed their own interest in fossils, turning them into charms to protect themselves and their animals, or crushing them into powders which they thought gave them great healing properties. As far as the discovery of prehistoric tools of rough-hewn or polished stone were concerned, popular tradition claimed that the stone axes which were picked up off the fields after a thunderstorm had been produced by lightning, hence their name of 'ceraunia' which means 'lightning stones'. Like the fossil teeth of sharks, flint arrowheads were called 'glossopetrae', meaning the petrified tongues of snakes. All these objects were carefully preserved since it was believed that whoever bore them would be safe from lightning, shipwreck and kidney complaints!

How fossils are formed It is fascinating to think that an animal which lived so many millions of years ago should have left behind some evidence of its existence; a sign buried beneath our feet and one which we are able to decipher. Even the smallest remains can help us to imagine the strange world in which they lived. A world where everything was different: the plants and animals, the climate, the course of rivers and the shape of land and seas – even the rhythm of the days and seasons.

The likelihood of such evidence surviving is, however, very slight. When a plant or animal dies its remains are nearly always

Left: After death this cerith (a type of sea-snail from the Eocene Period) was buried in the sand on the ocean floor. This may have happened at the spot where the animal actually lived or after it had been moved for some distance, in which case the snail may be associated with plants and animals from a completely different environment. Over millions of years the shell gradually altered as the sediment surrounding it hardened to form sandstone. Today these rocks would reveal the fossil just like an internal mould.

Below: We often find fossils of the same species together; but it is unusual for them to have all died at the same time. The ammonites are common fossils found in many places, but each species quickly evolved into another form. This makes them very useful as it helps us to date precisely the rocks they are found in. Each ammonite species has its own 'age label'.

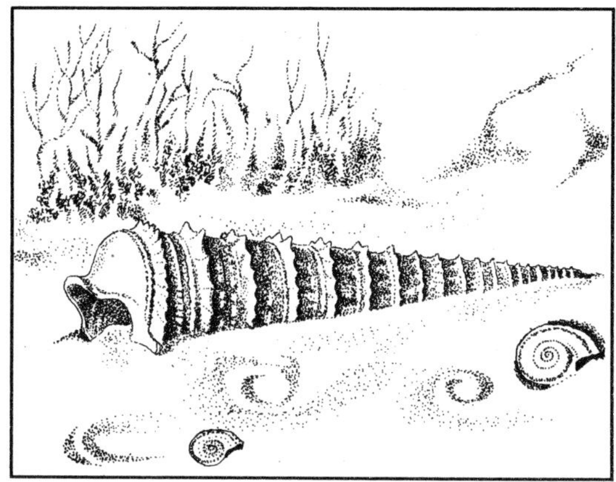

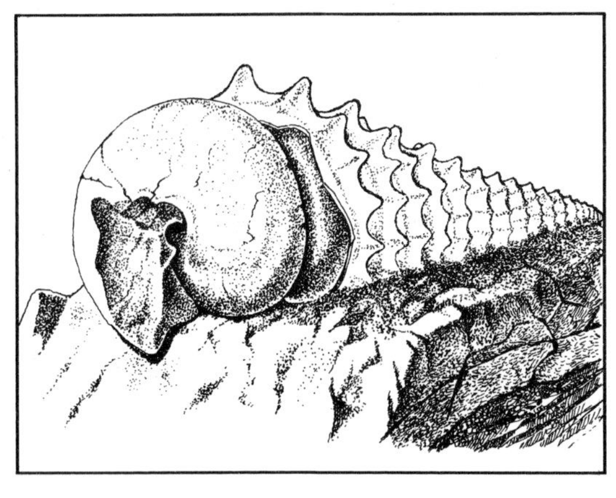

completely destroyed. It rots, and even its hard parts fall victim to bacteria. If conditions are right, though, some of the parts stand a good chance of being preserved. For this to happen a rapid burial of the remains is essential. That is why plants and animals can fossilize much more easily in aquatic surroundings, where mud covers them over quickly, than on dry land where, unless they are buried fairly rapidly by dust carried by the wind or by the ashes from a volcano, they must wait for the rising of rivers and lakes or even the filling of a rock fissure or cave to cover them. Drowning must have been the fate of many a land-dwelling vertebrate that has fossilized, which explains why it is relatively rare to find flying and tree-dwelling creatures in most fossil deposits.

Some fossils are, however, by no means rare. In some instances, especially in seabeds, the shells or skeletons of dead organisms have accumulated in such vast quantities that they have become cemented together to form rocks. The best example of this is chalk, found in enormous layers in Northern Europe. The microscope reveals that chalk is chiefly composed of tiny particles of the limestone shells of single-cell creatures (see page 64).

Once covered by the substance which can preserve them, such items as bones, teeth and shells are gradually transformed over a long period of time. Their organic substances disappear and are replaced by mineral

matter. Then the mineral parts of the skeleton are transformed as well. This means that the chemical composition of what must now be called a fossil is changed completely. The finest details of the creature's structure will still be visible, but the chemicals which originally made up its body will have been replaced by calcite, silica or iron pyrites. The fossil will often become harder and heavier, and will have lost its original colour. If this process continues it will lead over hundreds of millions of years to a progressive loss of the creature's detail and finally to the total disappearance of all evidence of the creature. Obviously, these stages will be more advanced in older fossils: certain shellfishes that are 'only' a few million years old have hardly changed at all, and still show their original colouring.

When the rock lies in the path of a current of water, fossils made of soft limestone may be completely washed away, leaving a cavity which corresponds to their external form. Later on, the external mould may once again be filled by a mineral carried along by the water. The overall result is an often very accurate reproduction of the original fossil, at least as far as its outside appearance is concerned.

The preservation of the soft parts of the body (such as organs) is very unusual, and happens only under exceptional circumstances. Very fine deposits formed in extremely calm surroundings can sometimes contain very delicate fossils. An excellent example of this is the Burgess Beds of British Columbia which date from the Middle Cambrian Period (see page 302). In the Burgess shale, organisms as frail as jellyfishes or worms have been discovered. The fossil worms of the Burgess Beds are so well preserved that scales, bristles and digestive tracts are still visible. Similar conditions must have existed in Germany at Holzmaden, Württemberg during the Lower Jurassic Period. Here, complete skeletons of great marine reptiles are still surrounded by a dark silhouette showing the outline of the body and flippers. In the fine Jurassic limestone of Solnhofen in Bavaria, even the feathers of the oldest known bird, *Archaeopteryx*, have been found.

Certain substances are able to conceal unusual fossils. Amber, found on the shores of the Baltic Sea and much prized for jewellery, is actually fossil pine resin. This resin can sometimes contain insects which, in spite of being thirty million years old, are very well preserved (see page 152).

You have probably heard of the frozen mammoths of Siberia. These animals, which lived during the last Ice Age, were always thought by local people to be giant moles which inhabited the depths of the earth, and they called them *Mammontokovast*. These creatures provided the main source of Russian and Chinese ivory for a very long time. The latest specimen, excavated in 1977, was a remarkably preserved newborn animal that had apparently fallen into a

Only the hardest parts of an animal stand a good chance of being fossilized. The belemnites (cephalopod molluscs from the Mesozoic Era, on the right), for example, normally only remain as the limestone guard (left) which these creatures possessed at the rear of their bodies.

In 1960 fossil-hunters in Spitzbergen found these traces (above) which were left by a great dinosaur of the Cretaceous Period on a slab of sandstone pushed upright by the geological folds of the Cainozoic Era. The dinosaur, an *Iguanodon* (below), walked on its hind legs and the imprint of its three-toed feet measured 68 centimetres ($26\frac{1}{2}$ inches) long. Similar trace fossils have been found in many parts of the world in Mesozoic Era deposits, and their identity has often been misunderstood. In the middle of the 19th century, an American priest called Edward Hitchcock made a huge collection of *Iguanodon* tracks, thinking that they had been left by gigantic birds and one of his predecessors, P. Moody, had even dug up tracks he claimed were those of Noah's raven mentioned in the Bible!

crevasse where it quickly froze. Fossils like these show us exactly what the living creatures looked like, and allow us to examine their internal anatomy and even reveal to us what they had for their last meal. Woolly rhinoceroses of more or less the same age have been discovered in Poland preserved in ozocerite, a wax-like substance. They were marvellously well preserved apart from their coats, which had disappeared.

In some instances it is the imprint of the animal's skin rather than the skin itself which remains. One example of this is the hadrosaur (a duck-billed dinosaur) discovered in 1908 in the Cretaceous Beds of Wyoming by Charles Sternberg (see page 195). Other exceptional fossils include the mummified snakes and frogs found in the Tertiary Period calcium phosphate beds of the Lot region in France.

A fossil may also be the trace or product of some biological animal activity, like dinosaur or birds' eggs that have fossilized in large numbers; 'coproliths', the fossil excrement of fishes, reptiles or mammals; and 'gastroliths', stones swallowed by certain reptiles to aid digestion and found among their skeletons, where they have been reduced to polished pebbles by the animals' gastric juices. Traces of other animal activities include burrows filled with sediment, birds' nests found still containing their eggs in very calm deposits of lakes which have dried up, and trails left on the sea-bed by crustaceans or in the hardened mud of ancient swamps by a dinosaur's foot.

The birth of a science

It is curious that palaeontology, the study of life-forms from the long-distant past, should be one of the youngest sciences. Apart from one or two early scientists, such as Leonardo da Vinci in the 15th century and Bernard Palissy in the 17th century, the true nature of fossils was not really recognized or admitted until the 18th century. It was a Frenchman named Georges Cuvier (1769–1832) who first undertook serious scientific studies of fossils. Cuvier is really the father of palaeontology, although he started out as a zoologist. One of his most important ideas was the 'principle of the correlation of organs', on which the science known as

You can see these fossil *Iguanodon* in the Royal Institute of Natural Sciences, Belgium. They were found in a coal mine at Bernissart, close to the French border, towards the end of 1877. The miners uncovered what they first thought were fossil tree-trunks but these turned out to be huge bones. Each bone was carefully dug up and its position marked. 130 tonnes of material were sent to Brussels, and it took over 30 years to put all the skeletons together. There were many different types of animals: turtles, crocodiles, amphibians, fishes and insects as well as the famous *Iguanodon*.

One day Cuvier recognized in a piece of Montmartre gypsum the small jaw-bone of a mammal similar to the modern opossum (below). To the amazement of a group of assembled friends he proceeded to reconstruct the rest of the skeleton of the animal.

Sure enough, two epipubic bones appeared on the reconstruction (right) just as Cuvier had claimed they would, these bones being characteristic of the group of marsupials to which the opossum belongs.

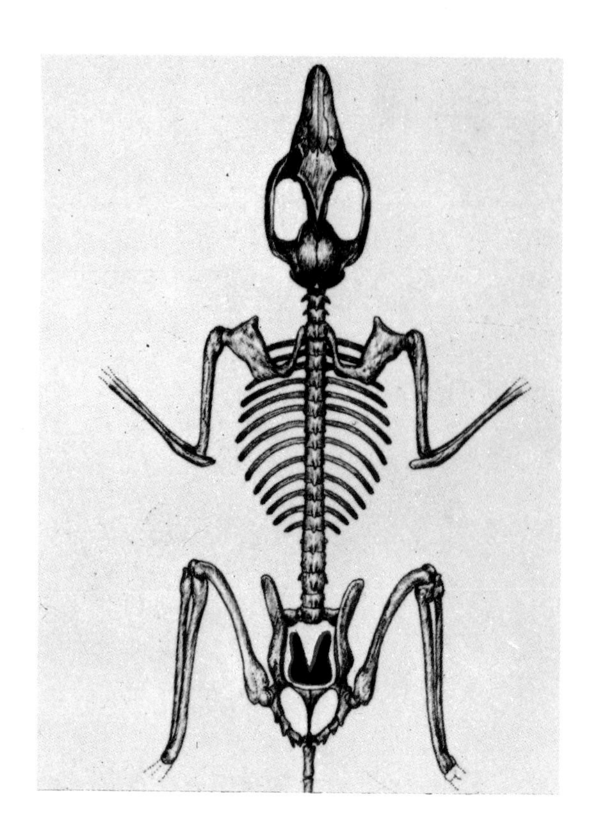

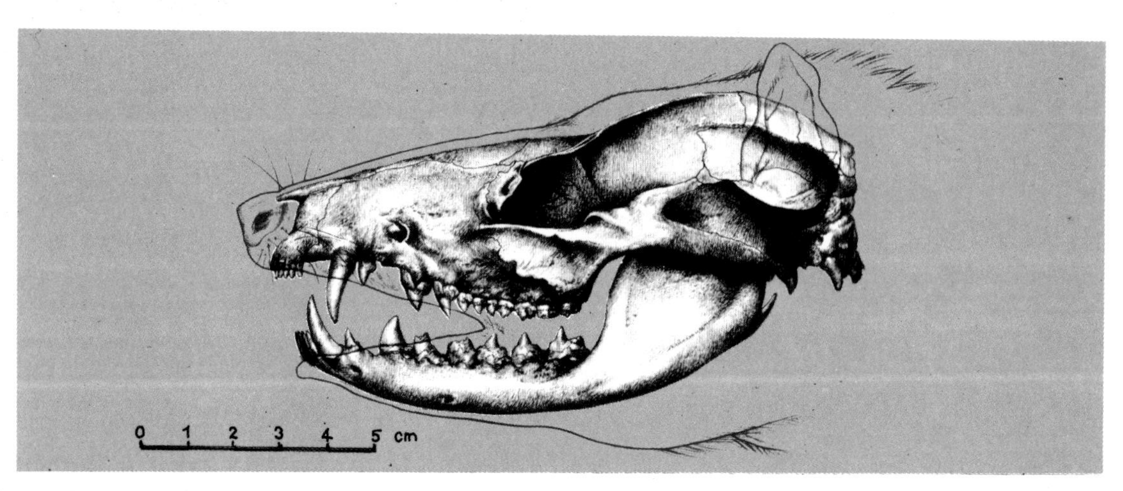

comparative anatomy is based. In a living being, the various organs are not independent, but interact with each other and work together. So much so that the anatomy of one element logically follows from the anatomy of the whole organism, and an analysis of one piece of the puzzle enables the missing pieces to be reconstructed. Cuvier demonstrated this brilliantly by studying the fossil mammals recovered from the gypsum deposits in Montmartre, Paris.

At that time Montmartre was one of the major centres for the mining of gypsum, and the product made from it, plaster of Paris, became so well-known that it gave its name to all very fine plasters. Abandoned underground quarries are today so numerous that the subsoil beneath the famous hill in Paris is extremely unstable, and the ground frequently subsides. The names of streets and squares in the area are evidence of this past industry, such as the *Place Blanche* which must have been covered in white dust. This gypsum, deposited in the Paris region towards the end of the Eocene Period, contained fossil remains that the workmen could pick up, especially since they were not using heavy destructive machinery.

Cuvier amazed his contemporaries by reconstructing the mammal fauna of Eocene lagoons and swamps. All he needed were a few bones to demonstrate that certain teeth had once belonged to a herbivore with hooflike feet, or that a clawed foot suggested carnivorous teeth and a body adapted to the capture of prey. Step by step, comparing these creatures with each other and with their modern relations, Cuvier succeeded in reconstructing the anatomy of each of them and learned to recognize each species just by looking at a single bone. This work, though normal routine to a modern palaeontologist, not surprisingly fired the imagination of the age.

The giant lizard of Maastricht

In 1780 the fragments of a huge skull, its jaws lined with sharp-pointed teeth, were dug up in a chalk quarry near Maastricht in Holland. It had a very strange history after its discovery, and was the object of much controversy. Opinions differed as to what it was: its sheer size – over 1 metre ($3\frac{1}{4}$ feet) long – obviously pointed to a gigantic animal, but which one? A whale? A fish? A crocodile? The experts who came to inspect it were very puzzled. There was also controversy about its ownership: did the extraordinary find belong to its discoverer or to the owner of the land on which it had been found? The case was finally brought to court and the owner of the land, Canon Godin, won, little guessing how useful the find would be to him some years later. In 1795, at the height of the revolutionary wars, French troops invaded Holland and laid siege to Maastricht. But General Pichegru's gunners had precise orders not to fire on the house of Canon Godin for fear of destroying the famous skull inside. Godin, warned of the envy of the armies of the Republic, took care to hide his treasure in the city. After the fall of Maastricht he demanded 600 bottles of wine before allowing the soldiers to carry off their prize.

Georges Cuvier was, like many palaeontologists, a failed churchman. He was a slight man with a huge head, red hair, a prominent nose and a modest nature. By the time that he died, he was a Baron and Councillor of State.

Faujas, then commissioner in Belgium in the wake of the armies of the north, had the specimen taken to the National Museum of Natural History in Paris where it still is today. By a series of deductions, Cuvier proved that it was in fact the relic of a gigantic marine lizard. The puzzle of 'the great animal of Maastricht' had been solved and the creature was named *Mosasaurus*, the 'saurian of the Meuse', the river on which Maastricht lies. It was one of the terrible creatures that lived in the seas during the Cretaceous Period (see page 175).

Cuvier's work had far-reaching effects. Despite his genius, however, he never acknowledged the accuracy of the theories of his predecessor, Jean Baptiste Lamarck, who proposed that animal species evolved through time and that this process of evolution formed the very basis of the great many different kinds of plants and animals which make up the living world. For his part, Cuvier explained the succession of different animal remains in the Earth's rocks by a series of creations and catastrophes. Not until the end of the 19th century was the idea of evolution finally accepted.

The dawn of the dinosaurs

One spring day in 1822, Gideon Mantell, a young country doctor of Lewes in Sussex, England, was visiting one of his patients. It was such a beautiful day that his wife decided to accompany her husband, who was a fossil enthusiast. While the doctor was indoors examining the patient, his wife walked around the house. Stopping in front of a pile of stones left at the roadside by a gang of workmen, she noticed something shining on one of them and took it to show to her husband, who immediately recognized it as the teeth of a large herbivorous animal, though he could not tell which it was. Mantell lost no time in finding the quarry from which the blocks had come, and there he was even more astonished. The sandstone being mined there was evidently from the Mesozoic Era. He thought it would be impossible for such ancient soil to contain the remains of large herbivores, and yet this was the only possible explanation for the teeth.

The learned men of the day, and Georges Cuvier in particular, all agreed that Mantell's collection of teeth had belonged to a large mammal, and could be the incisors of a rhinoceros – but none of them believed the teeth to be very old. However, Mantell carried on with his own investigations, convinced that he was right. He found that the quarry in the forest of Tilgate contained other fossil teeth as well as enormous bones. Now he was certain that it dated from the Cretaceous Period.

Searching among living creatures for a species which might be similar to his fossils, one day Mantell noticed a striking likeness between the teeth his wife had picked up and those of the iguana, the giant lizard of Central and South America. The shape was similar, though not the size. If the animal's dental proportions were any guide, Mantell reckoned that to arrive at the size of the prehistoric beast, the size of the iguana had to be multiplied by ten, and the iguana is 1.2 metres (4 feet) long!

So in 1825 Mantell published his description of the remains found in the forest of Tilgate under the name *Iguanodon* which means simply 'iguana tooth'. Even Cuvier admitted that he had been wrong and confessed that gigantic reptiles must have in-

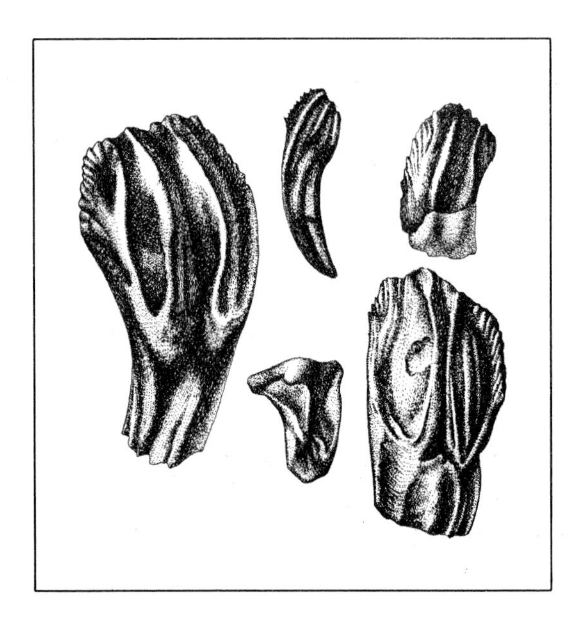

When Mary Ann Woodhouse, the wife of Gideon Mantell, picked up a few notched teeth from the side of the road, she did not realize she had stumbled upon the first evidence of the dinosaurs.

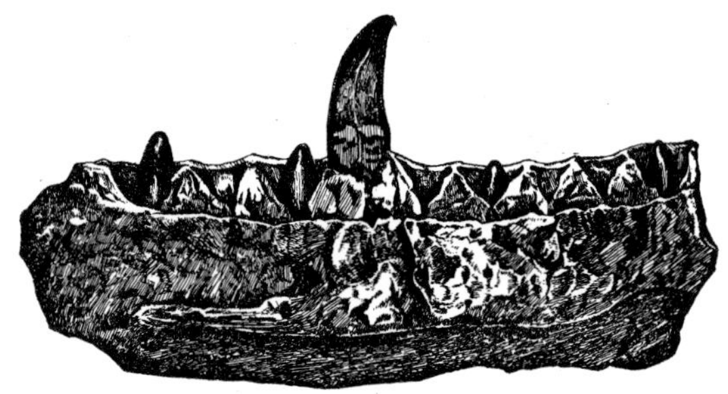

The very first dinosaur to be discovered was *Megalosaurus* (above), a few months before *Iguanodon*. But the British geologist Buckland who studied it did not recognize it as a dinosaur. Unlike *Iguanodon, Megalosaurus* was a carnivore with sharp, notched teeth like those shown on the left.

habited the continents before the mammals. Still, even Mantell's own reconstruction of *Iguanodon* was very different from reality. He saw it as a giant lizard 13 metres (42 feet) in length and moving on all fours, quite different from the way we now know it walked (see page 22). Mantell further believed that its thumb (also preserved as a fossil), used in the living animal as a defensive spur, was a horn borne on the end of the animal's snout. This explains why some early models of the creature show it walking on all fours and bearing a rhinoceros-like horn. *Iguanodon, Megalosaurus* ('great lizard') – described more or less around the same time – and *Hylaeosaurus* ('wood lizard') that Mantell discovered some years later were the three first representatives of a new group to which Richard Owen in 1841 gave the name of dinosaurs, which means 'terrible lizards'. The dinosaurs gave the study of palaeontology a popularity which has lasted until today, and Owen contributed much to the fame of these creatures. As the first director of the Natural History Museum in London, he employed the sculptor Waterhouse Hawkins to make life-size models in cement of these famous creatures. The huge statues were erected in a park in South London not far from the Crystal Palace. For the inaugur-

From left to right: *Stegosaurus*, *Apatosaurus* and *Allosaurus*, three species of dinosaur excavated from the Morrison Formation of Utah, U.S.A. – without doubt one of the most spectacular deposits yet discovered. Just before the First World War magnificent, complete skeletons were found, and today the visitor to the 'Dinosaur National Monument' can see the gigantic remains of these creatures, extracted from the rocks of the Jurassic Period.

ation of this exhibition, which was a tremendous triumph, Owen laid on a banquet on 31 December, 1853 right in the middle of the unfinished body of one of Hawkins' models of *Iguanodon*.

Ever since then dinosaur fossils have been hunted incessantly. The second half of the 19th century in particular was marked by a real 'dinosaur rush' in the west of the United States. Wyoming, Utah, Colorado and Montana were invaded by fossil-hunters who ran the risk of Indian attacks to exploit sites which were rich in finds. The two great American palaeontologists, Edward Drinker Cope and Othniel Charles Marsh, were deadly rivals. They both collected armed bands of men who nearly came to blows on more than one occasion, each spying on the other and trying to keep secret the more promising sites. Stimulated by this competition, however, Cope and Marsh organized the excavation of the fabulous finds that can be seen today in the great American museums. Since those days, Canada, Tanzania, Mongolia, Lesotho, China, India, the Sahara and Argentina have all provided fresh sites that let us reconstruct the world of the dinosaurs.

In search of fossil man

While many fossil animals were found during the first half of the 19th century, the origin of the human species remained a thorny problem. The first human fossil was found in 1830 at Engis in Belgium. But, like the discovery 18 years later in Gibraltar of the fossil skull of a human, this event passed almost unnoticed. In fact, the finding of prehistoric humans came about through the science of archaeology. The discovery of shaped tools in the same sites as the remains of extinct animals showed how old the human race is. This connection had been observed previously by the Englishman John Frere at the end of the 18th century, and by P.C. Schmerling whilst working around the excavations of Engis, but the persistence of a Frenchman, Jacques Boucher de Perthes, finally proved the point.

Director of Customs at Abbeville in the Somme, Boucher de Perthes was an eccentric character. After dabbling in economics, sociology and the natural sciences, and trying his hand at poetry, drama and novel-writing, he suddenly decided to tackle the problem of human fossils. His enthusiasm was sparked off by the apparently deliberately shaped flints found in the gravel quarries on the banks of the River Somme. A few looked like large almonds with sharp

edges, others just like daggers or spearheads. In his excitement, this customs officer-turned-poet visualized the remains of a civilization going back to the Tertiary Period. The scientists of the day were not convinced, but the general public were so enthusiastic about the idea that a scientific commission finally agreed to look at what Boucher de Perthes had found. Unfortunately he was ridiculed by the scholars. To begin with, the soil from which the flints had come was not Tertiary but much younger. And the fact that he consulted with spiritualists in order to 'communicate' with long-dead human beings led to his ideas being discredited.

For twenty years Boucher de Perthes battled to get his ideas accepted, and it was from England that this acceptance finally came. During a visit to his collections, the geologists Lyell and Falconer recognized the authenticity of the tools he had found so many years before. Fresh excavations were started, and the scientific world then admitted that the hard stones with the sharp edges which had been found in the Earth of glacial times were not all just unusually shaped stones of natural origin. Some bore definite proof of having been made by Prehistoric Man, and hence Cuvier's views of the matter were overthrown.

1856 saw the first discovery of Prehistoric Man that really drew the attention of the general public. In the small valley of Neanderthal near Düsseldorf in Germany, quarrymen emptying a cave of its clay uncovered a few bones. The natural history professor to whom they were passed, Johann Carl Fühlrott, saw at once that these were not the bones of cave-dwelling bears but of human beings. But how odd they looked! There was a low cranial dome; a receding forehead extended to the front by an enormous brow-ridge above the eyes, thick bone walls and extraordinarily massive long-bones. Fühlrott did not doubt that what he had in his hands was evidence of an intermediate being, some way between the great apes and Modern Man. The fossils seemed far older than those found at Engis, and the state of the deposits in which they were found ruled out the idea of a recent burial in the middle Quaternary layers.

Although the anthropologist Schaaf-hausen of Bonn, Germany took up the cause on behalf of Fühlrott, the rest of the scientific community rejected their interpretations, believing that the bones of Neanderthal had belonged to some malformed idiot with rickets, or even to a Cossack from the 1814 battles! For thirty years opinions differed; then in 1859 came the publication of Charles Darwin's *The origin of species by natural selection*. Darwin's ideas on the evolution of the living world and on the development of one species into another gave a theoretical basis to the apeman of Neanderthal. Despite the finding of a Neanderthalian jaw-bone at the 'Hole of Naulette' near Dinant (Belgium) a few years later, it was not until 1886 that the existence of a particular primitive human type was finally admitted. At Spy in Belgium, two complete skeletons with the same characteristics were unearthed from Pleistocene layers, and this time there could be no mistake. The doubters finally had to accept the evidence.

Back to the origins

The history of palaeontology is packed with amazing adventures, the most astonishing of which is perhaps that of a Dutch doctor named Eugene Dubois, another person who was obsessed with discovering human fossils. Towards the end of the last century this young man was a student of Ernst Haeckel who asserted that human beings could have descended from the gibbon, an idea which proved false later on. Dubois felt that he could perhaps find the traces of the 'missing link' between gibbons and humans in those regions which the former still inhabited. So he signed on as a military doctor and took ship for the island of Java, then a Dutch colony, where he was convinced he would find the remains of *Pithecanthropus*, the name Haeckel had already given to his apeman even before its discovery.

With its warm climate, Java had never suffered glaciation and formed an ideal environment, according to Dubois, for the evolutionary transformation he sought. Dubois' incredible story continues to amaze palaeontologists today. In November, 1890, after months of effort, Dubois actually found the mandible of a primitive human in one of his digs near Kedung-Brubus. One year later the sediments of the River Solo yielded up a calvarium (brain-pan) whose capacity was exactly half-way between that of the great primates and modern humans. When you consider the rarity of human fossils, Dubois' finds were a small miracle. The intensity of Dubois' joy made his return to Europe all the more painful: instead of the general enthusiasm he had looked forward to, he was greeted by disbelief from most of his contemporaries. Embittered by the attacks upon himself, he finally locked away his fossils, and though his collection continued to grow no-one was allowed to look at it for thirty years.

Towards the end of Dubois' life, *Pithecanthropus* was eventually recognized for what it was – an important link between modern man and his ancestors. However, by this time Dubois was denying its importance himself. In the meantime, however, milestones even further back along the road of human evolution had been discovered in South Africa: *Australopithecus*, the first link in the chain, whose discovery and study formed the chief contribution to 20th century human palaeontology, made the dreams of Boucher de Perthes, Dubois and all the others come true: humans had indeed lived in the Tertiary Period.

Discovery and extraction of fossils

Fresh deposits of fossil vertebrates have been found mostly by chance. Major earthmoving projects, quarries and construction sites of all kinds sometimes reveal treasures which have been locked away in the ground until then. For example, in 1977 the building of a new *Autobahn* at Kupferzell, in Württemberg, Germany revealed an extraordinary pile of bones containing literally thousands of fossilized remains going back to the Triassic Period – fishes, *Stegocephalus* (an amphibian) and early reptiles were collected in huge numbers by scientists from the museum of Stuttgart. In addition to these unexpected finds, there are sites which have been known for a long time but whose sheer size involves the continuation of digs over many years. In some quarries fossil bones are regularly found, but not often enough to warrant excavation. In these cases, the quarrying operations have to be closely supervised so that interesting specimens are not lost. The modern techniques of rock mining make such supervision much more difficult than in the past, and there is no doubt that numerous remains end up encased in concrete. In unknown, unexplored areas, systematic prospecting can reveal new sites; palaeontologists closely follow the

Professor Yves Coppens' team prospecting the Pliocene–Quaternary Period deposits in the Omo Valley, Ethiopia. Over the past few decades East Africa in particular has been a favourite area for many fossil-hunting expeditions by the French, British, Americans and Kenyans. These campaigns, often using complicated arrays of equipment, have uncovered many fossil hominids as well as the oldest known tools. They have also told us how to reconstruct in detail changes in the geology, climate and fauna of this vital region in Man's evolution.

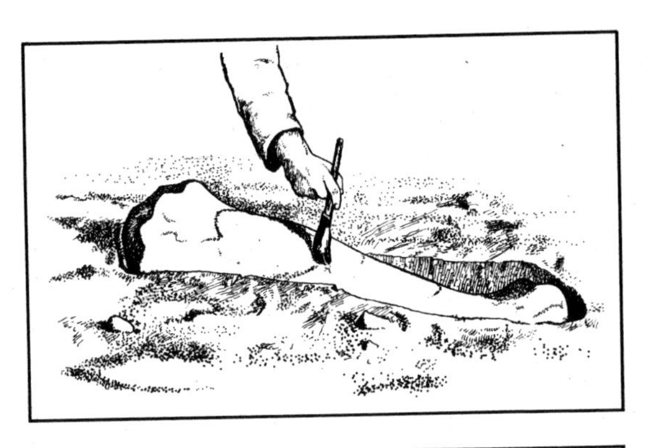

map-work done by geologists so as to select outcrops with soil conditions that seem to favour the preservation of fossils.

Sometimes whole expeditions are organized to recover important palaeontological material from remote areas. This happened during the 1920s when American expeditions to Mongolia found fantastic deposits of dinosaurs in the Gobi Desert. In 1969 a French expedition to Spitzbergen collected a number of fishes, amphibians and marine reptiles from the end of the Palaeozoic Era and beginning of the Mesozoic Era.

Fossil remains are often very fragile, and they must be separated with the utmost caution from the material which surrounds them. Large bones in particular must usually be protected in plaster before they are removed (see illustrations). Their position is first recorded by drawing (this might prove useful for helping in the removal process or to provide information on how they were deposited and fossilized). The latter problems are the subject of a separate science, taphonomy.

It is almost impossible to lift and transport a fossil, once carefully excavated, without certain precautions. As it is unearthed the fossil will be injected with substances to harden it when dry and then it is surrounded in a plaster cast much like a broken leg would be. The fossil is left on its plinth of rock, for the whole thing is taken back to the laboratory. The bone is covered in paper then encased along with its base in strips of fabric soaked in fresh plaster of Paris. If the fossil is very large, the protective shell must be strengthened with iron rods passed through the plaster strips. After a few hours this compact block can be handled without fear, since the fossil will be as safe and sound as it was in the ground.

This long crocodile jaw-bone is a very fragile specimen and has to be encased in firm plaster before taking it away to a museum.

Special techniques have to be used on prehistoric deposits containing large quantities of flint tools – and possibly human remains as well. The dig is accompanied by very careful map-work, since the distribution of rock and bone, together with the slightest traces of human activity, are of the greatest interest to the archaeologist, supplying him with the information he needs to tell him about the dwellings and ways of life of the people whose remains he is uncovering. A trace of ash on the floor of an ancient dwelling, or a pile of flint chippings, may indicate where a tent once stood, for instance, and may provide far more information than the most magnificent collection of Stone Age tools found at the same site. It is not difficult therefore to understand the great care taken when stripping away fragile archaeological layers. These are removed methodically, a small section at a time, after they have been drawn, photographed and even moulded in plaster. It is usually easier to unravel the story left by the remains of invertebrates, since they are without exception far more abundant than vertebrates. To collect the smallest of these – the 'microfossils' – you have simply to take a few handfuls of fossil-rich sediment and process it in the comfort of a laboratory. (This technique by the way is also used in deposits containing vertebrates; firstly to retrieve the most minute remains, and secondly to study any grains of pollen that may also have been preserved.)

In the laboratory In the laboratory, the isolation of fossils is usually carried out by mechanical means; the plaster surrounding the bones and their matrix is taken off, and the sediment removed by means of scrapers, chisels and needles. When the rock is dense, one must use drills not unlike those used by a dentist, or miniature pneumatic picks. As this work progresses, the fossil is prevented from falling apart by means of plastic solutions that enter the bone and make it much less fragile. Miniature sand-blasters can even be used to finish off the more tricky fossil extractions. These machines fire abrasive powders at very high speeds; by carefully selecting the right powder grade and grain size it is possible to clean the surface of certain specimens completely.

In light, loose sediments microfossils are normally found by washing: sediment samples are brought to the laboratory and then finely filtered under running water. The residue is then sorted under a binocular microscope. The final stage in this cleaning operation often takes place in an ultrasonic tank where the miniscule remains are suspended in a fluid and subjected to very high-frequency vibrations.

Chemistry is also of great help to the palaeontologist. It is occasionally impossible to extract a fossil from its surrounding matrix by mechanical methods, and so the

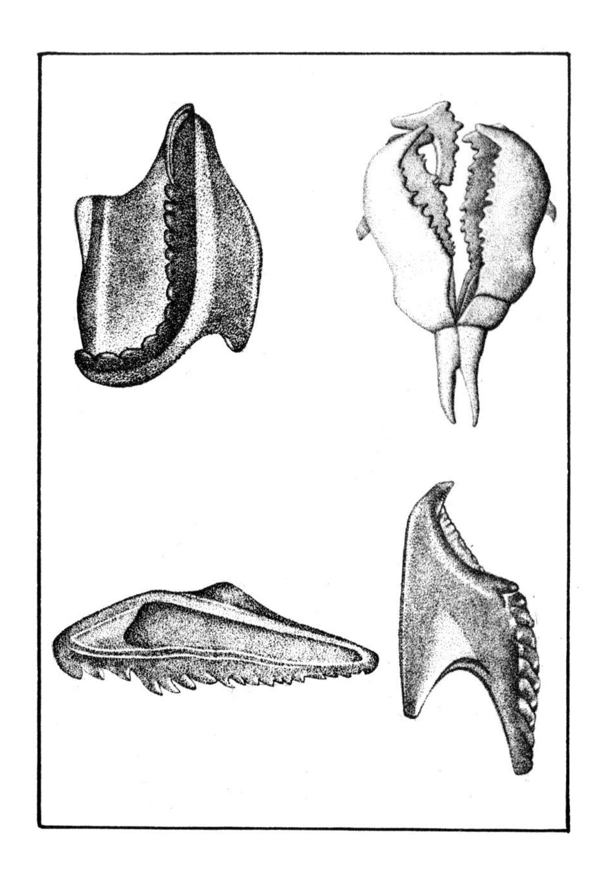

Scolecodonts from the Palaeozoic Era. These are the tiny mouth parts of marine worms that rarely left any other trace behind them. They are made of a substance called chitin and can be extracted from limestone by dissolving the rock away with hydrochloric acid.

matrix is subjected to chemical action. Provided it is not too fragmented, a bone which still contains phosphates can safely be treated with dilute formic or acetic acid that will dissolve the limestone in which it is trapped. Microscopic observation can reveal the great delicacy of this type of excavation. In the same way, microfossils composed of chitin will withstand the effects of hydrochloric acid which erodes carbonates, or hydrofluoric acid that destroys silica; this is the type of technique used to isolate pollen. Clay-based rocks are subjected to boiling in a weak solution of soda or potassium.

The fact that the fossil and its sediment are of differing densities is also turned to good use. In a heavy fluid such as bromoform, the teeth of small mammals will not sink but will stay on the surface. This separation by flotation is a lot faster than manual sorting under the microscope, especially when microfossils are very few and far between in the sample being examined.

How fossils are studied

To truly study a fossil it is essential to create, if possible, the appearance of the living animal from the preserved parts of its body. The hollow imprint of a bone may be filled with latex or a similar material in order to obtain a casting that can be compared easily with better preserved remains. It is sometimes easier to destroy a very fragile bone in acid so as to apply this technique to the tougher matrix that has so finely moulded the bone's smallest anatomical detail.

Radiography, too, can facilitate the study of fossils that are practically impossible to excavate from their surrounding rock. A large number of skeletons of reptiles of the Middle Triassic Period from Monte San Giorgio in Switzerland and Besano in Lombardy – like *Tanystropheus* (see page 174) or *Askeptosaurus* (see page 173) – were photographed in this way without any need to extract them from the slabs of bituminous schist in which they were imprisoned.

The technique of 'series polishing' has been successfully used on very primitive fishes of the Palaeozoic Era as well as on the shells of invertebrates. Although tedious and time-consuming, this technique has achieved some impressive results. It consists of eroding successive slices of the block containing the fossil, while drawing each section of the bone down to its tiniest detail. These sections are then enlarged and their outlines cut out from sheets of wax whose thickness is proportional to the depth of the successive sections. The wax silhouettes are then placed on top of each other and gradually build up a larger-than-life model of the animal's internal anatomy.

Microfossils are often studied on thin sections of rock only hundredths of a millimetre thick and which can be examined under an ordinary microscope. The same process is used to observe the structure of fossil bones or wood. The electron microscope and the colossal magnifications that it provides open up a whole new field of investigation to the palaeontologist. Examination and photography under infra-red and ultra-violet light are also frequently used. Ultra-violet rays in particular turn many fossils phosphorescent and reveal details invisible in ordinary white light. A shell that is perfectly plain to the naked eye can thus show unsuspected ornamentation when looked at under special lamps.

Reconstruction of fossils

Using all of this apparatus we can repair the fractures and deformations in most of the bones we find. The next step is to reconstruct the soft parts of the animal which covered, or were contained within, this skeleton. To do this, we must compare the parts of fossils with those of animals (or plants) about which we know more. This can provide us with precious information about the functioning of the joints, the working and size of muscles, the significance of the shapes of teeth or why the brain is a particular shape.

The conodonts are microfossils which can be extracted using diluted acetic acid. Palaeontologists are not sure exactly what they are. They are made of phosphate like bones and teeth and may have belonged to fishes. These microfossils have been the subject of much research since their identification enables us to date the deposits surrounding them with relative accuracy.

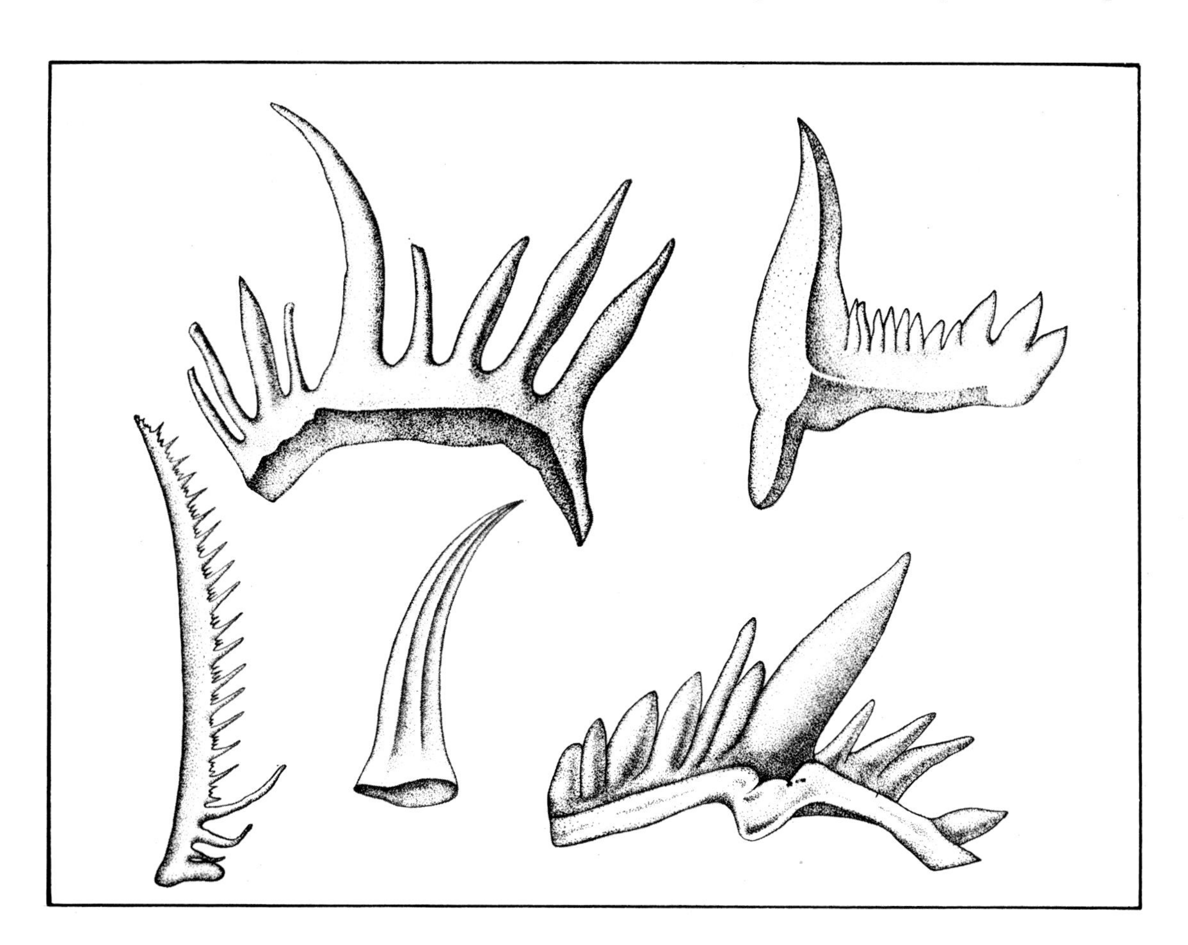

Triceratops (the 'face with three horns') was a North American dinosaur that lived at the end of the Mesozoic Era and sometimes grew to a length of 8 metres (26 feet). Because the skeletons are so complete (like the one illustrated here) we can get a good idea of what the animal must have looked like, especially since the imprints of its flesh are also known. But once the muscles have been put on the bones, it is hard to fill in the details. Was this reptile's mouth as wide as a lizard's (below) or did it have cheeks? We can only guess at its colour. By dipping a scale model of the animal in a water container we can find out its volume, and also its weight.

The skeletons of vertebrates are especially rich in detailed information, so much so that it is possible to make very reliable reconstructions of the appearance of animals that ceased to exist tens or even hundreds of millions of years ago. It may on the other hand be extremely difficult to form an idea of the appearance of an invertebrate animal such as a mollusc whose shell alone remains, or of a sponge in which only a few isolated parts of its body have become fossilized. In cases like this we have to rely on the discovery of specimens which have been very well preserved in order to solve the puzzle. Belemnites were creatures closely related to cuttlefishes (see pages 20 and 21) and the only part of their bodies easily preserved was the hard internal skeleton known as the guard – much like the present day cuttle bone. However it is most unlikely that we would ever have managed to reconstruct accurately these cephalopods, with their short tentacles lined with horny hooks, without the magnificent specimens which were discovered from the lower Jurassic Period of England or from the Portlandian region of Solnhofen in Bavaria. These specimens were preserved complete with their heads, arms and even traces of their ink sacs.

If we are not fortunate enough to stumble upon good specimens, however, and when living animals cannot provide even a distant cousin of our mystery fossil, it can stubbornly resist all attempts at interpretation. One well-known example of this riddle is that of the 'conodonts' (see page 35). These small, serrated fossils, abundant in certain sediments from the end of the Palaeozoic Era, have been thought to belong variously to fishes, worms, molluscs and even to arthropods – a guessing game that so far nobody has won. Similarly, mysterious traces left on the sea-bed or on the wet clay of riverbanks by organisms we should probably recognize elsewhere threaten to mystify palaeontologists for many years to come.

The creature called *Chalicotherium* (see page 254), and the various disastrous attempts made to reconstruct it, illustrate the extent to which science can sometimes be baffled by an animal that has no single equivalent in the modern world. For half a century its huge, clawed feet and horse-like skull were attributed to two completely different animals, one whose head had never been found, and the other whose legs defied discovery. It was not until the end of the last century that a complete skeleton, dug out from the Miocene rocks of Sansan in the département of Gers in France, made it clear that these head and feet belonged to the same animal.

Chalicotherium was so extraordinary that it was indeed hard to imagine its appearance and way of life. For a long time it was believed to have been some sort of tall, very bulky horse, but recent work done on the *Chalicotherium* of Czechoslovakia has shown it to be more like an animal with the outline of a gorilla, using its long clawed arms to seize the foliage on which it fed.

Understanding evolution

Once the anatomy and way of life of an animal or plant have been established, the palaeontologist's next task is to try and understand the relationship between this species and other forms of life similar to it. Even if the other species lived at different times, the chief aim is to find out the extent of their kinship.

The species is the basic unit used in the classification of the living world. Put simply, it is any group of creatures capable of reproducing among themselves and producing fertile offspring. It is customary to give each species, whether living or fossil, a Latin name comprising two words. These words are used internationally. For example, the *Iguanodon* of Bernissart (see page 23) is called *Iguanodon bernissartensis*. Now it is evident that the 'sterility factor' which separates the various species cannot necessarily be claimed for extinct species, and so their separation is based on physical features. The inability to produce fertile offspring outside the confines of an organism's own species is the method by which Mother Nature controls the types of creatures which exist on Earth. Nevertheless, even under this control great changes can occur. Just look at the contents of a bus, for instance, and see how variable the human race is. Apart from these obvious variations, all species slowly change during the course of evolution.

It is to J.B. Lamarck (1744–1829), and above all to Charles Darwin (1809–1882), that we owe an understanding of the process called evolution. The advance of genetics has provided us with a better understanding of the intimate mechanisms of this phenomenon. The characteristics that each individual inherits from its parents are contained 'in code' in its chromosomes which in turn are transmitted from one generation to the next inside the nucleus of reproductive cells. Sometimes, unusually arranged chromosomes give rise to what are known as mutations. In mutations, the information determining one or more of the organism's characteristics may be modified, and a new hereditary feature may appear. These alterations in the genetic heritage are not necessarily always beneficial to the mutant and its descendants, however. Sometimes they produce a characteristic of great advantage to the organism and which enables it to be more successful than others of its species, whilst at other times some mutations can be a great disadvantage to the organism. For instance, a mutation causing a loss of the

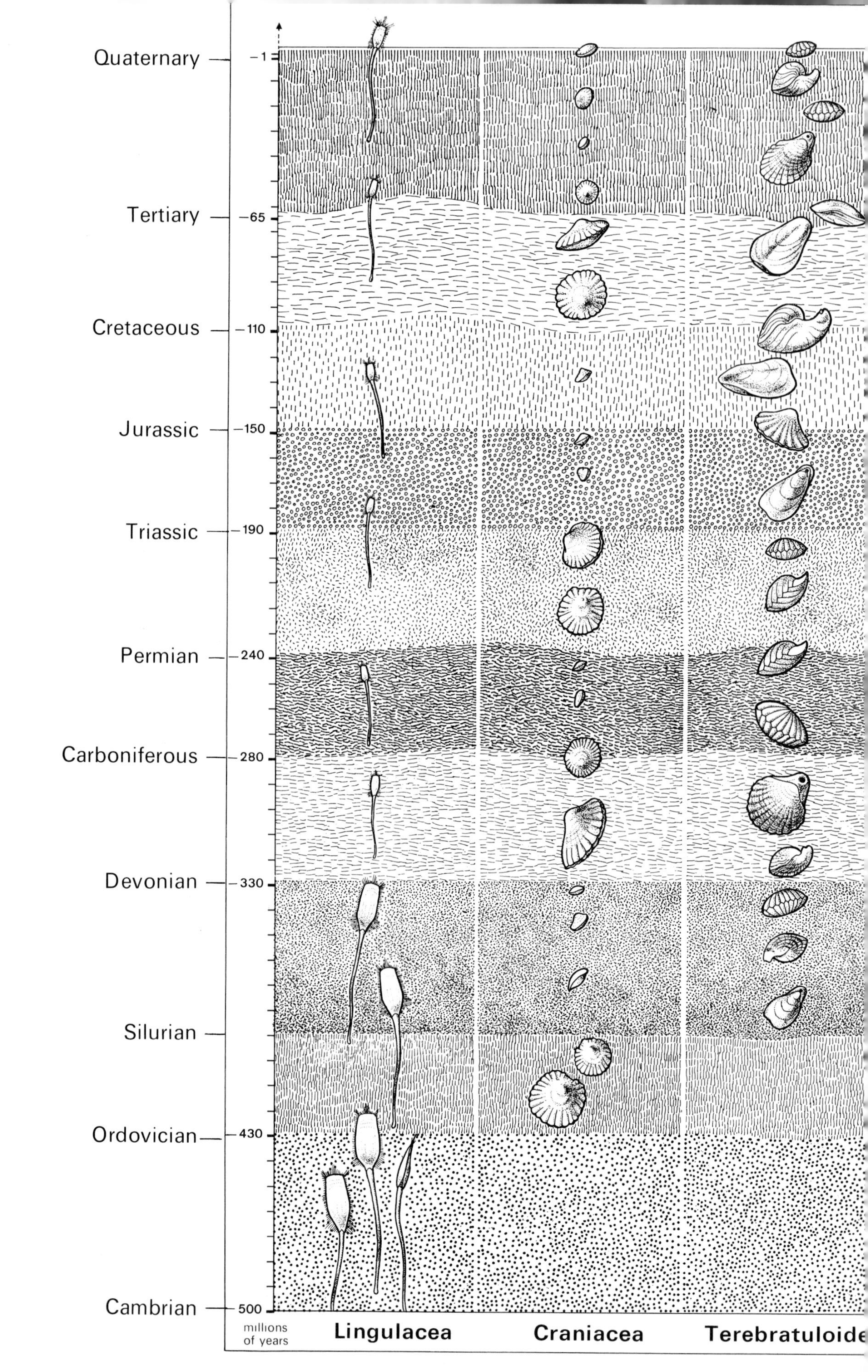
Quaternary
Tertiary
Cretaceous
Jurassic
Triassic
Permian
Carboniferous
Devonian
Silurian
Ordovician
Cambrian
-1
-65
-110
-150
-190
-240
-280
-330
-430
500
millions of years
Lingulacea
Craniacea
Terebratuloide

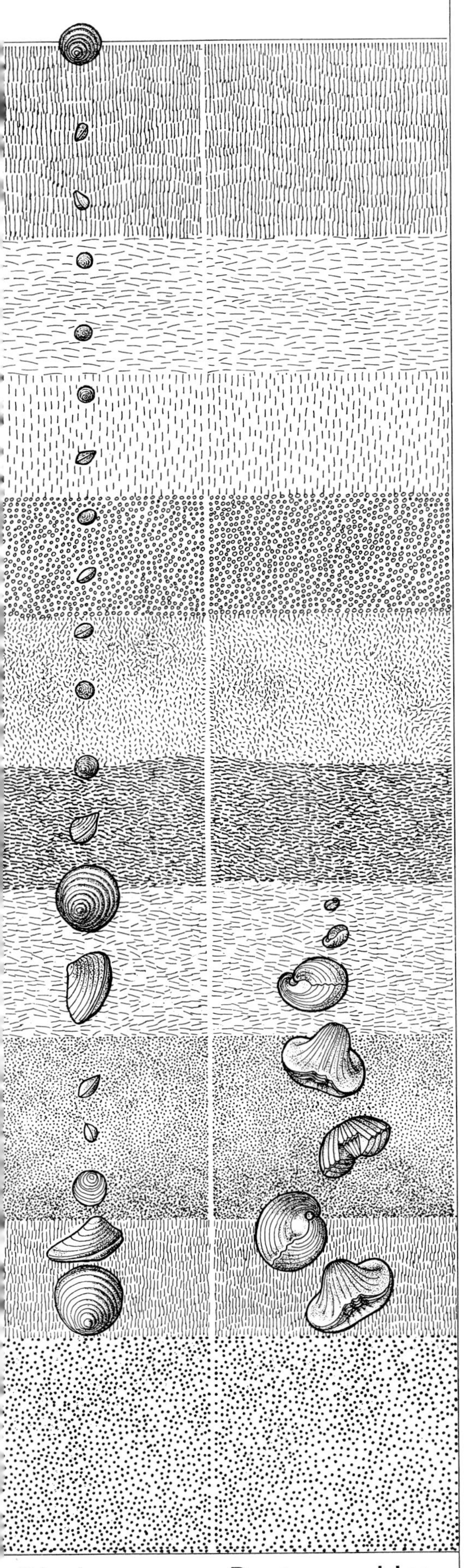

body pigment which usually enables an animal to remain camouflaged would mean that predators could soon spot it.

Darwin's great achievement was that he understood the importance of evolution and how it affected the species which he studied. He succeeded in underlining the fundamental role played by natural selection, a process which each individual undergoes. Each single representative of a species 'fights for its life' with a degree of success that depends on its own characteristics. Some will benefit by their greater ability to capture their prey, or to resist disease, or even to reproduce in large numbers. In a hostile environment they will be less frequently eliminated than those which are weaker, and it is these stronger individuals which will transmit their chromosomes to future generations. Above all, the process of selection exerts a stabilizing effect, causing the removal of any creature that is too unlike the rest of the species.

When the environment changes, however, a creature with different characteristics is often able to give rise to a new species. Let us imagine a population of rodents spread over a huge continent. Geological upheavals cut the continental mass in two. If the climate changes in one of the two resulting parts, becoming colder and wetter, the animals that inhabit it will, because of variation and natural selection, grow an increasingly longer coat. In time, as conditions in the two regions become more and more different, other characteristics will make them even more distinct. After several million years their chromosomes, bearing all these characteristics, will have become so dissimilar from those of the original population that reproduction between the two groups will be impossible. Thus from the original species a new species appears.

For the purposes of classification we have

In the course of geological time some groups evolved faster than others. In the brachiopods, sometimes known as living fossils, the lingules have hardly changed at all in 500 million years, while other brachiopods changed shape and size a great deal and sometimes suddenly, after long periods of stability. Some have survived up to the present-day, while others died out quickly and are characteristic of a given epoch.

created divisions larger than the species: genera, families, orders, classes and sub kingdoms – groupings that are all interrelated. So far as it is possible, these divisions are designed to show the degree of relationship among the species which they contain. One of the tasks of the palaeontologist therefore is to unravel the complex web of evolution, looking for characteristics acquired from a distant ancestor and reflected jointly in a group of species, in order that we may understand more about the past.

The fossil record

For many years fossils were the sole means of geological dating. From the beginning of the 19th century, before the idea of evolution had been accepted, the early palaeontologists quickly realized that the fauna hidden in the various layers of the Earth varied with their age. Except where land upheavals had occurred, the most recent layers of sediment covered layers of older sediment, and it was obvious that the further one went back into time the more primitive were the animals and plants found in the fossil state, and the more distant they were from their present-day cousins. The geological layers were divided up into successive Eras, each made up of several Periods (see page 302). By using groups whose evolution is rapid, it has been possible to narrow down these divisions even further. Certain fossil species with a relatively brief period of existence, like the ammonites of the Mesozoic Era, are characteristic of very shallow geological zones. The fossils used to establish the geological history of a cross-section of Earth are almost always invertebrates (see page 19), since these are far more abundant than vertebrates. In sequences deposited on dry land, however, it is frequently necessary to consult the remains of the latter creatures. This is amply demonstrated by the fact that our knowledge of the sequence of events which took place during the Triassic Period in South Africa is entirely founded upon the succession of remains of mammal-like reptiles. Thanks to their huge abundance, microfossils are the geologist's favourite aid in this task of unravelling this particular part of evolution. Since these organisms abound in certain types of rock the oil prospector, with his very costly boreholes, is able to date each layer he drills through from just a small amount of sediment. Our knowledge never ceases to grow, and we are increasingly able to study entire groups of creatures rather than merely isolated examples.

Palaeontology provides a scale of relative time and age, enabling us to compare one particular rock unit in time with another. But it cannot tell us exactly how old the rock is, and it is only recently that special techniques have been perfected which can solve this problem. In fact what it comprises is a whole series of methods, each of which is applied to a specific 'slice' of time. As we shall see in due course, most of these methods are based on the radioactive properties of rocks.

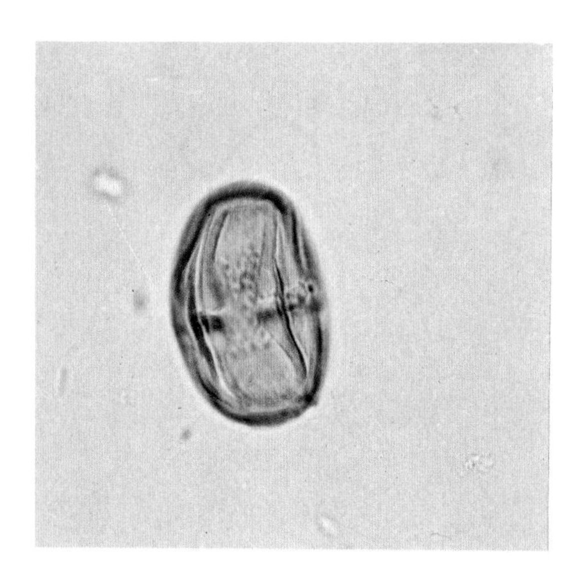

The study of fossil vegetation tells us a great deal about ancient environments. On the left is a perfectly preserved grain of pollen from the Oligocene Period photographed under a microscope. On the right is a leaf from the Miocene Period deposits at Oeningen, the famous site of Scheuchzer's 'Victim of the Great Flood'. Some fossil plants are quite similar to the plants we see today and by looking at where modern plants grow we can get an idea of what fossil climates must have been like.

The changing face of the Earth

All these patient efforts are succeeding in retracing the successive scenes in the evolution of the Earth and of life itself. A fossil coral found in a limestone bank tells us first of all that the sea once invaded the region, for corals are sea creatures. But it also reveals that this sea was warm and shallow at this particular spot, since the familiar algae that lived in the coral needed strong sunlight to survive. Its position within the rock mass will indicate if this area was once a trench into which the block of coral tumbled, or a stable part of the reef. Thus the ancient environment is reconstructed, with each fossil supplying fresh data. The nature of the rock itself can tell us about the chemical conditions of the environment and its relationship with the continent. The growth lines on the shells of bivalve molluscs will tell us that the duration of the lunar cycle was a little longer than it is today, for instance.

On Earth, the distribution of the various animal species at each epoch reflects a changing geography. Palaeontology has demonstrated the drift of the continents as they dislodge and move further away from each other. Let us look at one example of this: fossils of a giant river crocodile from the beginning of the Cretaceous Period, *Sarcosuchus*, and a land crocodile of the same period, *Araripesuchus*, are found only in Nigeria and Brazil – witnesses of an age when Africa and South America formed but a single continent.

With a complete knowledge of the geology of any region of the World we can retrace each episode in its history: the advances and retreats of the seas, the raising and erosion of the mountains, the changes in climate from the cold glaciations to the hot and humid periods in between. In each of these landscapes we can look at how the living world adapted to the changing environment.

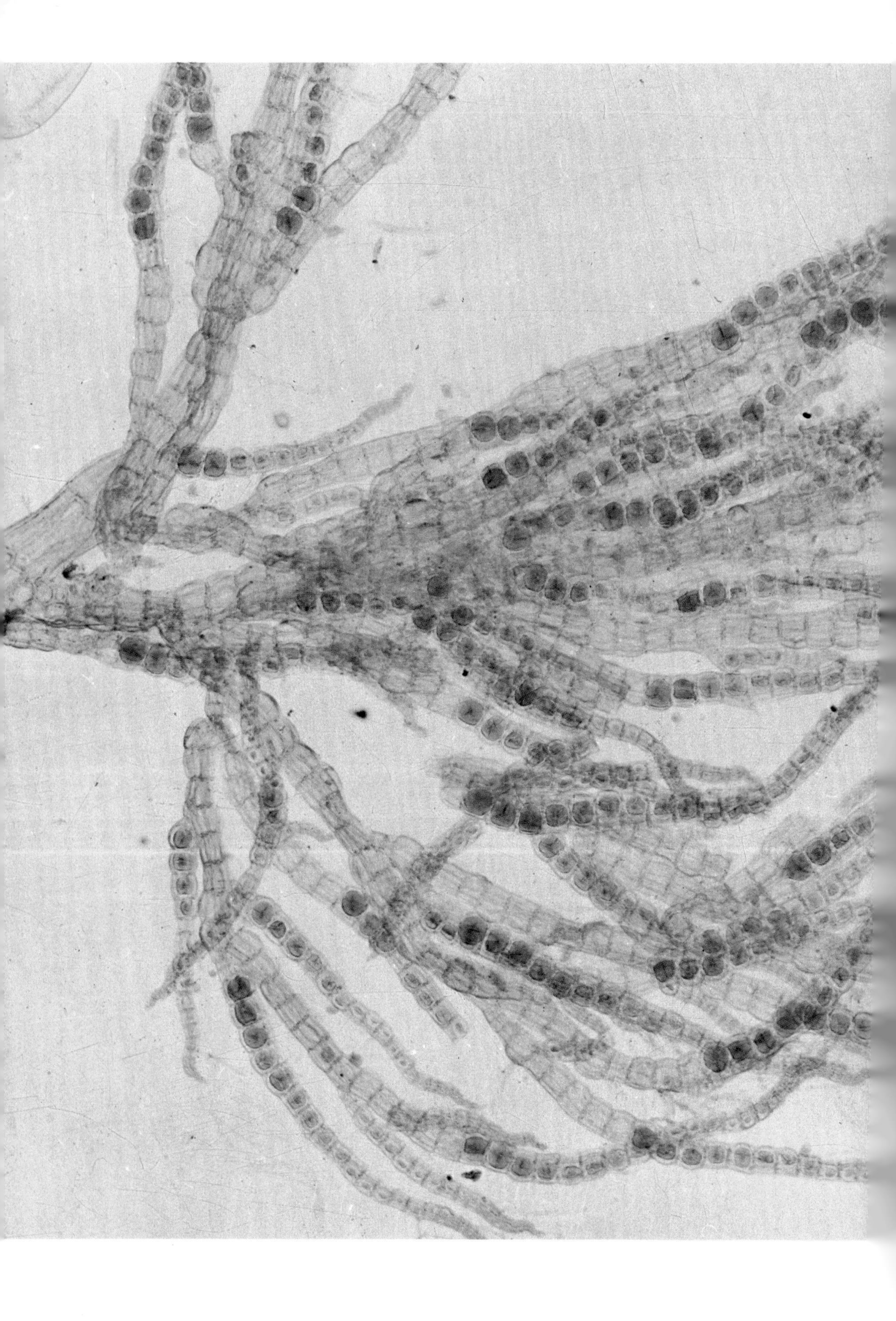

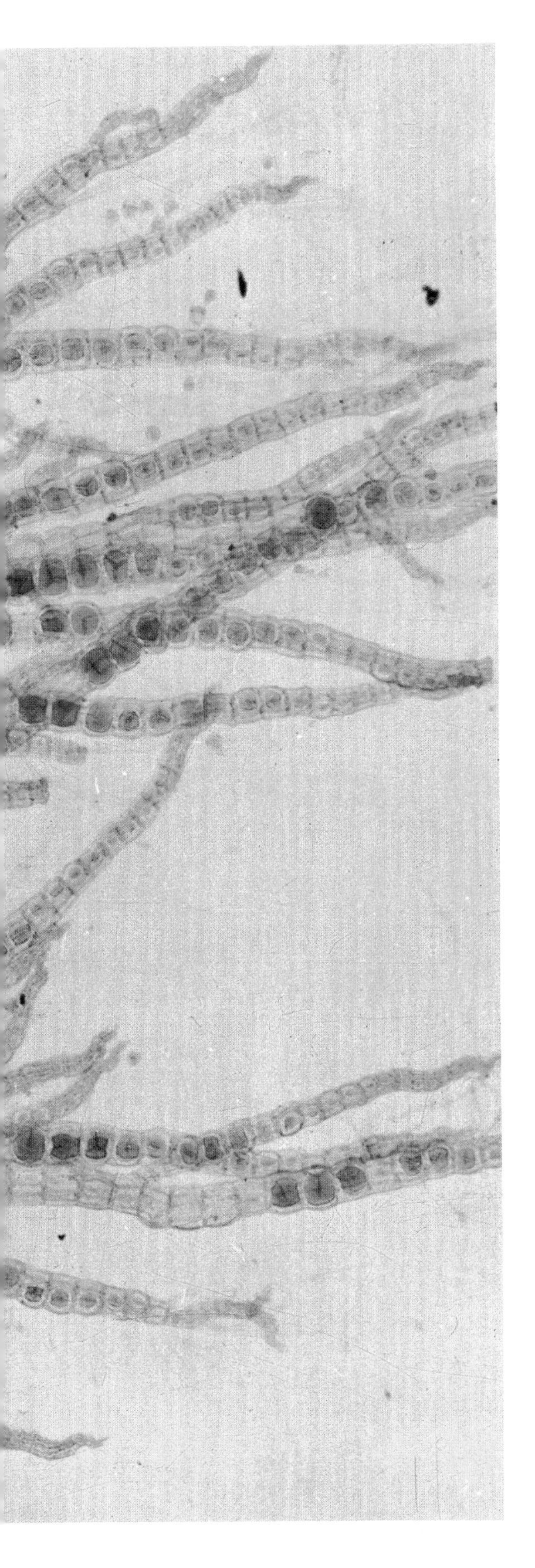

Chapter Two

First Forms of Life

First forms of life

Our planet

The Earth is one of the nine planets of the solar system, orbiting quite close to the Sun, between Venus and Mars. By no means a large planet, its diameter is 11.18 times smaller than that of Jupiter which is, of course, the giant planet of our solar system. In the centre of the system a yellow star, our Sun, accounts for 99.8 per cent of the total mass, and is a furnace burning at a temperature of some 20 million degrees centigrade, where nuclear fission transforms hydrogen into helium, releasing vast amounts of energy, a very small amount of which reaches our Earth in the form of light and heat. It is this small amount, however, that is vital for all life on Earth.

The solar system itself is a mere speck of dust in the arm of a spiral galaxy containing perhaps as many as 100,000 million stars, and this galaxy is lost in the vast reaches of an infinite Universe, of which the part that is observable with our instruments contains 100 million such galaxies.

The 'birth' of our Universe took place some 10,000 million years ago. Many astrophysicists believe that at the beginning all the matter of the Universe was concentrated into quite a small space, and an initial explosion, the 'Big Bang', dispersed this highly condensed mass which consisted for the most part (if not entirely) of hydrogen. The progressive dispersion of matter was accompanied by a gradual cooling and condensation of elements with a high melting point. Under the influence of the forces of gravity, vast clouds of dust and gases were formed. Within these whirling masses, the so-called 'proto-galaxies' – small aggregates of matter – formed in their turn and, contracting in upon themselves gave rise to the stars within which thermonuclear reactions were triggered. It appears that the stars were frequently formed in groups, some of which have remained so closely linked that today they represent twin, triple and even quadruple stars.

The study of the galaxies has shown that they are constantly rushing away from each other – to put it another way, the Universe is expanding. This phenomenon could go on indefinitely, or it could end in a contraction that would recreate the original conditions.

To explain the formation of the planets, astronomers make use of a theory put forward as early as 1796 by the brilliant mathematician Pierre Simon de Laplace, who believed that the solar system had probably separated itself off from an interstellar cloud turning faster and faster upon itself. The resulting disc of gas then split up into rings encircling the 'proto-star' in the centre, and the condensation of these rings into increasingly large masses formed the beginnings of the planets. As the weight of the Earth increased and the forces of gravity compacted the elements within, our globe became extremely hot, and the molten metals sank towards the Earth's centre leaving the lighter elements on the surface.

The age of the Earth's crust (the outer surface) is approximately 4500 million years, and special techniques of dating the oldest rocks have provided us with a fairly accurate account of the events that followed its consolidation. These techniques use the properties of radioactive particles which decompose in the course of time. Uranium 238, for instance, transforms very slowly into lead 206, a very different isotope (chemical form) from the usual, which is lead 204. Since we know that it takes 4250 million years for half of an amount of uranium 238 to decay into lead, all we have to do is measure the quantity of these two elements in a rock to find its age.

How life began

Life can be thought of as the ability of a group of molecules to maintain their individuality and structure, to assimilate matter, to consume energy and finally to reproduce. This concept of life seems obvious, but there are nevertheless certain structures on the very borderline of the mineral (non-living) and the living worlds that continue to perplex the scientist. Viruses, the cause of so many infectious diseases, are microscopic

An artist's impression of what the Earth must have been like in the very earliest days of its history, some 4500 million years ago. Life began in a very hostile environment. The poisonous atmosphere was shaken by violent thunderstorms and bombarded with deadly radiation. The Earth's crust had only just started to solidify and there were many violent volcanoes (these may have helped to make some organic molecules). The primaeval oceans, formed by condensation of water vapour in the air were hot and shallow, but over millions of years they provided the food for the first organisms to live on.

structures capable of multiplying within a host organism by making their own molecules. Once outside that organism, however, they revert to being totally inert crystals. Viruses in fact probably represent that rather poorly defined border which certain molecules have crossed in order to become living.

Once Darwin had demonstrated the mechanism of evolution, the next step was to determine at which point in time life began. In the middle of the 19th century when the microscope was already in use, scientists still believed in the idea of 'spontaneous generation' that had been fashionable ever since Ancient Greek times. This suggested that the worms had been born spontaneously out of mud, flies out of rotting meat and mice from piles of refuse. Although the notion had been up-dated since Aristotle's time and was now only applied to micro-organisms, people still firmly believed it. After lengthy arguments, it was finally the French chemist Louis Pasteur who succeeded in showing the extent to which almost everything in our environment is polluted by all sorts of microbes. The hands of a surgeon, even the very cleanest of receptacles, were all covered in microbes.

The initial appearance of microbes remained a puzzle, unless one believed in a supernatural origin of even the most primitive life-forms. For a time at least, the theory of 'panspermia', first formulated at the beginning of this century by a Swedish chemist called Svante Arrhenius, provided an answer. He thought that the Earth was an environment that had been engulfed by germ-laden dusts and micro-meteorites from outer space. However, this idea did not explain the origin of the microbes as it would be hard to imagine organisms capable of withstanding the radiation and extremes of

This diagram shows the apparatus designed by Stanley L. Miller for his historic experiment, which made simple organic molecules from ingredients present in the primaeval oceans and atmosphere. The air was pumped out of the flask and Miller introduced a blend of methane, ammonia and hydrogen, the components of the primaeval atmosphere. He used steam to push the gases into the spark flask and gave them electrical discharges of 60,000 Volts. After several days the condensed water contained some organic molecules.

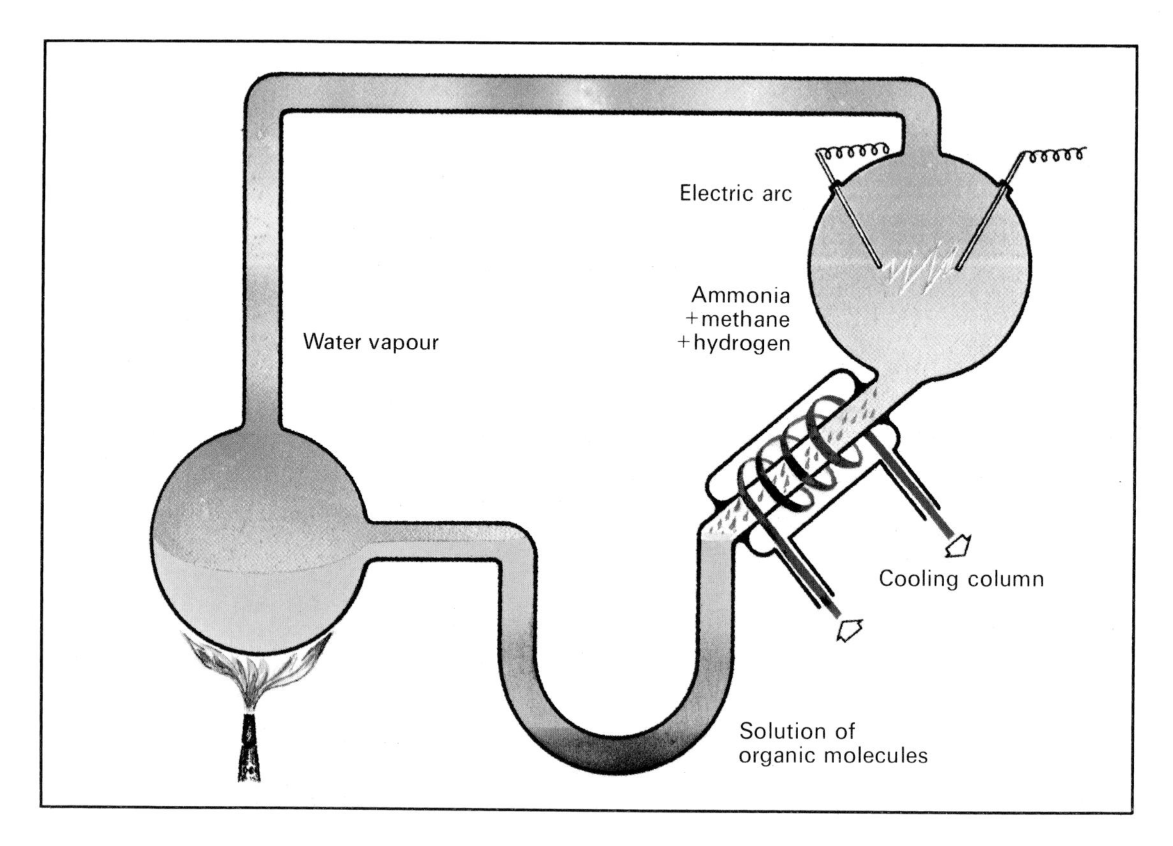

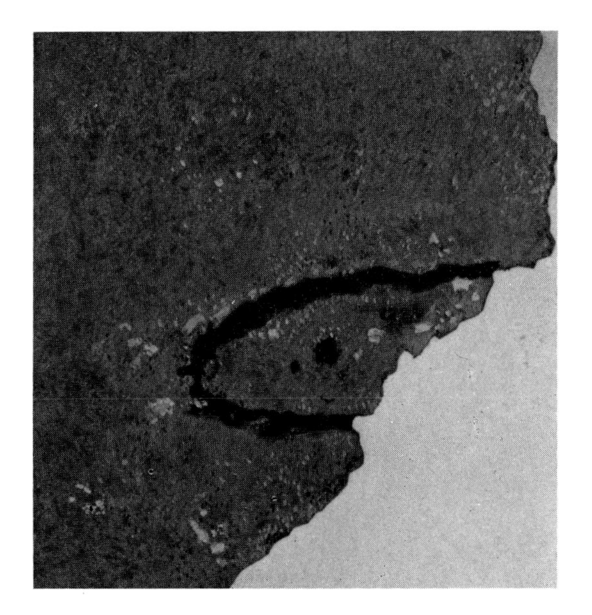

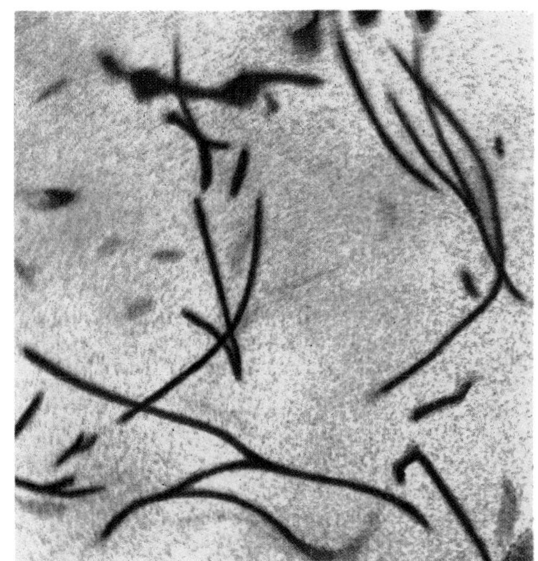

temperature found in interstellar space. Some were discovered in certain meteorites, like the famous meteorite of Orgueil in France, but it was found that these came from the Earth, not outer space.

Recently scientists have detected cellulose, a chief component of vegetable tissue, in the tails of comets. Some experts now think that a comet's tail is the original cradle of life and even go so far as to claim that the great epidemics of the Middle Ages were the result of the periodical passage of these heavenly bodies close to the Earth. Since 1968, water, alcohol, ammonia and some forty other molecules have been found in space. The existence in interstellar clouds of amino acids (the basic 'building' bricks of animal and vegetable proteins) is suspected, and have in fact been recovered from meteorites that have crashed to Earth.

Although the detection of organic molecules in outer space does not prove the concept of panspermia, it does show that chemical compounds essential to life are far more commonplace in the Universe than was once believed. Extra-terrestrial life has yet to be discovered in our solar system, however. The Viking Probes 1 and 2 failed to find any trace of life on Mars, a planet which would seem to provide the most favourable conditions for at least primitive life-forms. However, according to Laplace's theory such conditions must be fairly frequent throughout the Universe. Life-supporting planetary systems might even exist around stars quite close to our own Sun, such as Epsilon in Eridan and Barnard's Star. When one thinks of the countless numbers of stars in our Universe, then space seems quite capable of holding planets inhabited by living creatures.

Corycium enigmaticum (left) is an extremely ancient fossil from the Upper Precambrian Period of Finland. This little dark sac, shown here more or less life-size, is generally thought of as organic. By looking at its carbon we find that it was produced by a living creature – though not everyone agrees with this conclusion. For a long time, blue-green algae and bacteria formed the only life on our planet, and the blue-green algae have left their mark as microscopic filaments in Precambrian rocks (right). The bacteria were more numerous and some, called the siderobacteria, were responsible by their activities for huge deposits of iron ore.

The primaeval soup In the very earliest days of its existence, the Earth was vastly different from today, and the first germs of life developed in a most inhospitable environment. The primitive atmosphere contained no oxygen at first (this is confirmed by the large deposits of unoxidised Precambrian iron). The main gases were hydrogen, a light

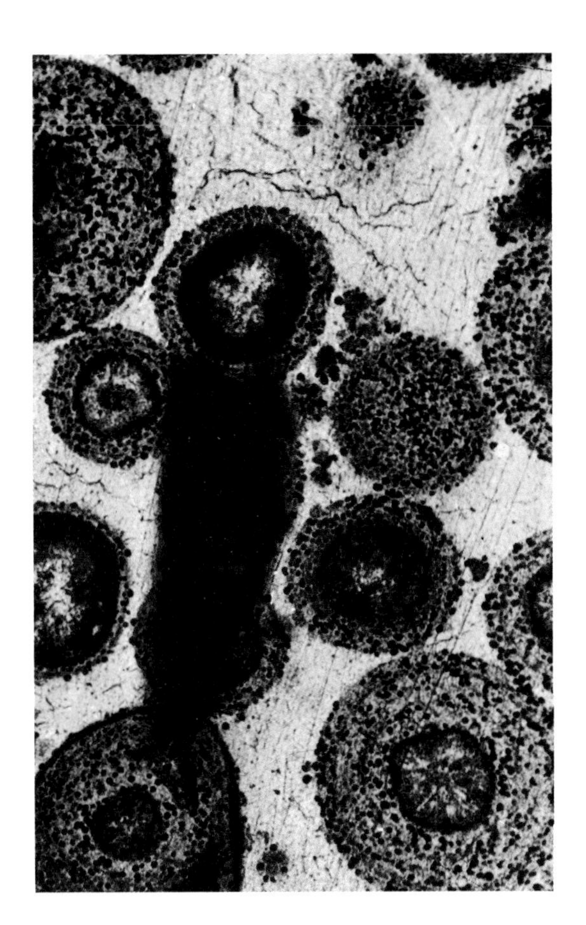

Asterosphaeroides darsii from Korera (Mali). According to D. Boureau, these spheroids, only tenths of a millimetre in diameter, are cellular forms representing a very primitive stage in the evolution of life. However, they date from the end of the Precambrian Period only, which would seem to point to several successive appearances of live cells during the Precambrian Period as a whole.

element, much of which escaped the Earth's atmosphere as it formed, methane, ammonia and water vapour. The latter condensed to fall as rain upon a still-hot terrestrial crust to form the first oceans: merely stretches of shallow water.

In 1924, the Soviet biochemist A.I. Oparin gave a description of the processes which, in the early days of the Earth, could have formed the first organic molecules. The primitive atmosphere had no protective ozone layer (which shields us today) and so the molecules were intensely bombarded by radiation from space. This formed new, more complex particles and these accumulated in the newly formed seas, constituting in the words of J.B.S. Haldane a 'primaeval soup' – the food of the first living beings.

Nearly 30 years later a young chemistry student at the University of Chicago decided to artificially recreate the supposed conditions of primitive Earth in order to verify the reasoning behind Oparin's theories. He developed an experiment which is now famous but at the time it seemed so naive that he kept it a secret (see page 48). He took a balloon flask containing a mixture of hydrogen, methane, ammonia and water vapour, and into this mixture generated an electric arc whose discharges simulated the electric storms which must have been present in the skies above the primitive Earth 4.5 thousand million years before. After a week the water had condensed, and had taken on a curious orange hue. He found that it contained a number of organic molecules, including the famous amino acids. The synthesis of simple organic molecules – the primaeval soup – had been proved beyond doubt! Miller's experiment was developed by other scientists; the effects of electric sparks were replaced by those of ultra-violet rays, ionizing radiation and even the heat from an electric oven. Various mixes of gas were tested, as were blends of the substances that had appeared in Miller's bottle, in order to reconstruct the primitive world in miniature. All the experiments led to the same result: a number of organic components formed spontaneously as well as some more complex elements like sugars, urea and molecules close to proteins.

Even the genetic material of our own chromosomes was made in this way. Sheltered by water from ultra-violet rays, without any oxygen to oxidize them, and without any living being around to destroy them, these substances enriched the primaeval oceans in ever-increasing concentrations. This slow, gradual process was accelerated as some of them gained the ability to make copies of themselves. They thus proliferated using a mode of reproduction which was the forerunner of that used by living organisms.

The molecules of life We all know what happens when we pour a little oil into a pot of water: if we shake the mixture tiny droplets of oil will be formed which will refuse to dissolve in the water. If we stop shaking the pot the droplets will float to the surface and form a thin film. This phenomenon is due to the particular structure of the molecules of fatty substances. These molecules possess a hydrophilic (water-loving) pole and a hydrophobic (water-hating) pole. This is why, in water, oil has a tendency to form small globules each of which has its hydrophilic pole turned outwards and its hydrophobic pole turned towards the centre. This characteristic, which is shared by the proteins, explains the composition of cellular membranes in which two layers of protein are separated by two layers of lipids.

Cells are the fundamental elements of living tissues within which they are highly integrated but they may also be isolated, as is the case with the simplest of creatures. Protected by its membrane, vital reactions take place within the cell. The minute drops of protein, isolated from the outside by their 'double membrane' are the first stage from which the large organic molecules of the 'primaeval soup', and eventually the cell, arise. Experiments have shown that the walls of these tiny spheres can let in certain ingredients of the 'soup' but keep out others. High concentrations of these chemicals inside the 'proto-cells' trigger off specific chemical reactions, and since each droplet has its own special characteristics, a type of selection is produced which precedes the natural selection which we have already mentioned in connection with living beings. Many of the droplets are destroyed by their own reactions, while others can develop using the energy released from their chemical reactions. Having reached a certain size, they split up into several identical droplets. Those whose internal structure makes them more efficient at this very simple metabolism, necessitating above all water, sugars and amino acids, will multiply at the expense of

The most famous Precambrian deposits are found at Gunflint in Canada. Layers of chert sandwiched between the slates and conglomerates contain not only bacteria and the filaments of non-septate algae (those with no dividing walls) but also curious micro-organisms such as *Kakabekia*, shaped like an umbrella, and *Eoastrion* with its branched outlines, which lived some 2000 million years ago (approx. × 1500).

the others. These successful 'proto-cells' contain substances called enzymes which are capable of greatly speeding up the chemical reactions inside them.

We must assume that some of these droplets which burst when they reach a critical size acquire chains of complicated molecules closely related to those found in our chromosomes. Starting with simple molecules selected from the environment, these chains have special properties which allow the production of all the elements that make up the droplet. This is the point at which our drops become true cells which live and reproduce. From sugars and primitive proteins they were able to copy themselves: one droplet can easily produce two identical ones. This extraordinary property gave rise to many of these cells and the more primitive aggregates gradually disappeared. As we have seen, mutation and natural selection subsequently made these still very basic organisms evolve towards increasingly elaborate types.

The first animals and plants

In time, some cells evolved which were capable of feeding off carbon dioxide gas generated by the fermentation of sugars in the 'soup'. This process needed the power of solar energy and was all made possible by a vital complex protein called chlorophyll which trapped this energy. This new step is called photosynthesis and was to have enormous consequences, since the products of the process were chiefly oxygen gas and glucose. Hence the newly evolved single-cell plants were able to supply the atmosphere with oxygen, and the marine world with sugars.

Ionizing rays reacted with oxygen and formed a layer of ozone high up in the atmosphere. Ozone is the gas that filters out many of the Sun's harmful rays and makes our sky look blue, but it also has the very important property of arresting the more

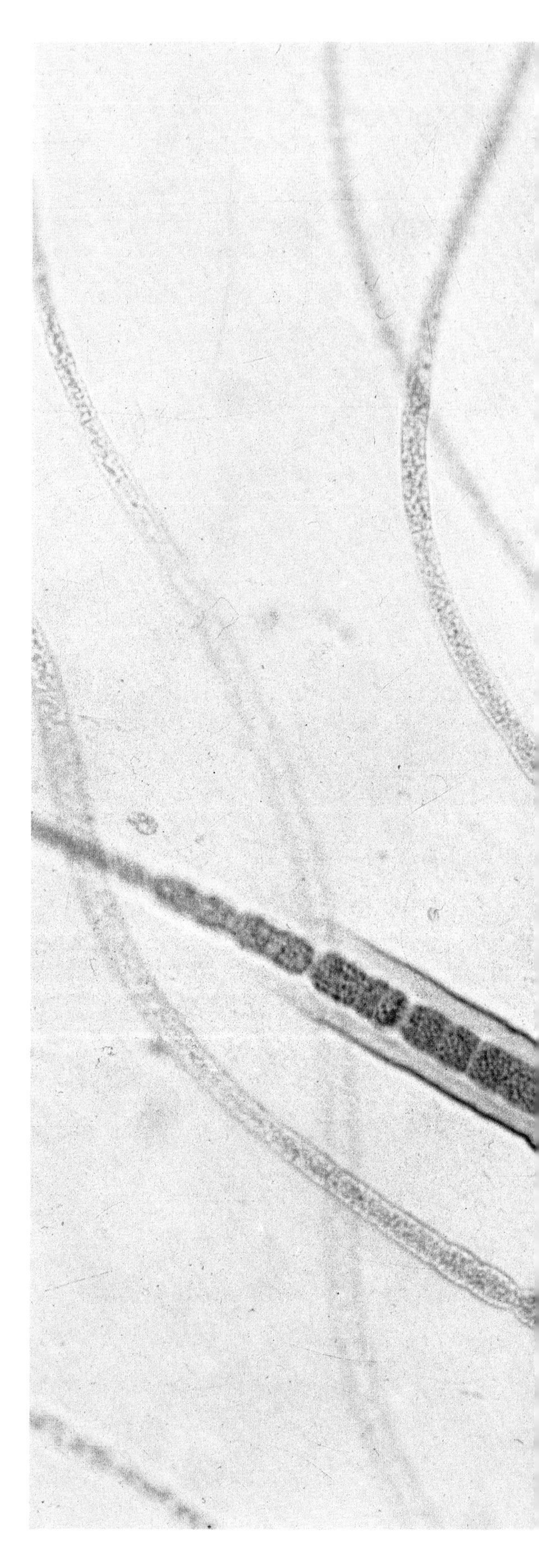

Rivularia (approx. × 640) are members of the Cyanophyceae (blue-green algae) and form long filaments that stick together in jelly-like masses. This modern genus is similar to fossil algae found in some Precambrian Period layers, like the ones at Gunflint, Canada.

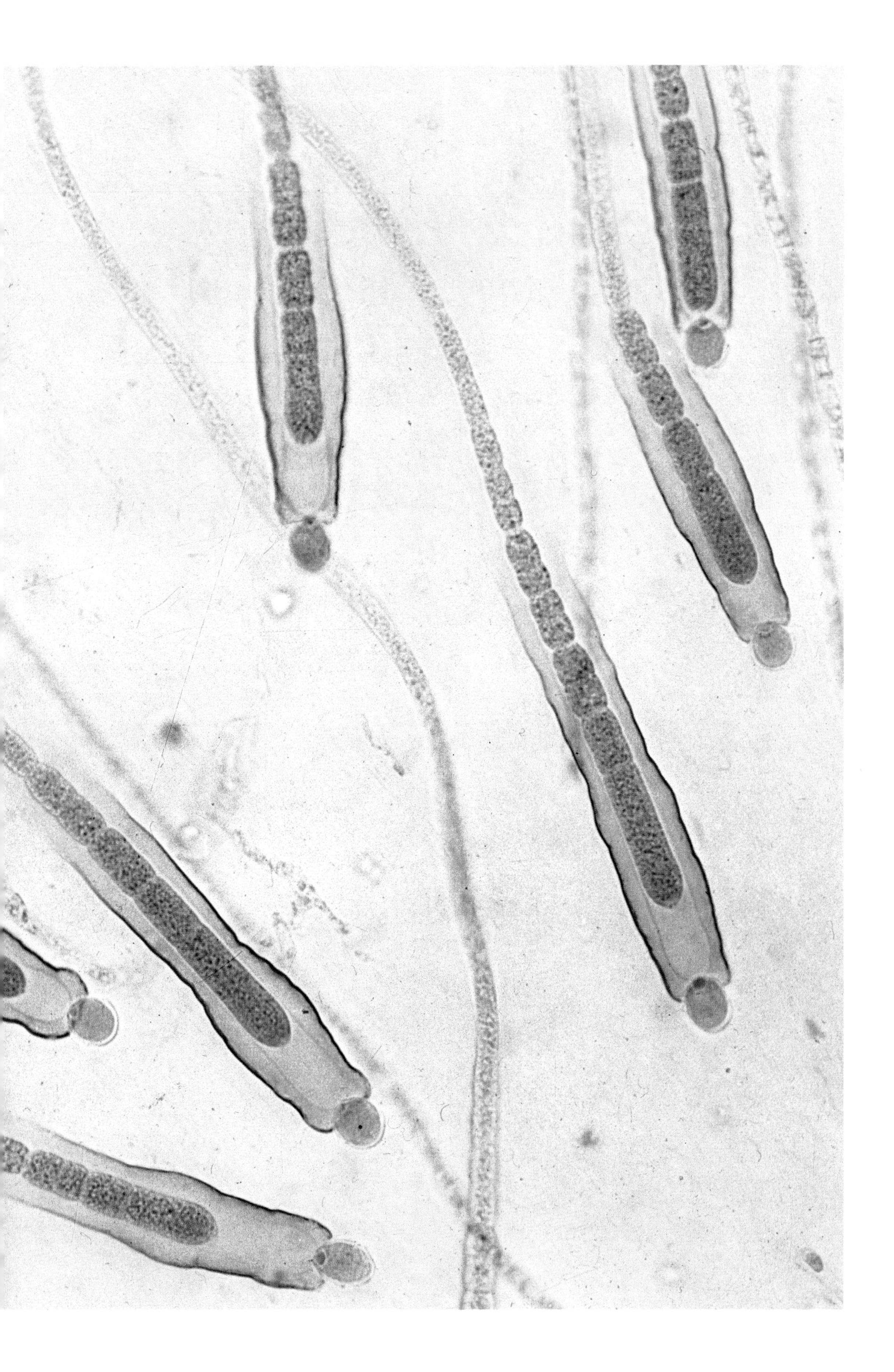

On the west coast of the island of Andros in the Bahamas, we can observe in temporary lagoons and mangrove swamps covered by the tides the development of stromatoliths. In these waters of variable salinity the algae grow on limestone muds, forming carpets which the Sun blackens when the tide goes out.

powerful ultra-violet rays. These rays could not penetrate the surface of the seas but made the air and land fatal for all organisms. Glucose and oxygen also assisted in the nutrition and respiration of cells which were incapable of photosynthesis. These were the very first animal cells. We can now see that animals have depended upon plants from the very beginning.

Looking for signs of the origins of life needs a knowledge of biochemistry, not fossils. Traces of the very earliest living forms are extremely rare for two chief reasons. The first is that these microscopic creatures had no internal or external skeletons suited to fossilization. Secondly, those deposits that are most likely to hold the most primitive fossils are of course the oldest, and they have frequently been distorted and their fossils destroyed.

The layers of sediment which are deposited in the oceans as a result of the erosion of continents or of chemical precipitation are gradually transformed from muds and sands into rocks. Once compacted they undergo many upheavals. Often thrust up above the ocean's surface, and deformed by folds, the structure of these rocks is modified in the course of geological time. Subjected to the high temperatures and enormous pressures in the depths of the Earth's crust, limestones turn to marbles, clays to slates and shales, and sandstones to quartzites. Eventually, all these rocks crystallize into the granite of the continental land-masses. In the course of this metamorphosis, the fossils gradually fade and disappear completely. We cannot therefore expect to recover remains that are older than a few thousand million years except in regions that have enjoyed much stability during the Earth's history.

The so-called 'Precambrian' deposits (those more than 570 million years old – not very much compared with the age of the Earth) contain very few fossils. Nevertheless, much effort and perseverance have revealed very ancient organic traces. The very oldest fossils have frequently been mistaken for quite lifeless forms (for instance the folds produced by currents and formations of metamorphic origin). Even today, the discovery of organisms in the Precambrian

Period can always be interpreted in several different ways.

In 1979, some tiny spheres only three hundredths of a millimetre across and similar to modern yeast were found among the cherts (a sort of flint) of south-west Greenland, some 3800 million years old. Some were grouped in colonies or filaments, but it seems hard to imagine them as yeast cells, as these were already highly complex at this time. Although their origin is organic, these *Isuasphaera*, as they are called, could be a stage preceding the appearance of the first cells, and if so would be the oldest fossils discovered so far.

'Stromatoliths' make up the majority of fossils known from the Precambrian Period. These are structures made from limestone layers piled up to form knolls or pillars (see opposite and below). Some attain gigantic proportions, such as *Conophyton* of the Sahara whose lamellae are conical and are as much as 30 centimetres (1 foot) high. Their continued existence down to our own time in tropical seas tells us much about their origins. The limestone of which they are formed was laid down by the activity of blue-green algae, plant cells without a nucleus. Stromatoliths have recently been detected in deposits 3400 million years old in Western Australia, which would make the oxygen-

Stromatoliths are particularly spectacular from the Precambrian Period, but they are found in more recent rocks. This is a section through a stromatolithic encrustation of the Alpine Triassic Period included in an iron-based sediment that shows signs of emergence (magnified about 5 times).

This is what the sea-bed not far from Ediacara might have looked like during the Upper Precambrian Period. Jellyfishes swam through the shallow waters, while worm-like creatures wriggled across the sandy floor (on the left are two *Dickinsonia*; on the right a *Spriggina*) among the *Charnia* (see page 58). Algae and sponges probably lived there as well.

producing phenomenon of photosynthesis very ancient. Certain stromatoliths could be non-biological in origin, but the fact remains that as long ago as 2700 million years, stromatoliths definitely showed the presence of algae or bacteria – especially those found in Canada.

More cherts, this time 1900 million years old from Gunflint on the Canadian banks of Lake Superior, have provided us with quite an assortment of flora including bacteria, filaments of blue algae and strange shapes that remind us of microscopic fungi.

The iron oxides in Precambrian sediments tell us that at that time quite large amounts of atmospheric oxygen must have begun to develop under the action of algae, but the concentration of this gas at the end of the Precambrian Period was still only 1 per cent of its present-day value. Around 1000 million years ago the first cells with a nucleus appeared – these are green algae. They are found at Bitter Spring in Australia and are still associated with bacteria, blue-green algae and – no doubt this time – fungi. In this deposit we have even found green algae at varying stages of cellular divisions. The analysis of the 'organs' of the living cell coupled with certain fossil evidence (see page 50) would seem to indicate that bacteria played a vital part in the development of the first nucleic cells, and perhaps these were formed by groups of several bacteria which constituted the ancestral type of these 'eucaryotes' (cells with a nucleus).

The hills of Ediacara

The next stage in the evolution of life saw the grouping together of the numerous cells in a single organism into distinct tissues and organs such as muscles and heart. All such animals are called metazoans. Apart from the lowliest forms of animal life such as protozoans and sponges, all living animals are metazoans. Early metazoan animals have been detected in Precambrian rocks. In

most instances, however, these structures have been wrongly interpreted. A supposed jellyfish found in the Grand Canyon of Colorado turned out to be the natural cast of a salt crystal surrounded by concentric fissures. On another occasion – more extraordinary still – a kind of giant scorpion was 'discovered' whose carapace consisted of plates of mud that had cracked and rolled up before being fossilized. This imaginary beast was also surrounded by false sponges and bogus corals.

In 1947, however, a geologist named R.C. Sprigg found a bed of Precambrian Period fossils in South Australia that was truly exceptional. Among the hills of Ediacara, Sprigg discovered a series of jellyfishes magnificently cast in the sandstone. At first he thought that the layers containing these specimens had been deposited at the beginning of the Palaeozoic Era, but it was later realized that they date back to the Upper Precambrian Period. In the absence of useful radioactive substances in the rocks, it is impossible to determine directly the age of these strata. However, their position indicates that they are about 600 million years old.

Sprigg's jellyfishes were followed by equally spectacular finds. Fossil hunters at Ediacara also found types of algae that resembled large feathers, and segmented worms, as well as their tracks – indicating that these creatures had been fossilized where they had lived. The strata also revealed other organisms having no association whatever with known fauna. The feature common to all the animals was the total absence of any limestone or silica-based skeleton except for a few tiny needle-like spikelets. All these soft creatures had owed their preservation to a very special 'burial' process. Imagine a shoreline where dead worms and jellyfishes sink into the mud in shallow water. A layer of sand covers them quickly, moulding their every detail. Hundreds of millions of years later their imprints are found at the base of sandstone banks.

It was not because they lived in very deep water that fossil metazoan animals were not found prior to the explosion of forms at the base of the Palaeozoic Era, but quite simply because their consistency made fossilization practically impossible. The palaeontologist Martin Glaessner even speaks of the 'age of the jellyfishes', when describing the animal world which existed at the end of the Precambrian Period. Without the unique fossil beds found at Ediacara we would be almost completely ignorant of Precambrian metazoan animals. This veritable mine of extraordinary fossils allows us to fill the enormous gaps in our knowledge about the early stages of evolution, but since the animals of Ediacara are already quite complex creatures, they must have been preceded by much simpler creatures of which palaeontology has so far found not a trace.

The fauna of Ediacara is rich in many species, with the jellyfishes representing several orders that have now completely disappeared. Some, like *Conomedusites*, had a body which is arranged in four-fold symmetry. The kinds of feathers which Sprigg took to be algae were in fact associated with *Rangea* and *Pteridinium*, both of which were discovered before the First World War by German geologists in South-west Africa. Later on, the same type of animal, *Charnia*, was found in England. These forms (see page 57) are like modern soft corals.

Various segmented worms were also found. *Spriggina* had a flattened body some 4 to 5 centimetres (1 to 2 inches) long, and a massive head bearing a large, pointed protuberance on either side. Its appearance makes one think it could be a potential ancestor of the trilobites, a group of arthropods that we shall discuss in the next chapter. While *Spriggina* was not unlike the marine worms of today, *Dickinsonia* was far more intriguing (see pages 56 and 57). It was shaped like an oval disc, subdivided into a number of thin segments, and the furrows which separated them converged on the central axis of the creature. The general appearance of *Dickinsonia* resembled that of a worm, although it is generally thought of as being a very special type of jellyfish.

Creatures like *Praecambridium* (see page 59), *Parvancorina* and *Tribrachidium* resist all comparison with modern animals. Apart from the fact that *Parvancorina* is frequently found folded or deformed, and must therefore have been fairly soft bodied, we have nothing to indicate the precise nature of this creature which was, perhaps, the larva of a trilobite. Some experts thought they saw in *Tribrachidium* a kind of very primitive echinoderm (a group of invertebrates that includes the sea-urchins and starfishes). But the echinoderms have a five-fold symmetry and not a three-fold symmetry, and so cannot be related to them. The species remains a mystery.

At Ediacara, tracks and burrows point to the existence of additional creatures that are

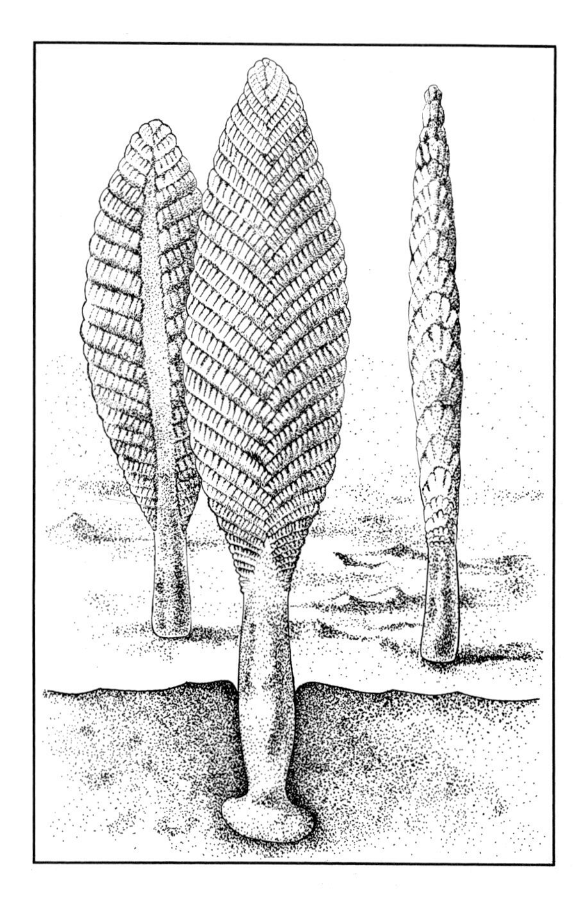

Rangea found in Australia and Namibia, like *Charnia* of Great Britain, reminds us of modern sea-pens whose arms were joined together. *Charnia* had a small disc at the end of its peduncle. These creatures may have evolved from a small, free-swimming jellyfish which became anchored to the sea-bed (approximately one-quarter size).

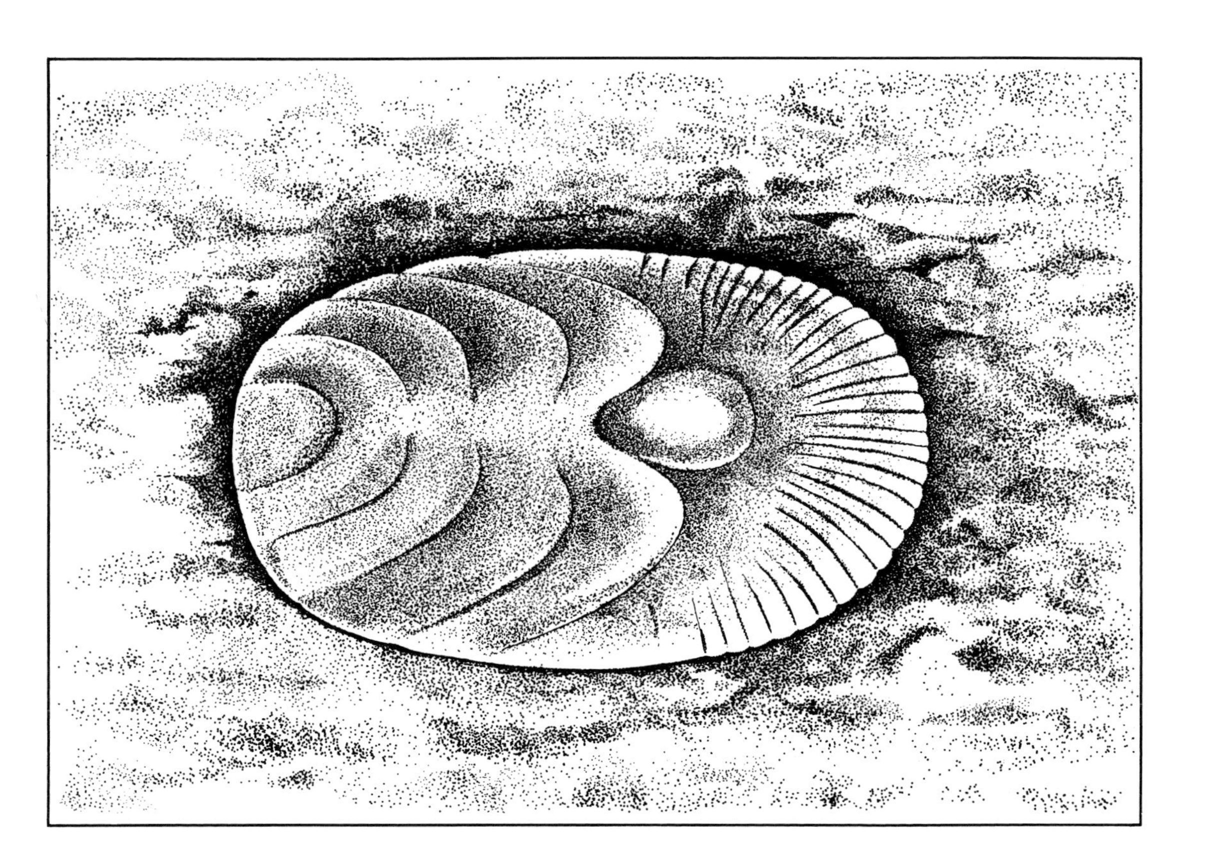

Praecambridium sigillum from Ediacara never grew more than 5 millimetres in length. A closely related form, *Vendia sokolovi*, has been found in Upper Precambrian Period rocks of northern Russia. They are both sometimes thought of as primitive arthropods (segmented animals) still without shells and related to the trilobites.

not themselves fossilized. For instance certain isolated spicules (lime or silica particles forming the skeleton of a sponge) found in the sandstone indicate the likely presence of sponges. Finally, the picture is completed by the almost certain existence of both algae and single-celled creatures.

The early Cambrian Period was to see the sudden appearance of a whole multitude of new species, and these evolved in abundance. The most important feature of this was the development within many groups of shells and skeletons. This meant that a large number of creatures would now be preserved when dead, for these hard parts would remain. We do not know for certain why so many different creatures suddenly acquired hard protective body parts, but it may have been the result of a change in the environment – such as a drop in the acidity of the oceans. Whatever the reason, it is likely that this new protection gave their owners such an advantage in the face of natural selection that the new characteristic spread extremely fast through the animal kingdom.

AUGUSTANA UNIVERSITY COLLEGE
LIBRARY

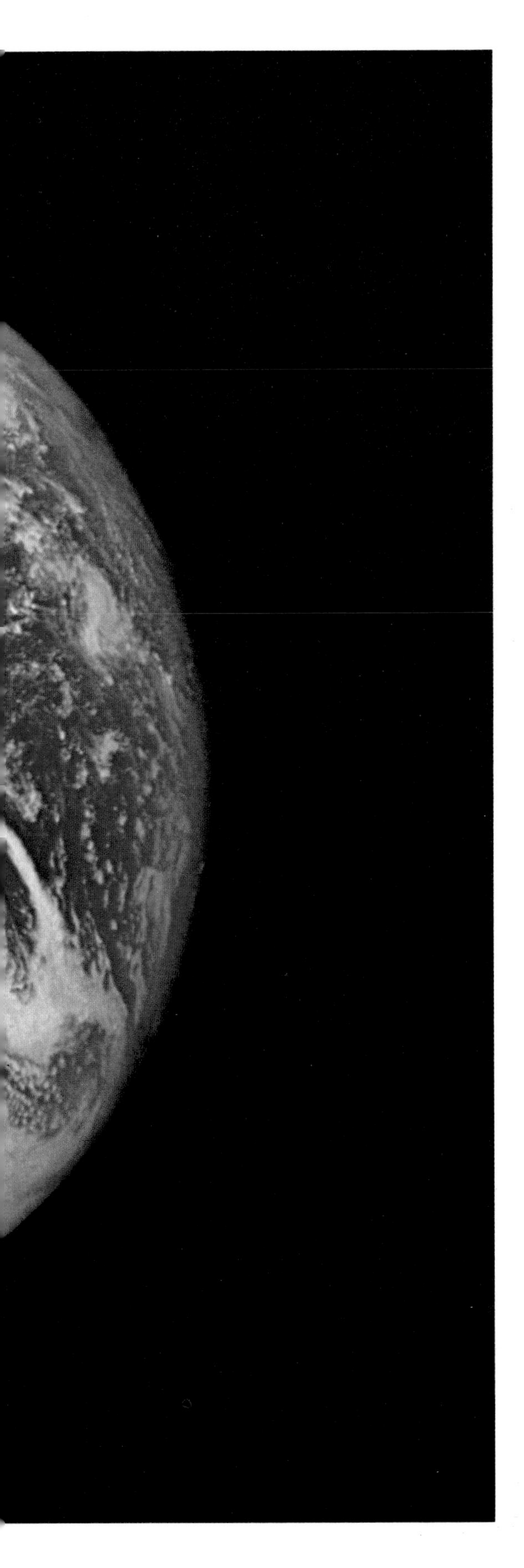

Chapter Three

The World of Water: Invertebrates

The world of water: invertebrates

The blue planet

Who, on seeing the very first pictures of Earth from space, can have failed to have been struck by its appearance: so strange and yet so like our classroom globes? While it was at times difficult to make out fragments of the continents through the cloudy whorls of atmospheric depressions and anticyclones, we were all aware of the immensity of the areas of water that give our planet its familiar deep blue colour. No less than 70.8 per cent of the World's surface is covered by water in fact.

We have already seen that the marine world formed the environment in which life started to develop, the environment which it first colonized and which was always most receptive to it. Life's progress under water was prodigious both in terms of the numbers and variety of the organisms born there. Furthermore, a long time after the conquest of dry land, most of the groups which had adapted to life on land saw some of their kind return to their original watery world. This was the case with a group of mammals which included the largest animal that the World has so far seen – the blue whale. Life has never completely cast off its origins; each one of us starts his or her own existence surrounded by the warm amniotic fluid of our mother's placenta, a fact which reminds us that life began in water.

The Cambrian Period marks the beginning of the time when fossils become relatively plentiful. It extended over some 570 million years, roughly one ninth of the World's age according to our estimations. Nevertheless, the presence of numerous groups of animals from the end of the Palaeozoic Era would seem to point to a very long period of evolution before that: the fauna of Ediacara proves this point. Although 570 million years may seem modest compared with the Earth's total age, it is still a fantastically long time, and difficult to comprehend. All the same, an idea of the enormous timescales involved is necessary if we are to begin to understand the mechanisms whereby geological changes brought about the upheavals in the Earth's crust and how, too, the process of evolution has been able to create quite complex creatures out of the most elementary forms. Let us imagine a walker on the road of time, and that each step he takes represents a human lifetime. Our walker would have to go back 20 paces to find Julius Caesar conquering the Ancient Britons. But the origins of humanity would be nearly 32 kilometres (20 miles) further back, while the beginnings of the Cambrian Period would be as far as the distance between Britain and Afghanistan!

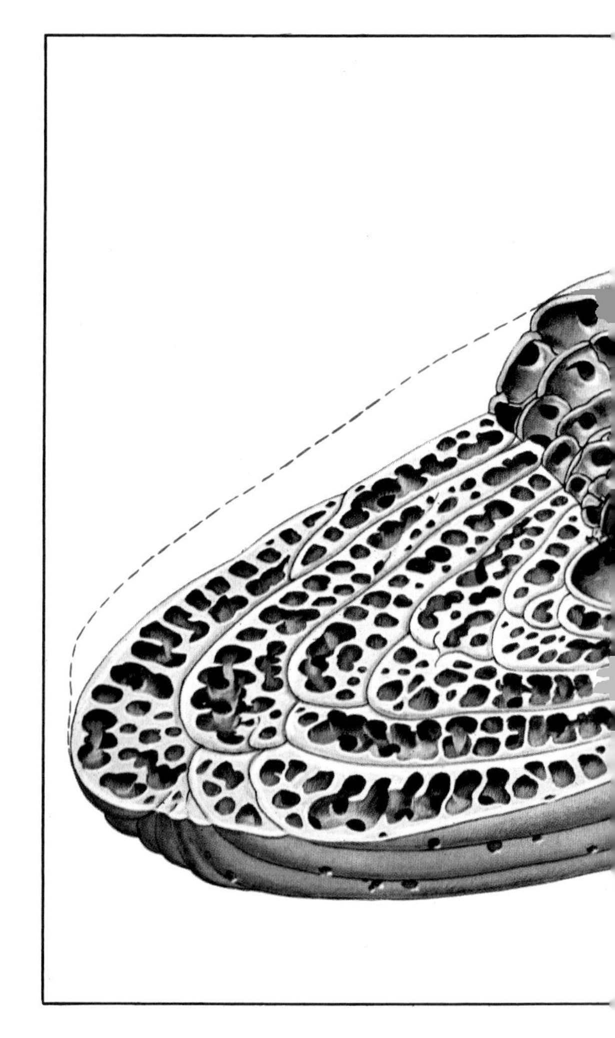

Single-celled creatures

The invertebrates were among the first creatures to colonize the marine world, the simplest of these being the protozoans, animals whose bodies consist of a single cell. The foraminiferans and the radiolarians were the two most important groups, at least in terms of the part they played in the formation of certain rocks, although a group of single-celled algae – the coccoliths – have also been discovered.

The foraminiferans have been the subject of much study owing to their immense stratigraphic interest; they are known from the Cambrian Period onward and are often most abundant within sediments which, if loose, have yielded up their fossils through washing and screening. In some cases they have been discovered by using the 'thin-slice' technique described in the first chapter. In modern nature these minute creatures are found almost solely in marine waters. Some are 'benthic', meaning that they dwell on the sea floor; while others, even smaller, are 'planktonic' or 'pelagic', in other words they float close to the surface or are suspended in the midde layers of the oceans. As shown in the illustrations on this page and page 65, these organisms are protected by a kind of skeleton called a 'test', and it is this test that is preserved by fossilization, presenting a huge diversity in size and shape.

There are some planktonic species that never exceed a few hundredths of a millimetre in length; at the other end of the scale are the benthic foraminiferans that can attain a size of several centimetres. The tests consist of chambers that frequently form a coil. Their external surface reveals an extraordinary variety of appearance due to the

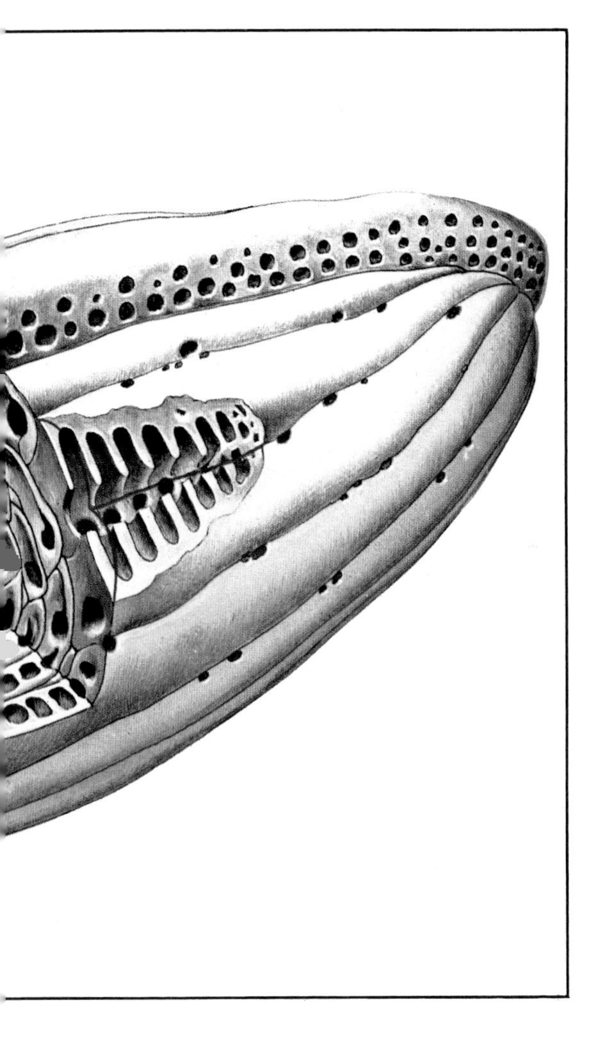

From this cutaway view of an alveoline test we can see that the foraminiferans had very complex skeletons. The test can be 8 to 10 centimetres (3 to 4 inches) long and is arranged around a central chamber in a series of larger and smaller cavities. Other protozoans like the amoebae may look primitive, but these organisms with their delicate shapes appear very specialized.

Who would think that a chalk mark magnified some 12,000 times under the electron microscope would look like this? Coccoliths, sometimes arranged in a sphere, are mixed with the remains of other organisms like the nannoconids to form nearly all of this rock. On the right we see some modern foraminiferans (1 to 3) and radiolarians (4 to 6). These creatures, especially the foraminiferans, also played a major part in the formation of sedimentary rocks. Today some 130 million square kilometres (5,019,300 square miles) of ocean floor are covered by muds containing globigerines, a planktonic foraminiferan.

different types of ornamentation (such as points, grooves, outgrowths and fluting), and due, too, to the very nature of the test itself. The latter is sometimes chitinous and sometimes made of mineral substances, in which case it may be made of silica, compacted masses of alien particles or even different forms of limestone: granular, fibrous, porcelained, and so on. Countless species have been found illustrating this huge range of shape and form.

One of the most original features of this group is its relatively complex way of reproducing – a peculiarity that was first revealed not by a biologist but by a palaeontologist. At the end of the last century it was noticed that a single species of foraminiferan had two distinct forms: one characterized by a large central chamber, the other by a small one. There were also other differences concerning the animal's overall proportions. Close observation of modern foraminiferans showed that these two types corresponded to two types of development: those with the small central chamber being the result of sexual reproduction, while those with the large initial chamber had come about by a sort of budding process. Even more curious, it was found that both types alternated their method of reproduction between sexual and budding down the generations.

The radiolarians on the other hand surround themselves with a silica shell, like an opal in shape and decorated with a magni-

1

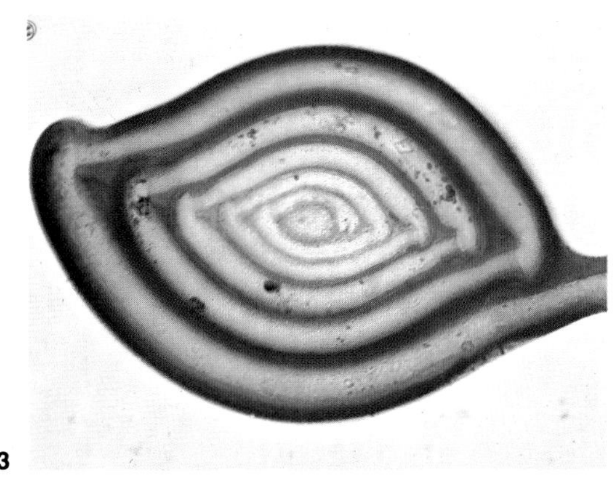

3

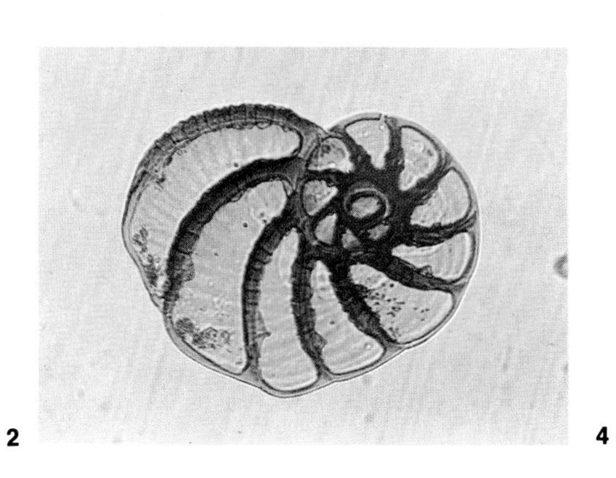

2

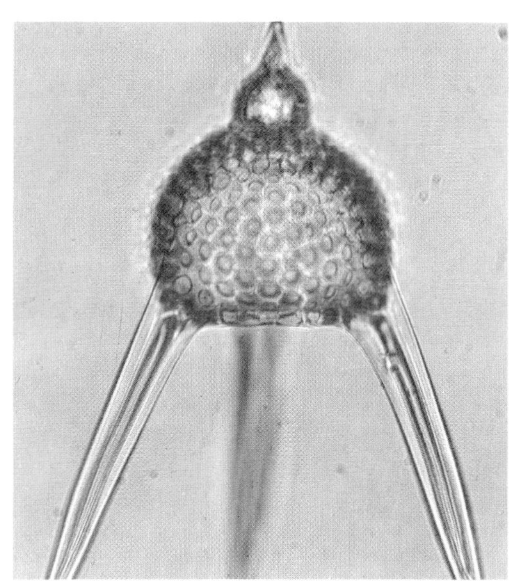

4

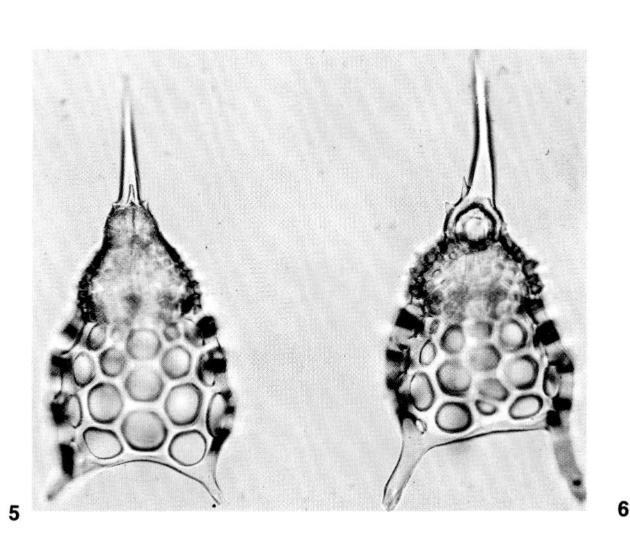

5

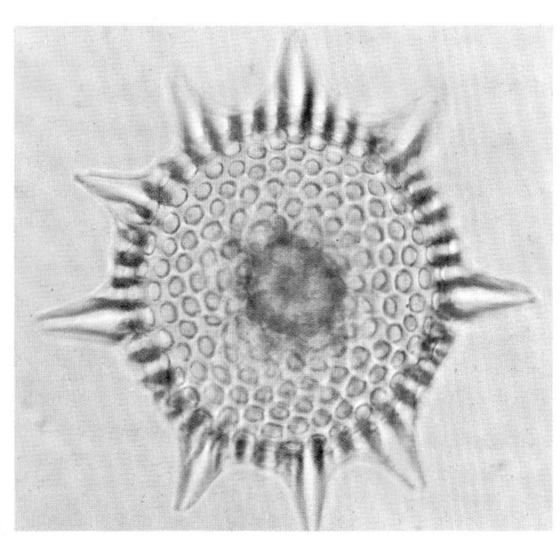

6

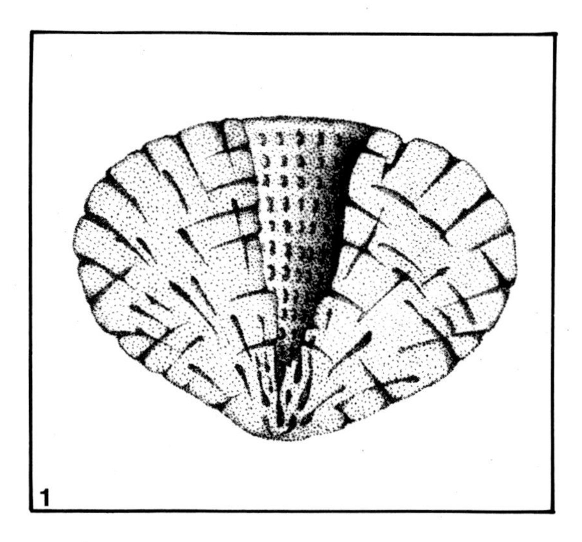

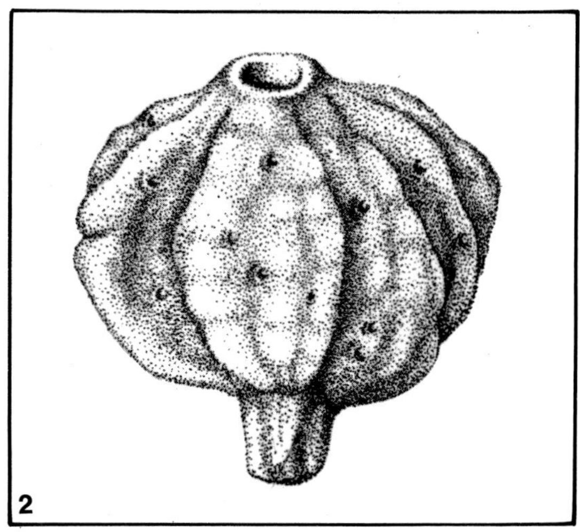

ficent array of points, hooks and perforations through which protrude the creature's long, arm-like body extensions, which are both rigid and viscous. The well-known German biologist Ernst Haeckel called them 'the miniature jewels of the deep' (see page 65). The multiplication of these splendid planktonic animals was evidently favoured by the enrichment of their environment with silica, and thus it was that the deposits in which the radiolarians are plentiful were frequently accompanied by submarine volcanic discharges. They appeared perhaps as early as the Precambrian, and their silica tests fossilized very easily thanks to their ability to resist becoming dissolved. These microfossils are found in abundance, especially in sediments deposited at very great depths.

The coccoliths are single-celled flagellate algae that possess a little 'whip' that they beat in the water in order to move around. They, too, are enclosed by a protective shell which is so small – a few microns only – that the electron microscope is its only feasible method of observation. Its limestone envelope is in actual fact an assembly of 'coccoliths' or minute rosettes of calcite crystals (see page 64) which become detached after death and are often found in the fossil state, at least from the Jurassic Period onwards, indicating warm or temperate seas.

As we stated earlier, the foraminiferans, the radiolarians and the coccoliths were active in the formation of certain rocks, and a large part of the most magnificent monuments of Paris are made of Eocene limestones moulded by milioles, which were millet-shaped foraminiferans. Another foraminiferan, but a larger species, is the basic component of certain limestones: this is the nummulite, whose appearance resembles a coin. These limestones are also called 'liard stone' from the name of an ancient currency, while the pyramids of Gizeh in Egypt have also given their name to a species of nummulite (*Nummulites gizehensis*) that is found in many of the massive blocks of those great structures. Strabo, the geographer and historian of Ancient Greece, criticized the common interpretation of these fossils as being the petrified remains of lentils fed to the slaves who toiled on the construction of the pyramids! It is also worth noting that the nummulites hold the record for size among the foraminiferans, since some Eocene types grow to as much as 12 centimetres ($4\frac{1}{2}$ inches) in length.

The coccoliths and other related forms constitute the basic material of chalk, but they are also present in other lime-rich deposits. The radiolarians contributed to the formation of rocks such as jasper, chert and, of course, the 'radiolarites', a basic list to which we could add petroleum, whose origins lie in the sedimentation and transformation, within an almost oxygen-less environment, of vegetable plankton, foraminiferous debris, minute crustaceans and larvae of all kinds. In the form of liquid hydrocarbons this organic matter – stockpiled within the 'mother rock' – migrates towards the porous 'carrier rock' where it may be trapped.

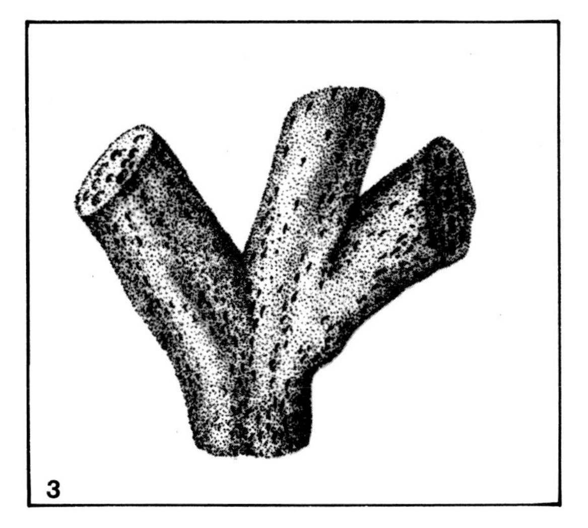

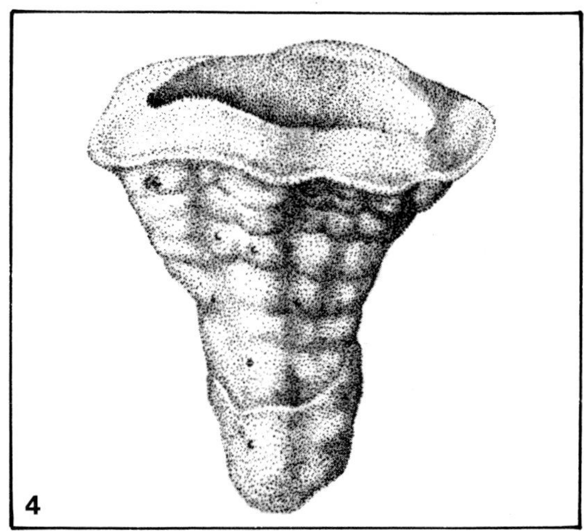

Because the lithisid sponges had mineral skeletons they fossilized in complete units. Sometimes they were rather bowl-shaped like *Aulacopium* of the Silurian Period (1). Some were bulky and pedunculate such as *Hallirhoa* of the Upper Cretaceous Period (2), or branched like *Doryderma* of the Carboniferous to Cretaceous Periods (3). *Chenendopora* (4) looked like a thick-walled trumpet. *Plocoscyphia* (5) is yet another type, a meandrosponge hexactinellid of the Cretacous Period. Most fossil sponges belong to these two groups.

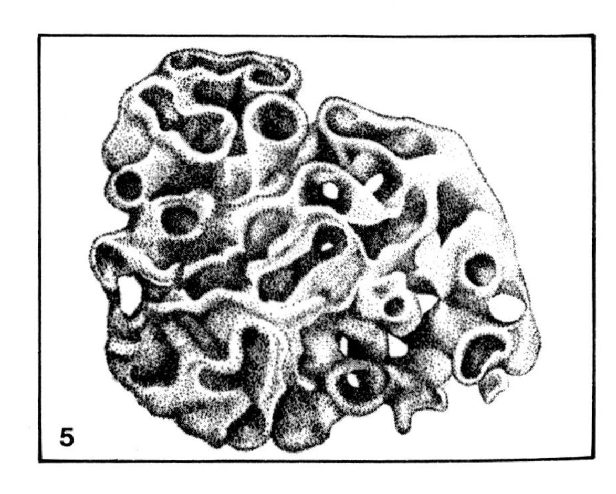

The sponges

The sponges occupy a rather special place between the protozoans and the more advanced forms of animal life. Right up to the 19th century sponges were thought of as plants although the way in which they are built is rather similiar to a sort of colonial protozoan. Many scientists set them apart from the higher animals (the metazoans) and place them in their very own subkingdom: the Parazoa. The porous nature of sponges is well known, their other scientific name being Porifera. A living sponge can be regarded as a sort of fleshy bag supported by a skeleton and perforated by thousands of little holes, or pores. Using special cells provided with a whip-like flagellum, the sponge draws in water through its small pores, takes from it tiny food particles, and then expels the remainder through its larger pores. This is a very simple way of breathing and feeding. This living filter has to process enormous quantities of water in order to survive and develop, and some of the sponges found in the Bahamas get through a massive 2000 litres (440 gallons) of water every day!

The most primitive types of sponge have a central chamber covered with feeding cells, while in the more complex forms these cells are distributed among a huge number of small cavities. Our bathroom sponges belong to the latter type, and they represent an exception in so far as their skeleton, composed of a material called spongin, is completely horny and non-mineral (which is why we are able to use them). Generally speaking, however, the rigid lime-rich or silica-rich elements of the sponges are there to support the soft tissues, and the nature and form of this skeleton is taken as the basis on which the sponges are classified. We can distinguish between the lime-rich sponges with their triaxon calcite spicules (in other words those having a triple axis), and the

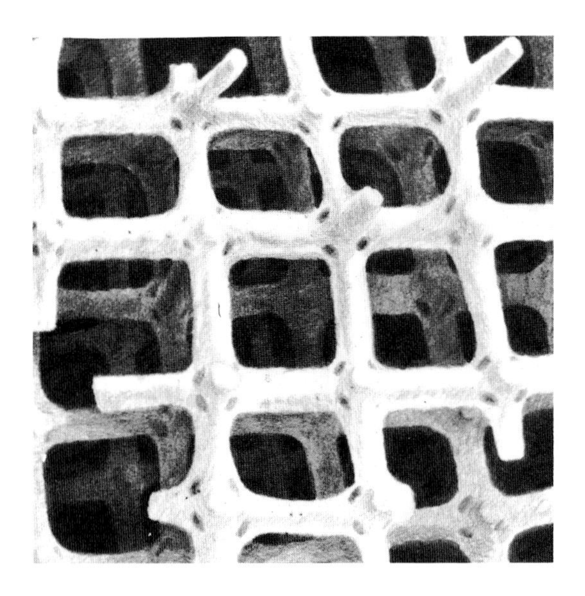

Two spicule networks of the hexactinellids: above, the cubic mesh of the Miocene Period *Craticularia* magnified some 50 times; below, the irregular pattern formed by *Proeuplectella* of the Cretaceous Period magnified about 25 times. These two types of arrangement – the one regular and tough, the other haphazard and fragile – separate the hexactinellids into two groups: the dictyonins and the lyssacins. The modern *Euplectella* ('Venus's flower basket') is a silica-based sponge from the Pacific Ocean and belongs to the second group. Its magnificent latticework structure is closed at the top by a perforated disc, and has been found with two shrimps inside this natural cage. Perhaps they entered this cage when young and could not escape later. This gave rise to an old Japanese custom where young married couples were given one of these sponges complete with its dried-up shrimps as a symbol of a lasting marriage.

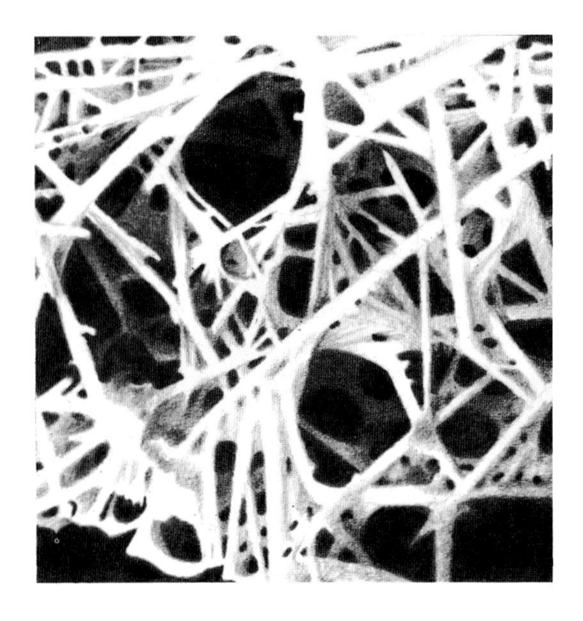

'siliceous' sponges that consist of spicules having a single, double or triple axis and spongin. Some of these sponges have only spongin or silica spicules.

With the exception of a few freshwater varieties, the sponges are most widespread in the oceans of the World, where they lead a static life, anchored to rocks or the drifting seabeds, and sometimes 'hitching a ride' on the shell of a mollusc or crab. Many species grow in large colonies where it is hard to tell where one individual finishes and the next begins. Sponges live in warm, fairly shallow waters where they present superb hues of bright red, orange and even green. They are also found in temperate and even cold zones and at depths of up to 7000 metres (22,960 feet). The major deeps covered in mud are the home of the hexactinellid sponges with their completely silica skeletons (see left). These sponges have a radial symmetry and look just like great cylindrical sleeves, funnels or bowls whose spicules form a magnificent network of finest 'lace'.

Those sponges with a mineral skeleton are obviously the only ones that stand a chance of being preserved through fossilizaton, and the most compact may appear in rocks in complete units. Often, however, only isolated spicules are found inside the sediment, and a large number of such items will probably never be identified. The outline of an entire sponge, whether living or fossil, is of limited interest only to scientists attempting to identify a particular species. These generally irregularly shaped organisms are actually very plastic; certain species are very variable, and some species may possess a large degree of external similarity. The palaeontologist is therefore most interested in the geometry of the spicule network and its mineral nature.

If we separate the archaeocyaths (see page 69) from the sponges, the first limestone sponges appear in the Devonian Period, while silica-based sponges are first found in the Cambrian Period. The latter are especially susceptible to treatment by acid in carbonated sediments and the finest, most delicate structures can be retrieved in this fashion. These structures are by far the most common to be found in the fossil state, in particular the lithistids which owe their relative abundance to their very robust skeleton. The spicules are encrusted with additional deposits of silica that often give the creature a stone-like appearance of great variety (see pages 66 and 67). Sometimes these lithistids, related to the hexactinellids, have formed veritable reefs like those, for example, found in the Upper Cretaceous of Provence, in France.

From time to time we have found isolated spicules from silica-based sponges that have accumulated in enormous masses forming their own individual rocks: the so-called 'spongoliths' and 'gaizes'. The sponges are considered to be fairly accurate indicators of the environment, and the succession of lime-rich sponges through lithistids to hexactinellids corresponds to an increase in depth. It would appear, however, that the last of these groups had, in the course of geological time, migrated towards increasingly deeper waters.

Cnidarians

The cnidarians are metazoans whose organizational plan is clearly more elaborate than that of the sponges, with cells tending to group together to form actual tissues. Cnidarians include the jellyfishes, hydras, anemones and corals. Cnidarians are typified by particular types of cells called cnidoblasts which are dispersed over the surface of the body. Cnidoblasts are defence organs which comprise a tiny capsule capable of injecting venom beneath the skin of its prey or of an aggressor. Anyone who has ever been stung by a jellyfish will know how painful this venom can be.

This group of creatures is of particular interest both in living nature where there are some 9000 different species, and in the fossil state. Cnidarians possess a radial symmetry; a single, central chamber leads to the outside by an opening that is both mouth and anus at the same time, and is usually surrounded by tentacles.

Jellyfishes, or scyphozoans, are adapted to a floating, wandering life and are found in all the seas. The largest among them, *Cyanea arctica* of the northern oceans, can weigh 1 tonne and may be 40 metres (130 feet) long. They have no skeleton and are largely composed of water, but as we saw at Ediacara they may sometimes fossilize. Quite apart from the Ediacara fossils, the most famous are without doubt those described by Walcott in the Middle Cambrian Period of Alabama and those found in the Upper Jurassic Period of Solnhofen in Bavaria. In deep waters a dead jellyfish is destroyed very quickly, so the conditions most favourable to its preservation are when the animal gets washed on to the shore and is rapidly enclosed by very fine sediment. Such fossils can be truly exceptional in quality. The other groups of cnidarians, on the other hand, are fixed at a single spot and may either live singly like the sea-anemones, or grouped into colonies. They are frequently mineralized and represent a substantial proportion of fossil invertebrates.

Some of the cnidarians inhabit fresh water. One such creature is *Hydra*, a small,

The archaeocyaths are a rather confusing group that we sometimes find with the limestone sponges. Basically, their skeleton consists of two cones, one inside the other, and connected by radiating partitions. The whole structure has many holes in it, but the walls have none of the spicules we find in the sponges. They are only found in the Lower and Middle Cambrian Periods, but even so they built some large reefs and banks – rather like coral reefs today.

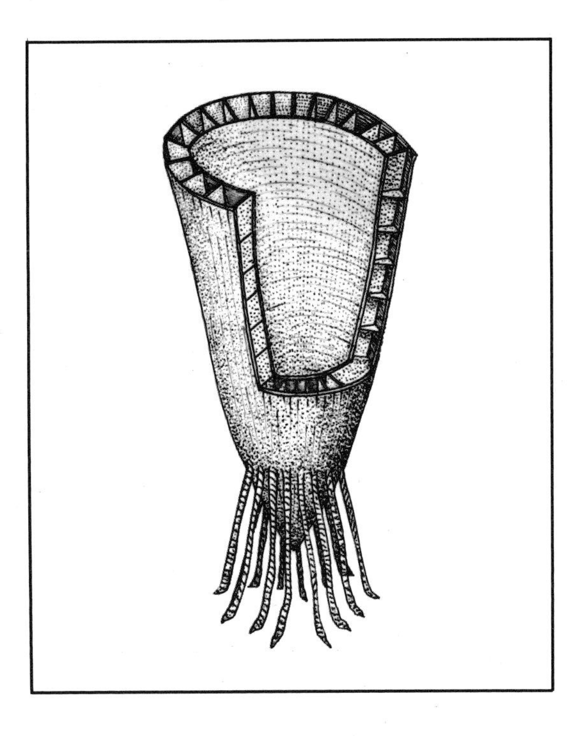

polyp-like animal that gets its name because of the astonishing power of regeneration that it possesses – just like the monstrous serpent of Greek mythology. It belongs to a group called the hydrozoans. Since they are rarely mineralized, fossil hydrozoans are known only by a small number of species; these are small forms living in colonies. The millepores, for instance, were built up by a stacking process of limestone layers interlinked by tiny pillars and perforated with pores that sheltered the polyps. Stromatopores, long thought of as belonging to the hydrozoans, are now linked with the limestone sponges. Known to us from the Cambrian Period onwards, their rounded forms have, over the course of time, produced whole reefs. They were especially common in the Silurian and Devonian Periods. During the Mesozoic Era the stromatopores became less common, to disappear completely at the end of the Cretaceous Period. The reason for their decline may well have been competition by other corals in the very same environment.

Mention must also be made of the Con-

In the Palaeozoic Era we find enormous reefs of tetracoralloids such as *Phillipsastraea* of the Devonian Period (below left) along with the stromatopores and tabulates. In the Mesozoic Era their place was taken by the hexacoralloids, an Eocene Period variety which is shown below right. The study of modern barrier reefs can help us to understand how fossil formations emerged. The hermatypical madrepore corals are very sensitive to depth and so a reef established at the foot of a volcanic cone will respond to a rise in sea level – or to the subsidence of an island – by rising higher and higher. The result of this process is that the volcano will eventually submerge leaving an atoll, a simple ring of coral surrounding an isolated lagoon in the middle of the ocean.

ularia, which are found from the Cambrian Period to the Lower Jurassic Period. These organisms have given rise to a number of different interpretations, some being similar in shape to jellyfishes but with an exoskeleton (external skeleton) made of calcium phosphate. *Conularia*, the most typical genus, was quite different, however, looking like an elongated cone with a radial symmetry of the order of four. Young specimens lived attached by the pointed end, the spines of which held it fast to the mud or even to the phosphate covering of adults, but these creatures are again not often found in the fossil state.

The third group of cnidarians, the anthozoans (the name means 'flower animals') underwent considerable development and are best represented in those geological formations where they were often capable of spectacular mineral production. Their body cavity is split up by radiating partitions, known as septa. Among the anthozoans, only the alcyonians and the stony corals have any interest to the palaeontologist.

The alcyonians include the red coral, seafan and the sea-pens still found today, and

The bryozoans were small organisms that lived fixed, colonial lives. As the illustration below shows, they have a more complex structure than that of the cnidarians and the sponges. They had a separate mouth and anus and their tentacles bore 'lashes' or cilia. Each individual animal lived for a very short time and was surrounded by a kind of gelatinous, horny or chalky envelope. They formed colonies of many shapes: encrusting, tree-like or even like a sail, as shown opposite. Some dwelled in fresh water but most lived in the seas and they have been there since the Cambrian Period, preferring choppy waters and firm substrates for a good foothold. Bryozoans have been found in water as deep as 6000 metres (19,680 feet), but the majority come from waters between 10 and 500 metres (33 and 1640 feet) deep. Sometimes they form 'meadows' on the sides of reefs where they retain particles of limestone and encourage the growth of the reef, building up considerable deposits. One of the best-known is the *Fenestella* limestone of the Franco-Belgian Lower Carboniferous Period. Some bryozoans tend to live with other animals filling the spaces that separate the calyxes of the madrepores, for instance. They are also sometimes found with molluscs and worms. The bryozoans are identifiable from small fragments and are important in the fossil record.

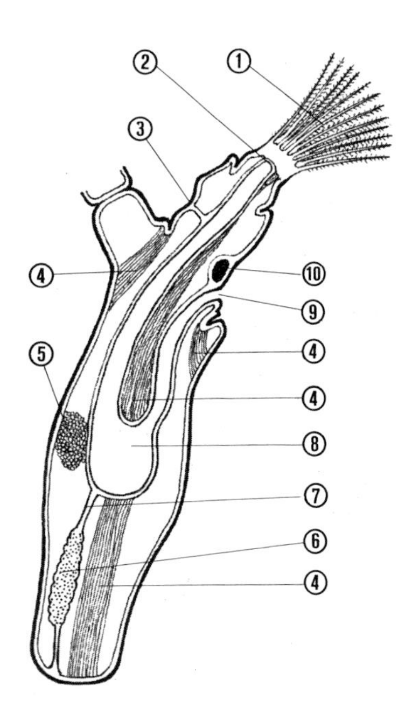

The parts of a bryozoan

1 Tentacle
2 Mouth
3 Diaphragm
4 Retractile muscles
5 Ovary
6 Testis
7 Funicle
8 Stomach
9 Anus
10 Nerve ganglion

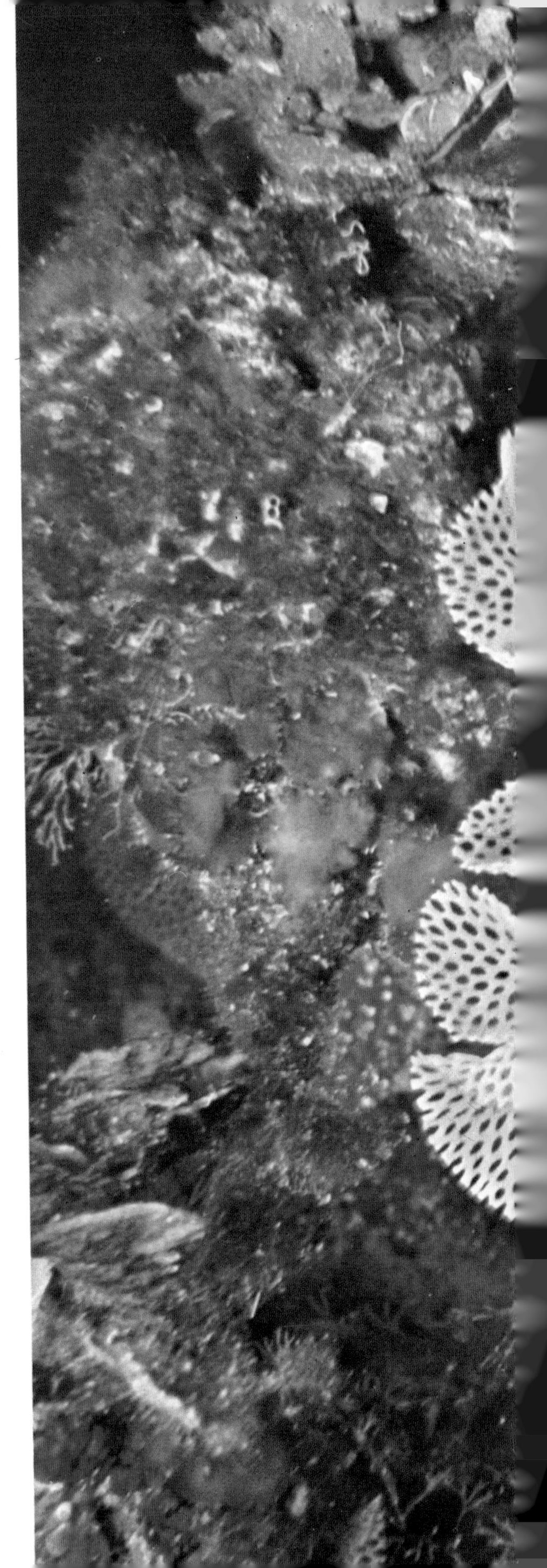

have been found from the Silurian Period, although their appearance among fossil traces is scant. One of them is *Cancellophycus*, which seems to have traced brush strokes across the surface of rocks. Long attributed to algae, these imprints conceal spicules that are closely related to those of modern sea-fans.

The stony corals, or Madreporaria, on the other hand, possess more substantial septa according to whose initial number we can make a distinction between the Tetracorallae, which appeared no later than the Triassic, and the Hexacorallae that followed them and which are also linked with modern corals. The Madreporaria are not only colonial; solitary species exist as well. The form of the colony can even vary within the same species depending on the depth at which they build up, being massive in waters whose surface is subjected to the action of the waves, and more delicate at greater depths. The Madreporaria are the most effective building organisms to have existed. Just take a look at any atlas: the Australian Great Barrier Reef, a 3000 kilometre (1864 mile) long structure of Hexacorallae that attains widths of up to 150 kilometres (93 miles). It has been estimated that one third of the weight of the limestone on the surface of the globe has been produced by these animals!

Present-day corals are divided into two groups. The first group – the reef-builders – live in symbiosis with the microscopic algae that inhabit their bodies and which are essential to them, probably due to their role as eliminators of waste products. The presence of the algae, which need light, imposes strict limitations on the environment in which the reef-building corals can grow, and they certainly do not survive at depths of over 50 metres (164 feet). They also require clear waters that are well aerated, of moderate salinity and very warm (22°C/71.6°F average for the coldest month). The second group of corals are less demanding and inhabit much deeper zones, down to 5000 metres (16,494 feet).

Brachiopods

Specimens of this group of creatures are among the most commonly found on any geological excursion into the marine terrain of the Palaeozoic or Mesozoic Eras. It was at this time that the brachiopods developed most abundantly, although they have continued to exist to the present day. Remarkably tolerant as regards their habitat, they have adapted to all kinds of environment, although they are found most frequently in temperate or warm and shallow waters.

Their shell, formed by the conjunction of two always unequal halves, or valves, is attached to the substratum either by the ventral valve itself or more frequently by a muscular stalk called a peduncle, which is implanted through a hole located beneath the narrow, arched part of the ventral valve (see below). In the most simple specimens,

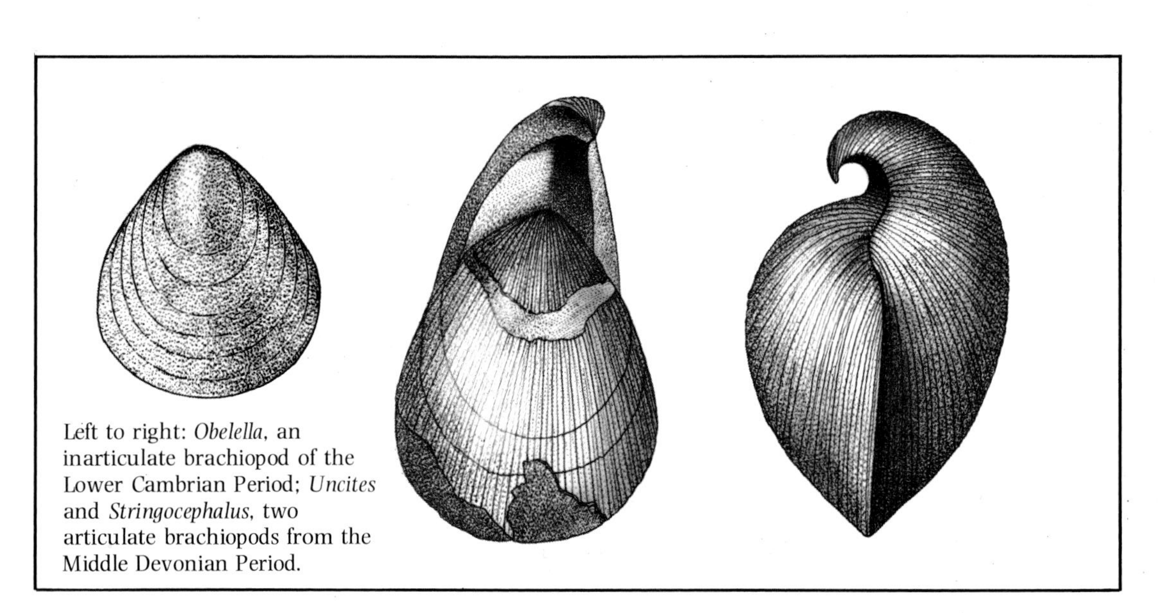

Left to right: *Obelella*, an inarticulate brachiopod of the Lower Cambrian Period; *Uncites* and *Stringocephalus*, two articulate brachiopods from the Middle Devonian Period.

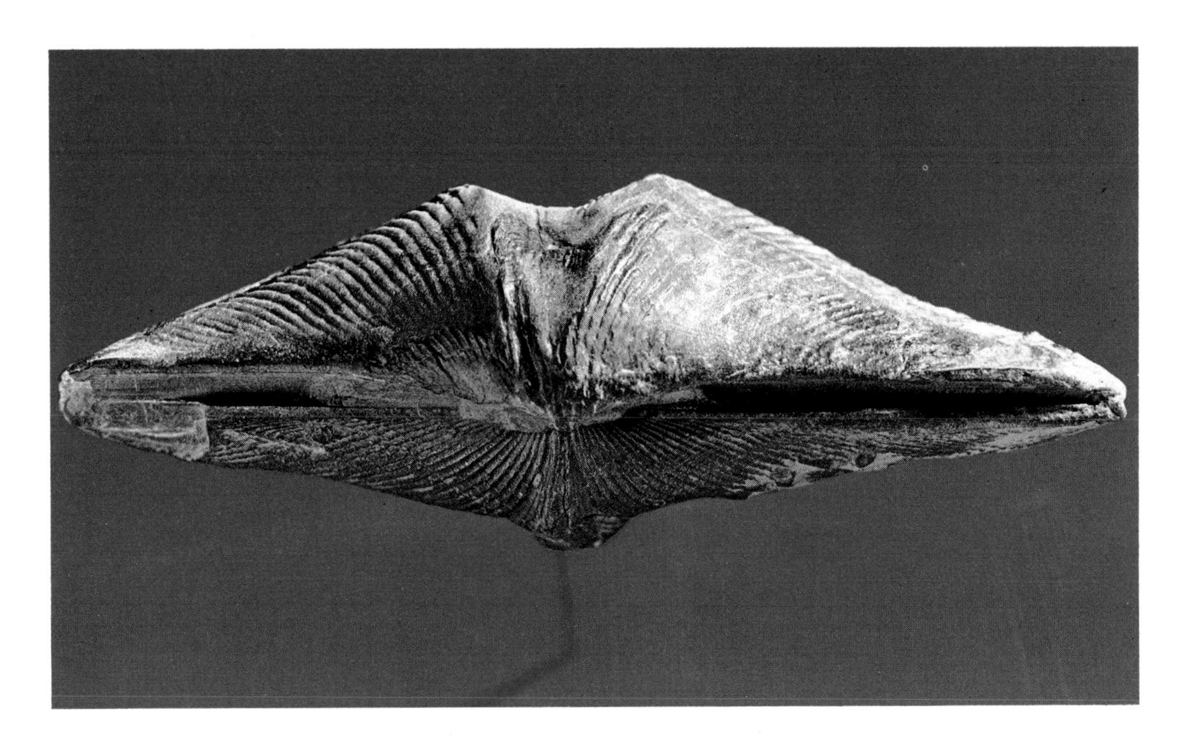

known as the inarticulate brachiopods, both valves are joined by a rather complex muscular system which makes them very independently mobile. This is the case with the modern genus *Lingula*, an animal that is often quoted as an example of a 'living fossil', since it has remained practically unchanged since the beginning of the Palaeozoic Era (see page 38). It possesses a chitinous-phosphate shell and burrows holes in the sand that can be tens of metres deep. Its retractable peduncle is around 20 centimetres (8 inches) long and is attached at the end of its hole to a plug of mucus and compacted mud.

The articulate brachiopods are much more numerous and have lime-rich valves interconnected by a system of meshed teeth and sockets, with the valves pivoting around this structure, while the muscles activate the opening of the shell. This cohesion explains why fossil articulate brachiopods nearly always have their two valves still joined. The brachiopods are provided with a characteristic organ called the lophophore, which is used for their nutrition and respiration. The lophophore consists of two hollow arms bearing tiny hairs called cilia. In the articulate brachiopods the arms are supported by skeletal components shaped like narrow bands or spirals. This internal skeleton, discovered thanks to the technique of series

The genus *Spirifer* was one of the most widespread brachiopods of the Upper Palaeozoic Era. It is identified by a skeletal support of the lophophore with well-developed spirals which stretch their tips either side towards the edges of the shell.

polishing or directly observed on exceptionally well-preserved specimens, has been used to establish the classification of the articulate brachiopods.

The brachiopods preferred to live on solid sea-beds or limestone muds; in the latter instance a branched peduncle was necessary for it to cling securely to the unstable substratum. Yet another method of attachment was adopted by *Productus* of the Carboniferous and Permian Periods, whose ventral valve was decorated with long spines that directly anchored the shell to the sediment. Next to this genus we find *Richtofenia* of the Permian Period, a deviant form whose fixed valve was transformed into an elongated lime-rich cone closed by a dorsal valve modified to form a kind of lid. The adaptation to reef-life shown by *Richtofenia* entailed an extraordinary morphological convergence with the madrepores and certain attached molluscs called the Rudista (see page 80). Finally, some brachiopods also attached

Helix, a terrestrial form, *Cerithium*, a dweller of brackish waters, *Patella* of the rocky coasts and *Aporrhais* of the sandy sea floors illustrate the variety of form and ornamentation of the gastropods and how they have become adapted to living in different places. The growth of the spiral occurred in such a way that it allowed them to grow while retaining their proportions. Their shells are mainly sinistral, (the opening is on the right when the creature is looked at with its tip upward). Dextral (opening on the left) gastropods are, however, known from the Palaeozoic Era.

themselves to other living organisms such as algae or crinoid echinoderm stalks (see page 97).

Molluscs

The term mollusc comes from the Latin *mollis*, meaning soft, and in fact the typical mollusc is a soft-bodied creature surrounded by a fleshy 'coat', which is capable of secreting a hard shell. Molluscs possess a basically bilateral symmetry which, in certain groups, is altered by the body undergoing twisting and coiling. A head may be discerned, also a visceral mass (the body organs) and a foot. The foot is a flattened ventral surface that allows the creature to move over solid ground. There are a number of exceptions to this general rule since the molluscs constitute an immense group consisting of tens of thousands of highly diversified species, occupying all kinds of ecological niches. Some have adapted to free swimming in water, some lost their shells, some have become burrowers, certain species have returned to a fixed life, and they have generally inhabited the marine world as well as fresh water. Several species, such as the slugs and land snails, have finally succeeded in living on land and breathing air. All these adaptations have been reflected by a group of animals which, if not represented by as many species as in their heyday, still boast a number of living species second only to the arthropods. Individually, molluscs can attain sizes unrivalled among invertebrates, even competing with some vertebrates on this level. The shells of the giant clam can weigh over half a tonne, while the great deeps of the oceans are the habitat of giant squids that can attain the size of a sperm whale.

As late as 1957, thoughts on the origin of the molluscs continued to be hypothetical. Then a small and completely unknown animal was discovered among the specimens harvested by a Danish ship, the *Galathea*, during a scientific expedition. Dredging operations at a depth of 3600 metres (11,800 feet) off the coast of Nicaragua had revealed ten complete specimens with a shell in the form of a Phrygian cap and which could only be associated with creatures thought to have been extinct since the Devonian Period! The internal anatomy of *Neopilina*, whose organs such as muscles, breathing and excretory system run longitudinally, demonstrate the original relationship between the most primitive molluscs and the worms, whose segmentation is very advanced. As we have seen, worms existed as early as the Precambrian Period, and so the separation of worms from their mollusc ancestors must be very ancient indeed.

Remains of molluscs abound at all epochs, since their shell lends itself particularly well to fossilization. Three classes are especially predominant: the gastropods, the lamellibranchs and the cephalopods.

Gastropods We have probably all seen these creatures, even if it be only the humble garden snails, which are very easy to observe. Here we have molluscs which have lost their bilateral symmetry, their shell being coiled about them in a spiral, deforming the visceral mass. During embryonic development their body undergoes a twist through 180 degrees which puts their anus and respiratory organs behind the head. But outside these phenomena of twisting and spiralling, they have remained fairly close to their ancestral forms and our earlier description of the typical mollusc applies just as well to them, too. A unique feature of the shell of the gastropods is that it is closed by a horny or calcareous lid when the animal withdraws inside. This lid cannot be thought of as a second valve (the two-part shell of the lamellibranchs) since it is not secreted by the shell but by a gland incorporated in the foot. Sometimes the spiral shell is coiled in a single plane only, and this is the arrangement found in the genus *Bellorophon* of Ordovician to Permian Periods, but more often a more complex spiral formation took place around one axis, sometimes with a secondary reversion to the flat coil, as with the planorbids found in our seas. Finally, there are instances where the shell has uncoiled, sometimes in huge individuals of a normally coiled species, or sometimes in a systematic manner throughout the whole group. This latter case is clearly illustrated by the genus *Vermetus* of the Tertiary Period to modern times, whose deformation corresponds with a sedentary way of life.

Other types reflect a very advanced regression of the limestone skeleton, for instance the slugs or *Aplysia* (the sea-hare), while in the nudibranchs the skeleton has completely vanished. There are on the other hand certain species that were prodigious manufacturers of calcium carbonate: the great Neritidae of the Upper Jurassic and

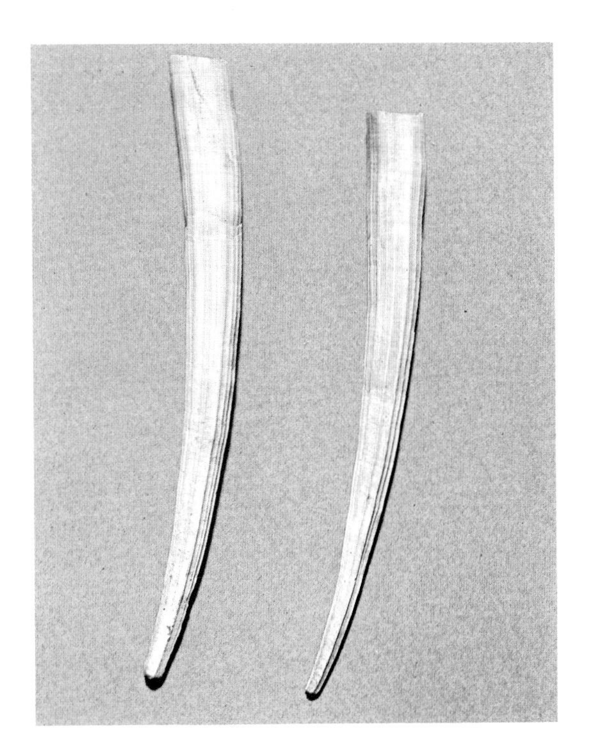

Left: The scaphopods make up a class of molluscs of which the typical genus is *Dentalium*. It only contains two families but this group is quite frequent in fossilized deposits of the Cainozoic Era.

Below: We can only rarely see the original colouring of a fossil – here is an exception: *Natica tigrina* of the Pliocene Period (Castell' Arquato, Piacenza, Italy).

Cretaceous Periods fabricated shells with such thick walls that they probably scarcely had the strength to move at all. A very large creature from the Eocene Period, *Campanile giganteum*, attained a length of 50 centimetres (20 inches) with its superb elongated cone decorated with tubercules. In the present-day two species exceed lengths of 60 centimetres (24 inches): *Pleuroploca gigantea* and *Megalotractus auruanus*.

The gastropods are the only molluscs that have succeeded in adapting to life on dry land; their system of breathing organs has enabled them to become air breathers. This development probably occurred towards the end of the Palaeozoic Era and is worthy of particular note since – as we shall see – only the arthropods and the vertebrates actually succeeded in colonizing the continental land masses. The acquisition of a lung very different to that of the vertebrates has been observed in various families of gastropods, in particular the pulmonates whose habitat is continental but not solely terrestrial. Like the slugs and snails, the planorbids and *Lymnaea*, found as fossils in lake limestones, use oxygen from the air and have to come to the surface of the water in order to breathe.

Lamellibranchs Mussels, oysters and scallops belong to this class, in which the head is no longer distinct from the rest of the body. The latter is still very compressed laterally, but has remained largely symmetrical. As with the brachiopods the shell is made up of two valves, but these are now lateral rather than dorso-ventral; in addition, the structure of the mineral layers that constitute them is very different in the two groups. In the lamellibranchs an internal layer superimposes sheets of organic matter and layers of aragonite (an unstable form of calcium carbonate). This internal layer forms the mother-of-pearl which owes its iridescence to the diffraction of light among the alternation of fine deposits; the pearls used in jewellery are mother-of-pearl lumps secreted by the irritated body of the oyster around some alien particle. The thickest part of the shell is built by prisms of calcite or aragonite arranged at right angles to the external surface of the mother-of-pearl layer. This prismatic coating can become highly developed, and in Jurassic and Cretaceous deposits prisms have been found in the free state which originated during the breakdown of shells of the genus *Inoceramus*.

There are other features that separate the brachiopods from the lamellibranchs. In the former, for instance, the opening and closing of the shell is activated by the muscles. As we have seen, these are active movements. In the lamellibranchs, however, an elastic ligament located near the hook passively causes the shell to open, while the creature's internal muscles resist this action. That is why it is enough to cut the large muscle joining the two valves of an oyster to make it open immediately. For the palaeontologist, the

The internal views of a valve of *Glycymeris* and *Venus* (below) clearly show the difference between the small teeth of a taxodont lamellibranch (*Venus*) and the more complex structure of a heterodont. On the right, a shell of *Trigonia*, a magnificent Mesozoic Era lamellibranch (× 5 approx.). These shells are distinguished by different zones of ornamentation at front and rear, separated from each other by a series of keels starting at the 'hook' at the top left.

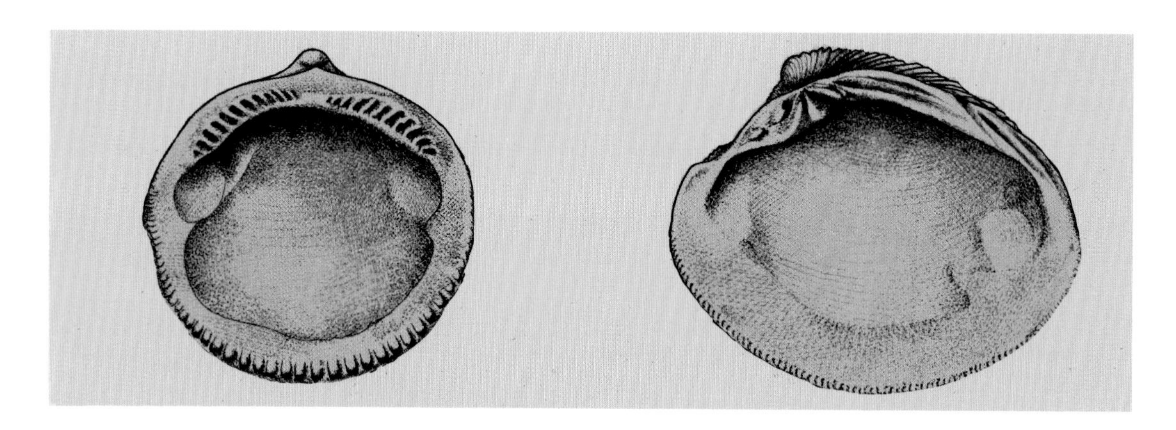

This Palaeozoic Era *Cyrtoceras* (left) has a more or less straight shell. The Mesozoic Era *Nautilus* (below) is similar to modern forms. Both show the arrangement of chambers divided by cross-walls or septa so typical of the tetrabranch cephalopods. The suture lines at the edge of each chamber are single in the nautiloids. In the example of *Cyrtoceras*, we can see the trace of the siphon that connected the animal's body to the end of the shell is near to the ventral edge of the top chamber.

implication of this anatomical device is that the putrefaction of a lamellibranch after death systematically dissociates the two parts of the shell which are subsequently found separated in the fossil state, contrary to what happens to the remains of the articulate brachiopods.

The lamellibranchs have been present since the Lower Cambrian Period, and today make up 62 different families. Their classification has been made possible owing to the variations in the number and form of the teeth of the hinge. The 'taxodonts' have numerous small identical teeth (see page 78). They are the oldest group but they still survive today, their relative stability eliminating the need for any rapid evolution. The genera *Arca*, *Cucullaea* and *Nucula* are among the types most widely found in fossil deposits. As their name suggests, the 'dysodonts' possess small, poorly shaped teeth, or even none at all, and it is this order which includes oysters (*Ostrea*) and their derivatives like *Gryphaea* or *Exogyra* – all adapted to a stationary life. This adaptation in fact involves a very advanced asymmetry of the shell, with the hollowest valve sometimes literally welding itself to its support. The 'heterodonts' for their part have only two or three cardinal teeth under the hook, plus a few elongated teeth front and back. Their evolutionary success has surpassed that of the dysodonts, and it is this order which contains the majority of lamellibranchs, both fossil and modern, such as *Cyrena*, *Cardita*, *Corbula*, *Lucina*, *Chama*, *Venus* and others.

The oddest of all lamellibranchs are without doubt the 'pachydonts' (thick teeth). Initially represented by heavy, individual shells (as in *Megalodon*), they gave rise at the end of the Jurassic Period to the first Rudista, *Diceras*. Rudistae are very specialized lamellibranchs – they, too, have adapted to a fixed way of life, but the modifications we saw in the oysters have been far surpassed here. In the Cretaceous species of the genus *Hippurites* the right-hand valve has taken on the appearance of a limestone trumpet with thick walls and standing on its tip, the inside of the chamber being sub-divided by vertical folds. The left-hand valve is held by elongated teeth and forms a circular lid. These creatures are notable for their similarity with the madreporite corals as well as for having erected substantial reefs consisting of enormous quantities of limestone. They disappeared very abruptly at the end of the Cretaceous Period at a time when they were at the peak of their development.

Cephalopods The cephalopods, a solely marine class, comprise the present-day squids, cuttlefishes, octopuses and nautiluses. Today, the latter have just five species in the genus *Nautilus* which has remained relatively similar to fossil forms. But in the Palaeozoic Era, and during the Ordovician and Silurian Periods in particular, they represented a most important group of nautiloids which without doubt constituted the original stock of the other cephalopods.

Cephalopods are molluscs whose body structure has become very complex. They retain the bilateral symmetry and their head is distinct from the body. Their eyes are as complex as those of the vertebrates and their foot is split up into tentacles which surround a mouth that is now armed with a horny beak that has sometimes fossilized. The shell of *Nautilus*, the only modern genus in which it has not regressed, is coiled in a flat spiral (see page 80), and is split up into successive chambers which have appeared as 'joint lines' on the internal moulds of fossils. These compartments are filled with gas to lighten the animal, which actually occupies the last chamber only. It is nevertheless linked to the end of the spiral by a siphon that passes through the mother-of-pearl walls which separate the various cavities. *Nautilus* is a gregarious creature, and is a swimmer and carnivore, like all modern cephalopods. It is found in the western Pacific and the Indian Ocean, and frequents the areas around coral reefs where it is on the lookout for crustaceans, burrowing during daylight to escape the strong sunlight.

The nautiloids have simple joint lines, and the very first among them were not coiled like present-day *Nautilus*, but had slightly incurved or even absolutely straight cones, like that of *Orthoceras*, for instance. Their shell was very elongated, and certain Ordovician orthocones attained lengths of up to 4.5 metres ($14\frac{1}{2}$ feet), while others were very short, their dwelling chamber having a very small opening suggesting an equally small number of tentacles (*Nautilus* has several dozen) as well as the ingestion of very tiny prey. *Ascoceras* of the Silurian Period is more complex, its shell, long and arched, ending in a bulging rear part. Coiled species rapidly developed, sometimes with disjointed coils like that of *Phragmoceras* of the Silurian Period. The spiral species of the nautilus type gradually predominated, and were the only

The suture lines of the goniatites are characteristic, tracing an acute angle towards the rear (this also gives them their name, gonia meaning 'angle' in Greek). The occupants of these shells probably looked like octopuses, squid or the nautiluses of modern times.

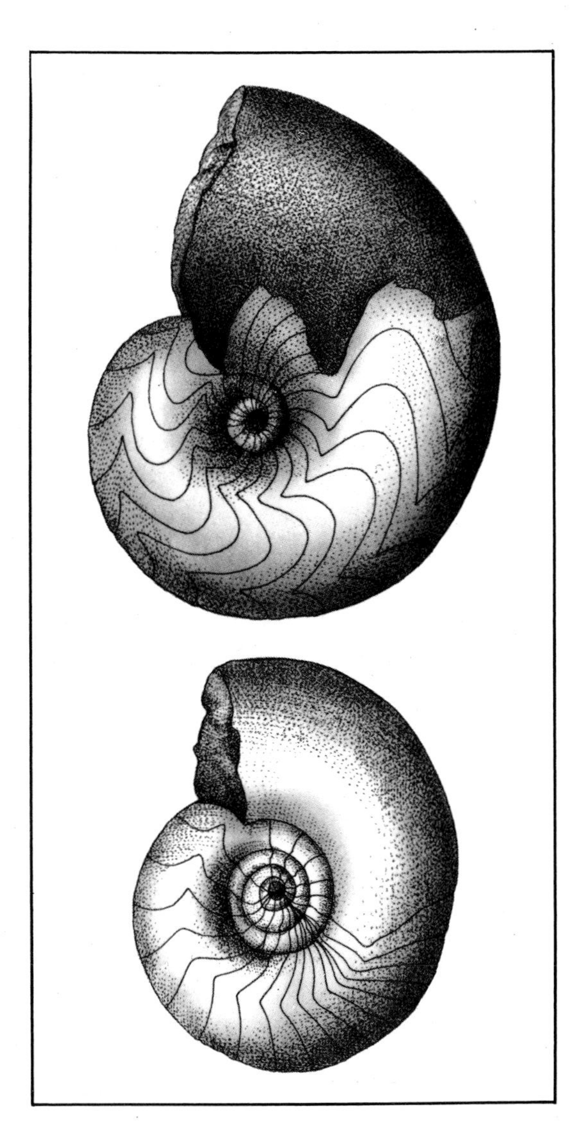

ones to survive beyond the Triassic Period.

The gradual disappearance of the nautiloids at the beginning of the Devonian Period was in all probability caused by the emergence of another group of cephalopods: the ammonoids. The ammonoids included the goniatites and the clymeniae of the Palaeozoic Era as well as the ammonites proper which succeeded them in the Mesozoic Era. The ammonoids were basically originally coiled forms which, unlike the nautiloids, showed a tendency to uncoil in later forms. Their siphon did not run to the centre of the shell walls as in *Nautilus*, but along the edge of the chambers and generally in a ventral manner (in other words, on the outside). Only the siphon of the clymeniae runs dorsally. The chief interest in the latter lies in their poor distribution in the fossil record, strictly limited as they are to the Upper Devonian Period where they evolved rapidly. In addition, the discovery of a clymenia enables us to date accurately the rock from which it is extracted.

The goniatites with their even simpler joint lines flourished during the Devonian and Carboniferous Periods, which were the peak of their development. During the Permian Period they were gradually superseded by the first ammonites (ceratites) that take their name from the god Ammon whose ancient statues bore the horns of a ram. They comprised six times more genera than the goniatites which they completely replaced by the end of the Palaeozoic Era. In the course of the Mesozoic Era, they developed most energetically, and the rapid succession of varied species makes them of prime importance stratigraphically. The periods in the Mesozoic Era are divided into narrow 'zones', characterized by the ammonite fauna, with shells frequently decorated by ribs and tubercules. This sculpturing is much more pronounced than that seen in the goniatites. Their joint lines became extra-

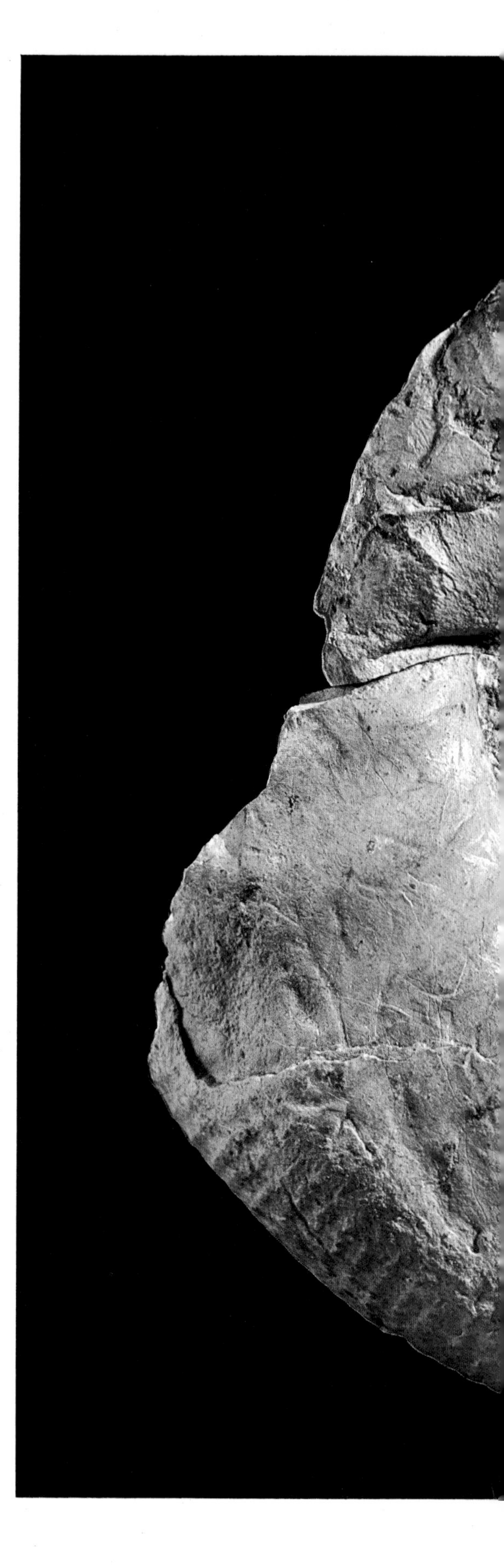

This huge ammonite belongs to the genus *Perisphinctes* of the Upper Jurassic Period. Looking at this specimen we can easily understand why the ammonites are named after the ram horns of the Ancient Egyptian god Amon and of Zeus-Ammon of Ancient Greece.

This beautiful ammonite (below) from the Jurassic Period of Britain (× $\frac{1}{2}$ approx.) has incredibly complex suture lines where the lobes have spread out to form lace-like 'leaves'. As ammonites evolved, the suture lines became more and more elaborate. *Crioceratites emerici* (× 1 approx.) (bottom) is an uncoiled ammonite from the Lower Cretaceous Period of Castellane in France.

ordinarily complex, engraving mottled images on fossils to beautiful effect.

These images are especially important for the palaeontologist since they characterize each species in a manner not unlike the way our fingerprints facilitate the positive identification of a certain person. The dwelling chamber could be closed by a lid system that sheltered the ammonite's body from the outside world. These 'lids', called *aptychus*, are generally found isolated among the sediments, and sometimes in large quantities. On the death of the animal they became detached, together with the creature's soft parts, while the shell with its extensive floating ability (nautilus shells have even drifted ashore on the coasts of Japan) were carried far away by the ocean currents. We have found deposits of *aptychus* without ammonites, and so it has been necessary to create a classification and a stratigraphy for these lids (or operculae) quite independently of the shells.

The size of the ammonites ranges from a few millimetres to 2.5 metres (8 feet) in diameter for the largest, *Pachydiscus seppenradensis* of the Upper Cretaceous Period. Dwelling within coastal margins as well as the open sea, they fell prey to the big sea

Turrilites of the Upper Cretaceous Period with its 'trochospiral' coils reminds us of the gastropods and is very different from the basic form of the ammonoids. Some nautiloids had adopted this type of spiral as early as the Palaeozoic Era.

reptiles (see page 168), and a specimen of the genus *Placenticeras* found in the recent Cretaceous of South Dakota had been bitten through by the sharp teeth of a mosasaur (see page 175). Attacking from above, the predator had bitten the shell 16 times before extracting the body of the mollusc.

From the Middle Jurassic Period onwards we see in certain families the deterioration of the regular spiral, with the coils becoming disjointed, as is the case with *Crioceratites* of the Lower Cretaceous Period (see page 84), or forming crooks (as in the case with *Macroscaphites* of the Lower Cretaceous Period) or even forming straight cones over almost their entire length (as in the case with *Baculites* of the Upper Cretaceous Period). Certain genera like *Turrilites* from the Upper Cretaceous Period in particular, adopted a gastropod-type coil, (see above). The majority of these shells must have belonged to bottom-living creatures, since their shells were poorly shaped for movement through the water. Like the Rudista and certain other groups of invertebrates and vertebrates, the ammonites became extinct at the end of the Cretaceous Period. A general cause capable of accounting for all these disappearances has so far been sought in vain. A cooling of the climate, cosmic phenomena, epidemics, geological upheavals – all these and other reasons have been put forward as possible explanations, but none is truly satisfactory, for how can we explain the simultaneous disappearance of animals living in completely different environments, while numerous species passed the Cretaceous Period–Tertiary Period frontier without difficulty?

Within the cephalopods we can distinguish between tetrabranchiate cephalopods (those with four branchiae) and dibranchiate cephalopods (those with two branchiae). The first include the cephalopods with the tough external shell that we have just seen, while the latter have a relatively reduced internal skeleton. In the world of the fossils the belemnites constitute the richest group of dibranchiates. In these molluscs, that portion of the skeleton divided into chambers ends in a small cone containing gas or fluid – known as the phragmacone – on which sits a calcite guard (see pages 20 and 21). This guard, located at the rear, often bears the traces of blood vessels which prove the presence of a layer of flesh at the surface, albeit a doubtless very thin one. We also find the perforations made by cirripeds, tiny parasitic arthropods whose orientation parallel to the axis of the guard shows that they were attached during the mollusc's lifetime, in order to take advantage of the current created by its motions. The soft parts of the body which we know from some exceptional specimens preserved in fine sediment such as that at Solnhofen in Bavaria are reminiscent of modern squid. They reveal the imprint of an ink sac used for defensive purposes and which certain curious minds of the last century had the idea of dissolving into writing ink.

The carnivorous – even cannibalistic – and gregarious belemnites inhabited coastal waters, serving as the basic nourishment for the great marine vertebrates like the Lower Jurassic Period shark found in Swabia whose stomach contained 250 of their guards. The latter are found in large numbers in Mesozoic Era deposits, with forms offering little variety. Nevertheless there are some that have been the object of extreme lateral compression, like *Duvalia* of the Upper Jurassic and Lower Cretaceous Periods. The belemnites, the largest of which attained some 2 metres ($6\frac{1}{2}$ feet) in length – *Megatheutis* of the Middle Jurassic Period – disappeared completely at the end of the Cretaceous Period.

The barnacles are very specialized arthropods adapted for a fixed life-style. They are frequently found on mollusc shells, like this *Flabellipecten* from the Miocene Period of Altavilla (Sicily) shown about $1\frac{1}{2}$ times life size.

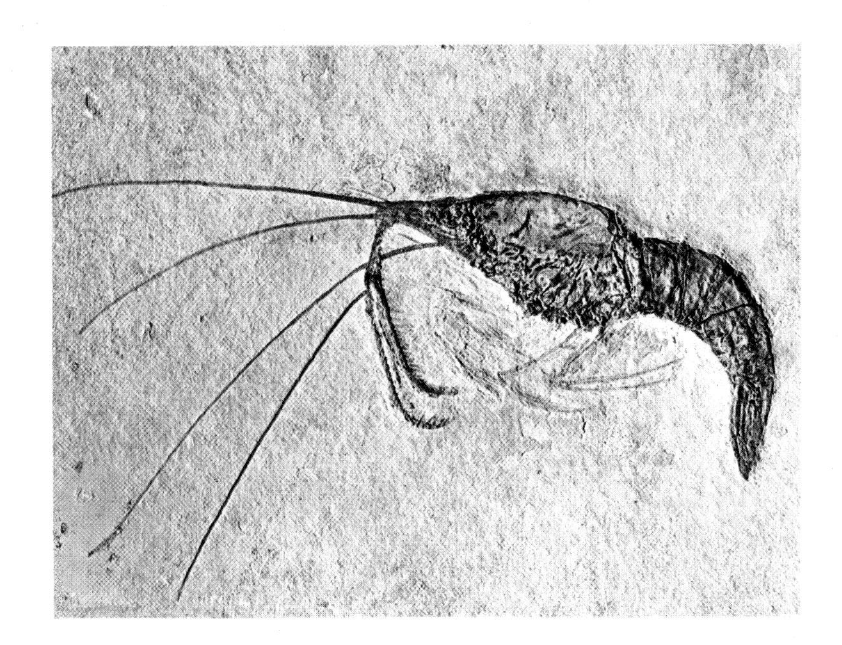

The fossilization of decapod crustaceans needed very particular conditions, and luckily they were present in the lake of Solnhofen (Bavaria), the home of this *Aeger insignis* ($\times \frac{1}{2}$) at the end of the Jurassic Period.

Arthropods

The arthropods constitute the most important phylum of metazoan animals. In particular, the class of the insects underwent a spectacular development: 900,000 species out of a total of 1,200,000 known animal species – not to mention the enormous number of insects that have not as yet been named. In the marine world the most familiar arthropods to us are of course the crustaceans (crabs, lobsters, barnacles and shrimps), but there are many others, both living and fossil.

The name arthropod means joint limbed – a feature possessed by all members. Also, their bodies are typified by an extensive metamerism, in other words the repetition of identical segments, which reminds us of their roots among the worms. This metamerism is, however, modified, and the body is divided into regions, with the head in particular having become quite distinctive. The circulatory and nervous systems are also quite complex. Another feature of the arthropods is the nature of the layer of skin which covers them. This stiff coating has a defensive as well as a supportive function. This external skeleton surrounding the whole body imposes an intermittent growth pattern based on periodical sheddings when it is abandoned before the animal produces a larger one.

Trilobites From the Lower Cambrian Period onwards throughout the Palaeozoic Era, trilobites represented the most remarkable group of arthropods with over 10,000 species in 1500 genera. *Spriggina* (see pages 56–57) and *Parvancorina* from the Precambrian Period of Ediacara, have given us an idea of the ancestral forms from which they descended in those far-off times, with their development culminating in the Cambrian and Ordovician Periods. During the first of these periods they became reliable fossil record indicators, but subsequently regressed down to the Middle Permian Period, at the end of which they disappeared.

These animals are generally 5 to 8 centimetres (2 to 3 inches) in length, although some species never exceeded a few millimetres, while giant trilobites are also known. One such creature was *Uralichas* from the Middle Ordovician Period. It attained 75 centimetres (30 inches) in length. Trilobite fossils are relatively plentiful, since each creature shed its exoskeleton between 20 and 30 times during its life, and each time the external chitinous skeleton that was left behind was capable of fossilizing. As their name suggests, the body of the trilobite was subdivided into three longitudinal lobes (see pages 88 and 89). At the same time, they can also be split up into three sections from head to tail: a head in the shape of a wide shield extended on each side by a point and usually with two visible eyes, a longish segmented thorax and finally a terminal region called the *pygidium* which shows up very clearly on the specimen of *Bathyuriscus* shown below. On the head, a line corresponded to the joint along which the carapace was shed to release the animal inside.

The eyes were nearly always of the composite type, that is, like the eyes of modern-day insects, they comprised a large number of small, light-sensitive facets arranged geometrically next to each other. Blind species have also been discovered, and these probably lived in the ocean deeps or led a burrowing life. The undersides of trilobites are rarely preserved owing to their being covered by a skin which was softer than that found on their backs. When threatened, these arthropods, the prey of the goniatites in particular, protected their soft bellies by curling up much in the same way that modern woodlice do, and death frequently 'froze' them in this attitude, which was subsequently fossilized. On the rare occasions when it has been found, the under surface shows the complex appendages on each segment that supported the branchiae and which were used to move the animal around. The absence of any chewing organs indicates that these invertebrates only took in tiny particles of food. Trilobite movements have sometimes left behind tracks imprinted in the once-moving sediment of certain rocky banks. These tracks are never more than 0.5 centimetres wide, which would indicate that the young of the species were basically swimmers. On the same rocky banks, oval imprints have been interpreted as the traces left by trilobites lying stationary on the sea-bed. Finally, at the end of some tracks, other marks seem to have been produced by these animals at the very instant they launched themselves off to swim. Small oval eggs have also been discerned among the puzzling structures found among trilobite deposits.

The evolution of the trilobites shows a clear tendency for the *pygidium* to grow. In the later genera their ornamentation was often very extreme, with the head covered in spines for example, and the body developing long points. Some of the most advanced forms, however, are very smooth.

Gigantostraca Fossil representatives of our present-day crabs, prawns and lobsters have also been found, as have orders of small crustaceans and marine arthropods of a less common kind, like the king crabs which still frequent the shores of America and Asia, and whose appearance is reminiscent of the trilobites. Of all these groups, the Gigantostraca have totally vanished leaving no equivalent in modern nature. They are classified as being related to the king crabs, the spiders and the scorpions to which they bear a

Some Cambrian Period trilobites: *Olenellus thompsoni* (× $\frac{1}{2}$) of the Lower Cambrian Period of North America; *Redlichia vernaui* (× $\frac{2}{3}$) of the Lower Cambrian Period of China; *Bathyuriscus formosus* (× $\frac{4}{5}$) of the Middle Cambrian Period of North America; *Paradoxides spinulosis* (× $\frac{1}{4}$) from the Bohemian Middle Cambrian Period; *Olenus truncatus* (× 4) from the Upper Cambrian Period of Sweden; and *Fremontia fremonti* (× 1) from the Lower Cambrian Period of North America.

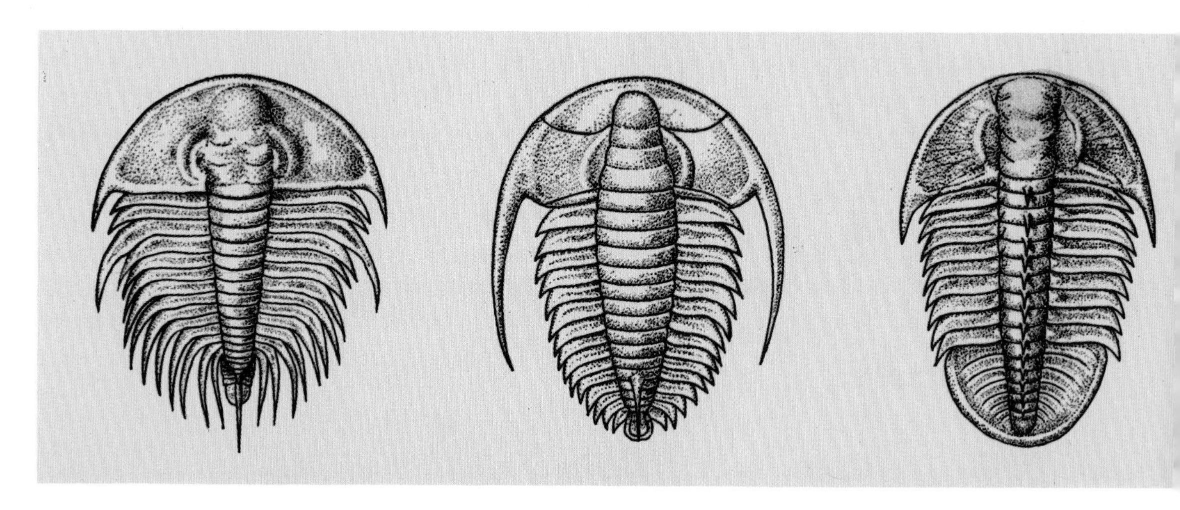

Trilobites are usually found broken up into their separate segments. This is because they used to shed their skins or carapaces (like insects today). Complete specimens are rather rare and nearly always appear with their back uppermost, like the one shown on the left, while the underside is seldom fossilized.

reasonable resemblance (see page 90) – but what scorpions! One of the largest, *Pterygotus* of the Ordovician to Devonian Periods, measured between 1.8 and 2 metres (about 6½ feet) in length. In the Silurian deposits of North America Gigantostraca approaching 3 metres (10 feet) have been found. These creatures appeared during the Ordovician Period and flourished until the Devonian Period, while during the Carboniferous and Permian Periods their size diminished prior to extinction. To begin with, Gigantostraca are found in marine strata only, so either they actually inhabited this environment or their remains were carried along by the rivers to be deposited along the coasts. They seem, however, to have gradually adapted to life in brackish lagoons and then to the freshwater swamps of the Permian–Carboniferous Periods, by which time the marine Gigantostraca had ceased to exist.

Their body, protected by a chitinous carapace, consisted of a cephalothorax bearing the visual organs: two large eyes plus a pair of ocellae in the middle, then a long, articulated abdomen. After the sometimes very powerful pincers came five pairs of legs – the front appendages were used for moving around on the bed of the sea or lake, and the rearmost pair were usually modified to form 'flippers'. At the tip of the abdomen, the 'telson' was a tapered structure which spread out to form a 'paddle' in swimming varieties such as *Pterygotus* (see page 90). In *Carcinosoma* of the Ordovician to Silurian Periods in particular, the pointed telson, curved like a scorpion's sting, may well have contained a poison gland with which it could have delivered a lethal dose.

The flattened shape of the Gigantostraca suggests a bottom-dwelling habit and way of life, and these arthropods probably burrowed in the mud in search of prey. Some must have been good swimmers, and it is possible

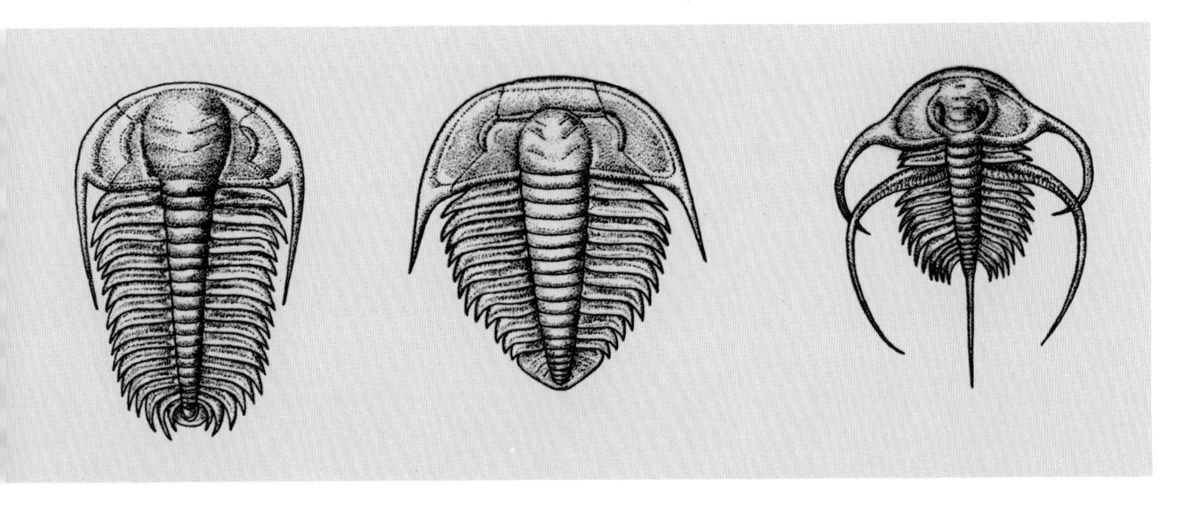

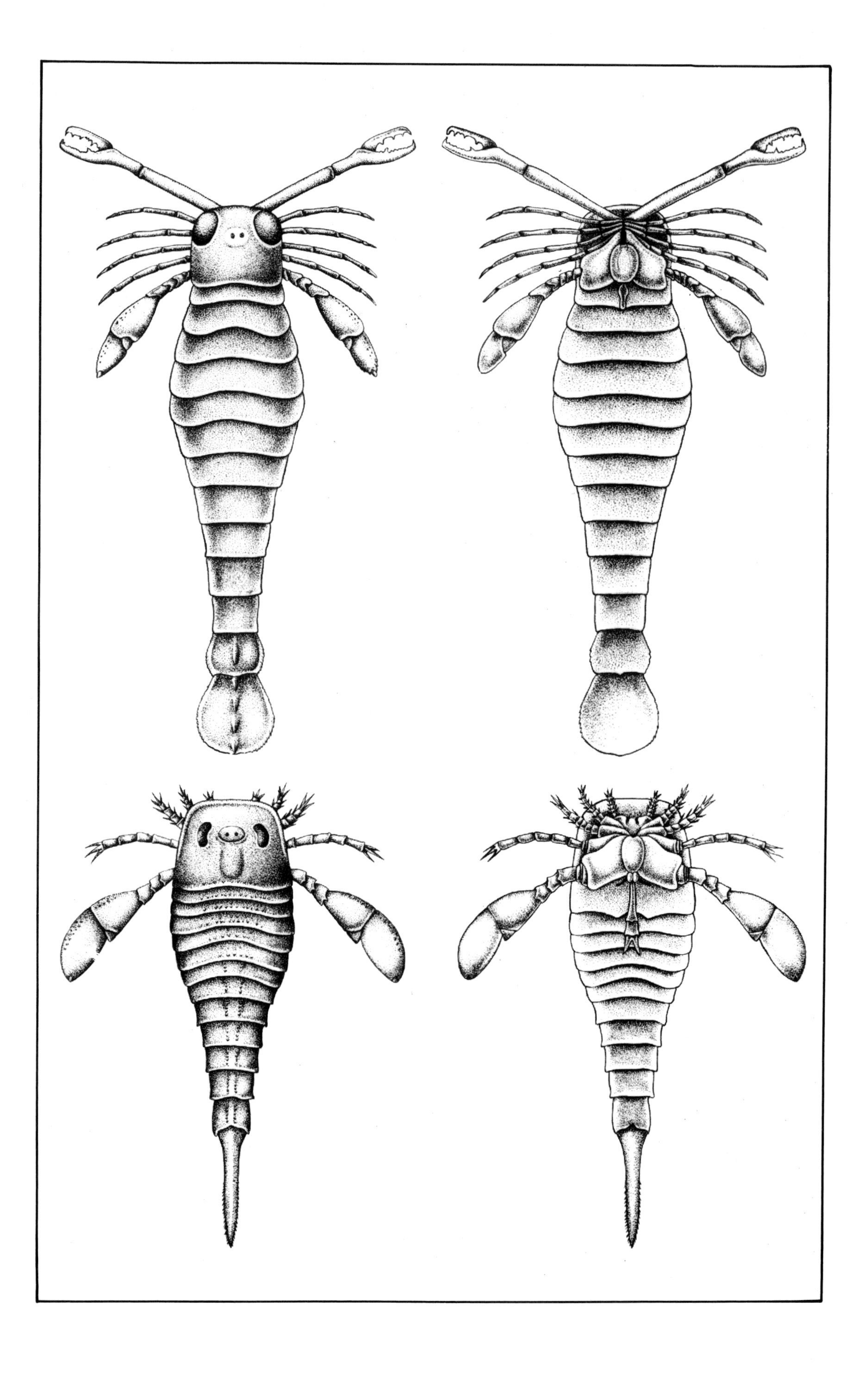

that the Gigantostraca swam on their backs like modern-day king crabs, undulating their abdomens. The huge claws of *Pterygotus* and their association in fossil beds with the most primitive form of fishes, the Agnatha (see chapter 4), naturally made them the predators of the latter creatures, a large number of which actually lived on the ocean floor. It is not hard to imagine these arthropods lying in wait for the Agnatha in much the same way that modern mantis shrimps patiently wait until a passing fish comes within reach of their terrible spined legs and chewing mouthparts, which are capable of cutting their victims in two with a single movement.

Ostracods This is a group of small crustaceans that enjoys a special significance in geology. The layman, observing a fossil ostracod through a microscope for the first time, would be forgiven for thinking that it was not an arthropod. The name ostracod means simply 'shell-shaped', and it does indeed look like a sort of small bivalve (see page 92). A living specimen caught in a lake or on the ocean shore is a curious creature that often never exceeds 1 millimetre in length. Very highly specialized from a morphological point of view, its body is compressed laterally between the two hinged valves. The indistinct head carries two pairs of antennae that serve as sensors and legs at the same time. The abdomen ends in a series of claws which enable the animal to burrow. The eyes, which are sometimes absent, are situated beneath a carapace which is translucent above them. The valves are formed by the superimposition of an internal layer of chitin and an external coat of limestone, and it is this mineral portion that is found in the fossil state, separated by a process of washing and screening of sediment. As with the lamellibranchs, the various genera are characterized by their overall outlines, the sometimes complex ornamentation of the valves and the nature of the hinge. The abundance of ostracods in certain formations and the ease with which they can be extracted from boreholes, for instance, explains the great interest shown in them by petroleum geologists who frequently use them as stratigraphic fossils.

Having appeared during the Ordovician Period, the ostracods attained their greatest sizes during the Palaeozoic Era. At this time some species reached 2.5 centimetres (1 inch) in length. They subsequently shrank in size and continue to evolve right up to our own time, both in the seas and oceans and in brackish and freshwater environments. Living ostracods are distinctly larger, with a finer carapace. They are omnivores that feed upon algae, vegetable debris and micro-

An original specimen of *Eurypterus remipes*. Like the trilobites, the Gigantostraca shed their carapaces and these are often found as fossils.

Two very different sized Gigantostraca: top, *Pterygotus buffaloensis* of the Upper Silurian Period ($\times \frac{1}{20}$); bottom, *Eurypterus remipes* from the Upper Silurian Period ($\times \frac{1}{3}$). Dorsal (back) and ventral (underside) views of each animal are shown.

organisms. Preferring calm waters that are rich in organic matter, they can be pelagic and planktonic or bottom-dwelling, crawling on the sea-bed at various depths.

Echinoderms

With the phylum of the echinoderms we come to another group of very elaborate invertebrates which, like the worms, brachiopods, molluscs and arthropods, have completely escaped from colonial life and whose origins go back a very long way. For instance, fossil traces found in Precambrian rocks have been attributed to them. Exclusively marine-dwelling, the echinoderms gave rise to many different forms, and were in general, organisms that were both fixed and free-ranging. They acquired a radial symmetry based on the number five – five arms in the starfish, five ambulatory (walking) zones in the sea-urchin – and the organs, too, are repeated according to this number or one of its multiples. The genus *Tribrachidium*, however, from the Ediacara Precambrian, presented a tri-radial and not a penta-radial symmetry. In addition, the most ancient groups of echinoderms are sometimes asymmetrical or of bilateral symmetry, and this represents a stage in the larval development of modern species through which they must all pass. The penta-radial symmetry is, therefore, a secondary process in this scheme. The embryonic development of the echinoderms is of great interest to the zoologist, since it is very close to that of the Chordata, the group to which the vertebrates (and Man himself) belong. The creatinophosphoric acid extracted from the muscles of the latter is found nowhere in the animal world except in the echinoderms, a relationship that is often invoked when tracing the roots of the phylum Vertebrata. Chordates and echinoderms are also classified together within the 'deuterostomes'.

The skeleton of the echinoderms is internal, and is based on calcite, but it also contains small amounts of magnesium carbonate. It is composed of spicules spread throughout the epidermis, as in the Holothuroidea – the group of echinoderms with the softest bodies – or formed by 'welded' plates, as in the sea-urchins for example. A highly original hydraulic system, known as the water vascular system, is used for respiration, locomotion and for sensing all at the same time. The nervous system is very simple, and is spread throughout the body, apart from a nerve ring around the mouth from which the other nerves spread out. This curious anatomical picture is completed by primitive features like the absence of a distinct head and excretory system.

External view of a valve of *Cythereis reticulata*, an ostracod found in the clay deposits at the end of the Lower Cretaceous Period of Glatigny in France (× 70 approx.).

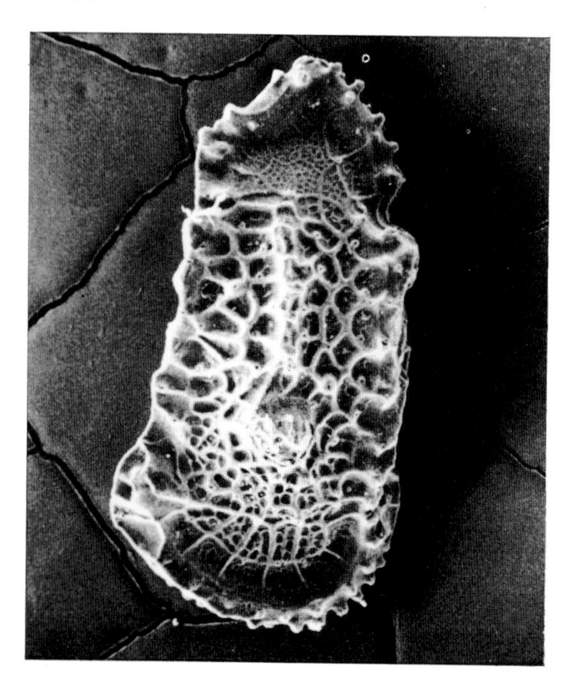

The primitive echinoderms From the start of the Palaeozoic Era several classes of echinoderms developed in the depths of the sea, and some of these strange animals have no direct descendants with the result that their anatomy and ecology are the subject of frequent controversy. A typical case in point is that of the ophiocystoids from the Ordovician to the Devonian Period. These animals are so rare and fragmented, that we know practically nothing about them.

The cystoids, including the species *Cheirocrinus walcotti* (see page 93), lived from the Cambrian to the Devonian, culminating in the Ordovician. There is little doubt that their preservation was difficult, for their fossils are rather rare. The main part of their

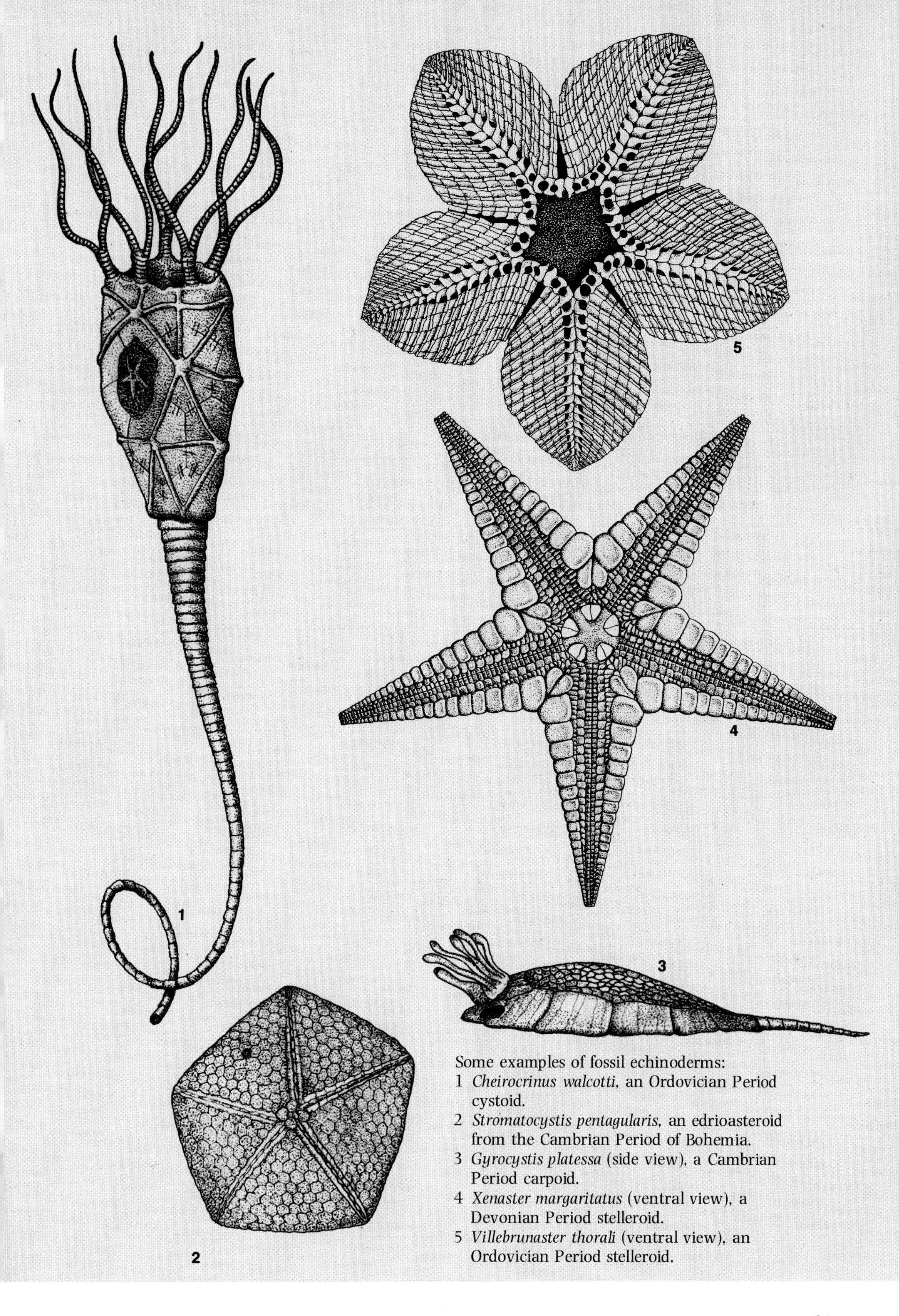

Some examples of fossil echinoderms:
1 *Cheirocrinus walcotti*, an Ordovician Period cystoid.
2 *Stromatocystis pentagularis*, an edrioasteroid from the Cambrian Period of Bohemia.
3 *Gyrocystis platessa* (side view), a Cambrian Period carpoid.
4 *Xenaster margaritatus* (ventral view), a Devonian Period stelleroid.
5 *Villebrunaster thorali* (ventral view), an Ordovician Period stelleroid.

body consisted of a sort of elongated box, the 'theca', armoured with jointed plates. Sometimes the latter were small and very numerous (up to 2000), sometimes they were larger and more regularly arranged as in *Cheirocrinus*, in which case they were consolidated by a network of ribs. Articulated arms surrounded the mouth towards which three feeding channels converged, and whose division raised the number to five, giving the animals a penta-radial body arrangement. Opposite the mouth at the other end of the theca the creature was attached to the substratum either directly or with the aid of a hollow stalk whose tip was anchored by means of roots or by wrapping itself around a fixed object.

The geometry of the blastoids (Ordovician to Permian Periods) is more regular – they strictly adhered to the penta-radial symmetry. Their theca looked like a rosette with five petals when seen from above and, like that of the cystoids, it was attached directly or by means of a peduncle or stalk.

Most primitive echinoderms were not free-ranging, and this applies to the class of edrioasteroids from the Middle Cambrian to the Carboniferous Period. Nevertheless, some of these latter were able to move, in which case their flattened, relatively supple theca was provided with a sucker that enabled them to halt temporarily. They have often been found attached to shells. The five ambulatory zones that radiated from the central mouth were straight in the primitive genera like *Stromatocystis* of the Cambrian Period (see page 93), but in later species they become curved, seeming to clasp in their arms a theca covered with small plates.

Out of all the groups of echinoderms one group in particular has intrigued palaeontologists for a long time. *Gyrocystis platessa* of the Cambrian Period (see page 93) gives us an idea of the group known as carpoids. This was a group which covered a period that ran from the Cambrian to the Middle Devonian.

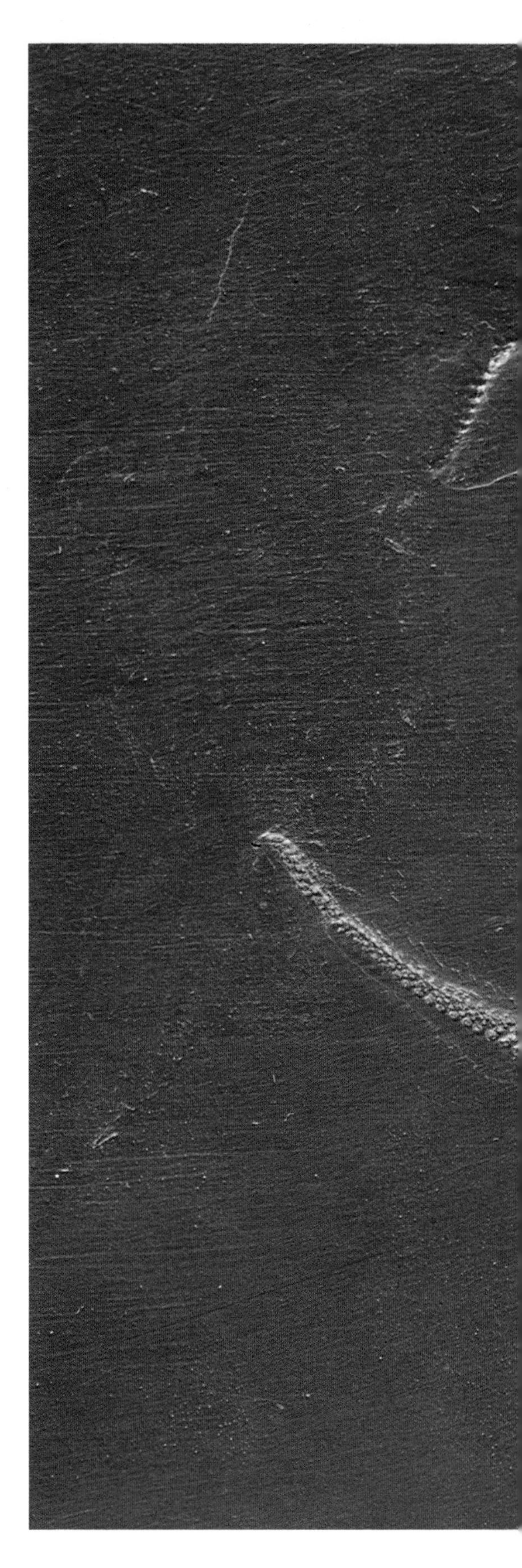

Like the brittle-stars and the sea-cucumbers, the starfishes do not have a rigid outer covering, and so their fossils are usually limited to spikelets and odd plates; specimens as complete as this one are very rare indeed.

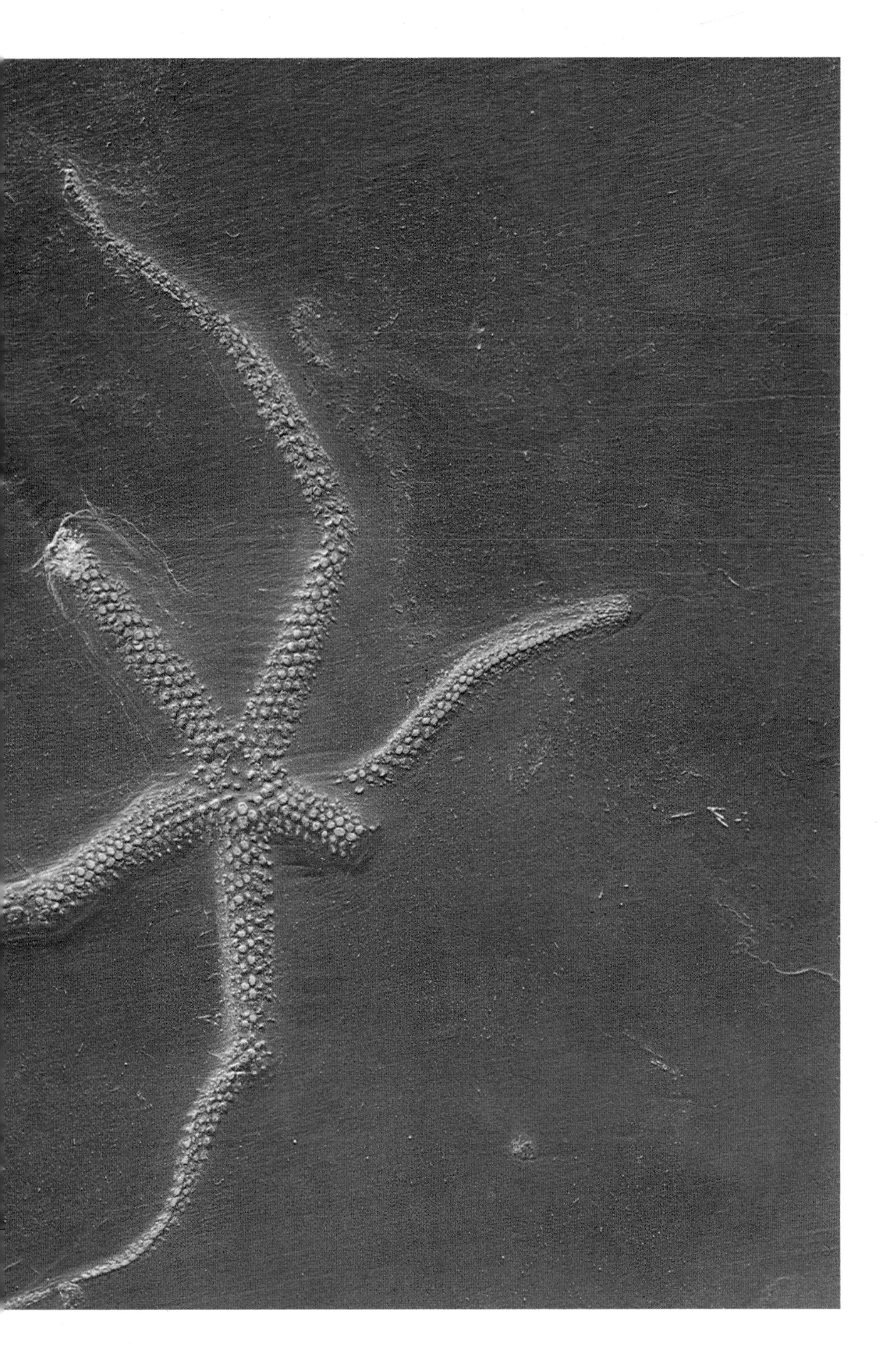

Left: Calyxes of *Uintacrinus socialis*, a pelagic crinoid of the Upper Cretaceous Period. It had no stalk, but extremely long arms over 1 metre ($3\frac{1}{4}$ feet) long.
Below: A group of crinoids from the Lower Devonian Period belonging to the genus *Bactrocrinites*. The discovery of complete crinoids is rare, because the pieces of calcite which made up each individual in life nearly always fell apart after death. The crinoidal limestones are rocks composed chiefly by the cementation of a huge quantity of this debris.

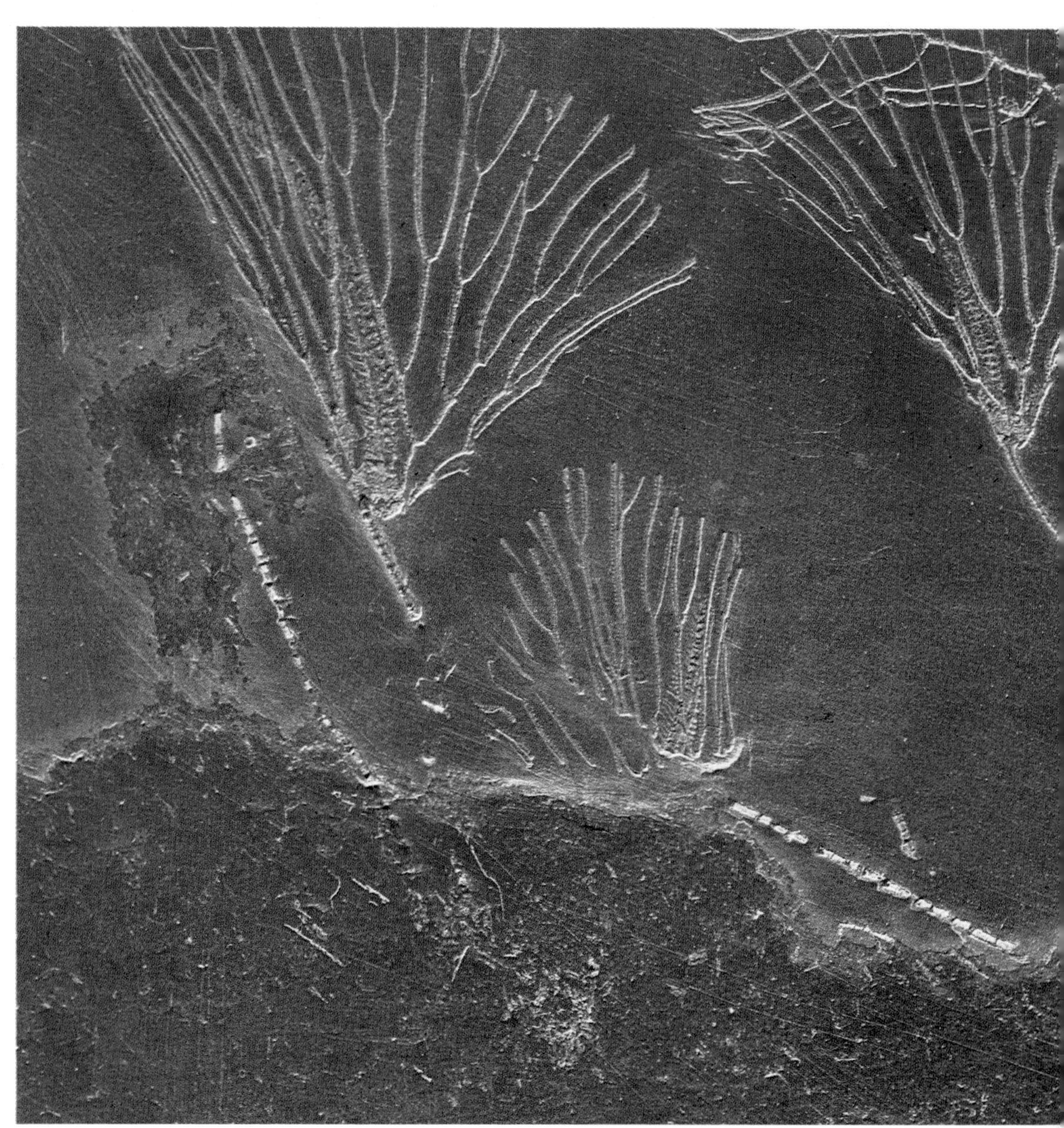

It is hard to describe them since they have been the subject of radically differing views, with the classical interpretation being that they were laterally flattened echinoderms that actually lived lying on their side. Far from being of penta-radial symmetry, they are sometimes completely asymmetrical and at other times bilaterally symmetrical. A single mobile arm would have been used to bring particles of food to the animal's mouth. The British palaeontologist R.L. Jefferies, however, asserts that at least some carpoids belonged to the Chordata and not to the echinoderms. In his view, the feeder arm became a sort of tail that superseded the peduncle which attached earlier types. With the mouth and anus in reverse position, their internal organization was totally different. Jefferies' theory once again points to the special links that seemed to unite echinoderms and the Chordata. It also has the merit of bringing to light the difficulties raised by an interpretation of remains which, although fairly complete, are very distant from any present-day forms.

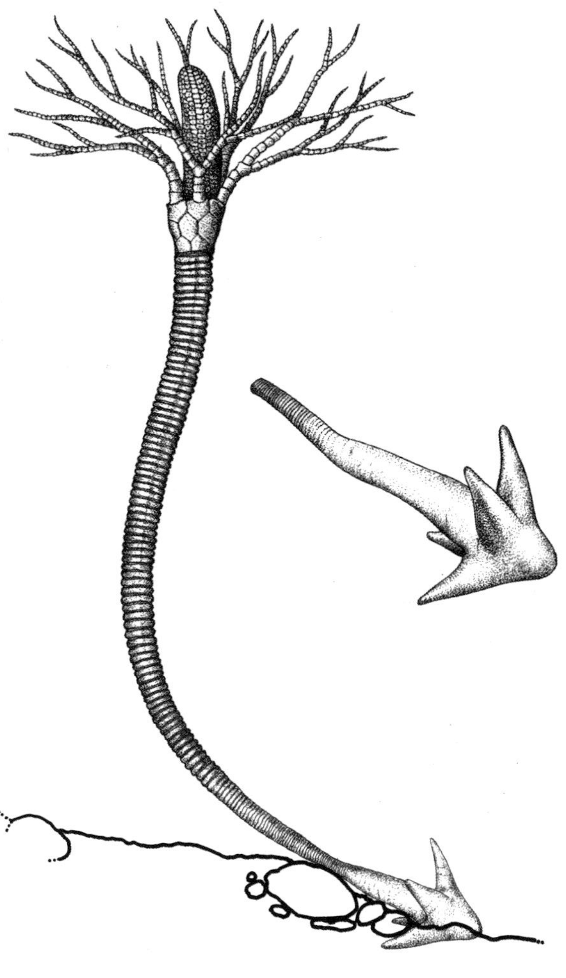

A reconstruction of *Ancyocrinus bulbosus* from the Devonian Period of North America ($\times \frac{2}{3}$ approx. for the complete specimen). It remained attached to the sea-bed by its anchor (like some present-day forms), but if it was torn loose it could move clumsily over short distances.

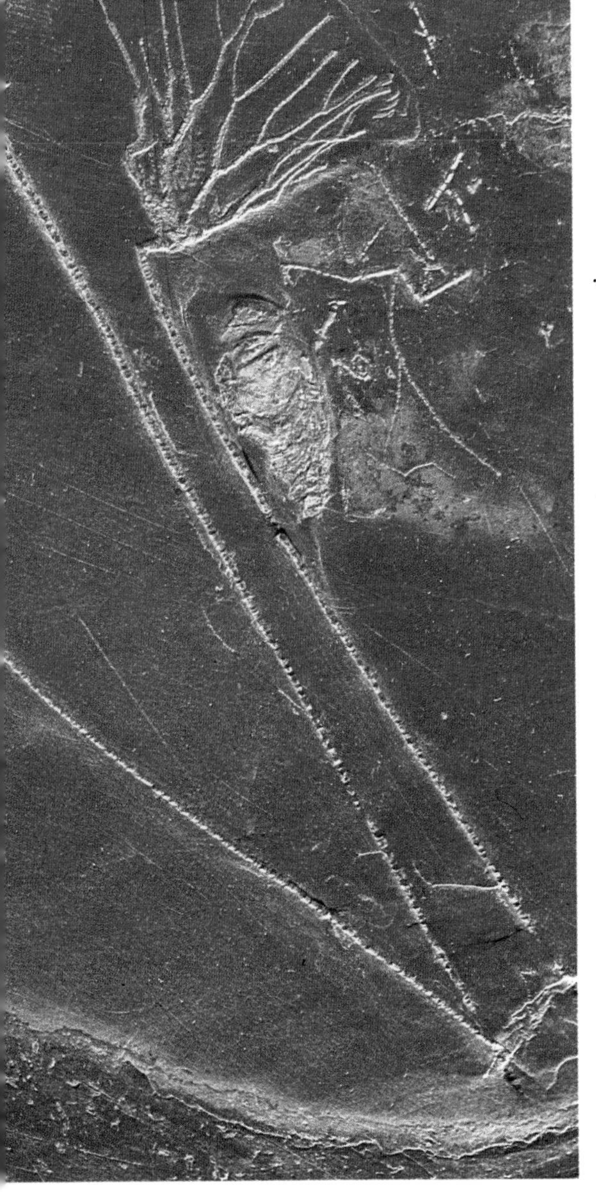

Crinoids Of the five living classes of echinoderms only two are truly important in the

Two very different types of sea-urchin: *Clypeaster pliocenicus* (above), a large irregular specimen with much-reduced spines and with a 5-rayed symmetry in strong contrast to the secondary bilateral symmetry; below, a cidarid showing tubercules bearing enormous radiolae and which has retained a perfect 5-rayed symmetry. This symmetry, a feature of many echinoderms, has been interpreted in different ways. One original explanation is mechanical: the pentagon is a simple polygon close in shape to a circle and without parallel sides, unlike the square or hexagon. A pentagonal configuration of plates is therefore very strong and will stand up well to certain forces exerted upon it.

fossil state: the crinoids and the echinoids. The very first crinoids have been found in the Lower Ordovician Period (see pages 96 and 97) and there were multiple variations on the basic theme. The Comatulae, for example, lost their peduncle in the course of their evolution in order to swim freely in coastal waters. In the fossil state this peduncle frequently possessed a branched root but may also have ended in a disc, a hook, an anchor (see page 97) or a bulb. Sometimes the theca was directly attached. The arms also reveal a great variety of different adaptations, the crown of *Petalocrinus* (of the Silurian Period) for instance bore broad and rigid fans. Some crinoids were minute in size, but the biggest, *Pentacrinus* from the German Lower Jurassic Period, consisted of two and a half million jointed pieces over a length of 20 metres (69 feet) and with a crown 1 metre ($3\frac{1}{4}$ feet) across. An underwater forest of these creatures must have been an incredible sight, especially if their colours were as rich as those of modern living crinoids.

The expansion and diversification of this group culminated during the Palaeozoic and Mesozoic Eras. In 1873, at a time when it was believed that encrines (fixed crinoids) had completely disappeared, the dragnet of the British vessel *Challenger* brought up living specimens from the depths of the sea. In modern times, encrines almost never inhabit coastal zones, but at the height of their development they colonized the agitated waters of coral reefs as well as the deeper, calmer regions of the oceans. If, as has been asserted, the hollow basal bulb of *Scyphocrinites* (from the Silurian Period) was indeed filled with gas, it must have lived a floating life like *Pentacrinus* (from the Lower Jurassic Period) which sometimes drifted with the waves, hooked on to a large piece of wood.

Encrine fossils can often form enormous accumulations, while the segments of their stalks, whose cross-section may resemble a five-pointed star, are frequently discovered alone, and are at the origin of numerous legends like the one from Central Europe that claims these stone stars fell directly from heaven.

Echinoids The other class of echinoderms –

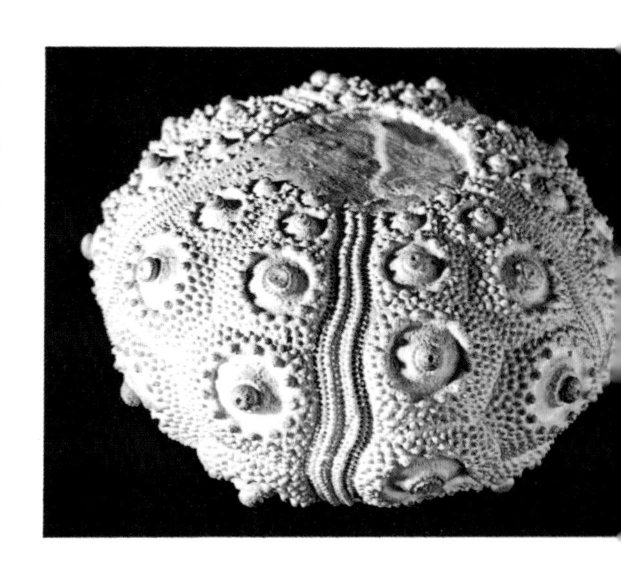

The greatly magnified section of a sea-urchin spine under the microscope reveals unexpected structural complexity. These geometrical patterns of calcite crystals are typical of each family of echinoids.

the echinoids – is much more familiar to us. These are the sea-urchins so commonly found along our coastlines. Many have been found in the fossil state, especially in the chalk deposited in shallow seas at the end of the Mesozoic Era. These creatures were totally free-moving, and are typically equipped with calcareous spines called 'radiolae' (see above). When transformed into small bristles, they are used to dig burrows by mud-dwelling urchins. They may, on the other hand, be as thick as a child's finger in a genus like *Cidaris* which is able to move around on the rocky sea floor with the aid of its strong 'stilts'. Fossil cidarids have produced spines of varying shapes – a swollen club, a bucket, a thorny stick or even a splayed bowl.

The sea-urchins' basic penta-radial symmetry is often superimposed on a secondary bilateral symmetry (see page 98). Originally the mouth opened downwards at the centre of the test, and a chewing organ, 'Aristotle's lantern', was situated behind it; the anus was located at the top of the theca. This type of organization, found among the more primitive forms, was frequently modified by the migration of the two openings, with the mouth shifting to the front and the anus moving down to the underside, sometimes occupying the original position of the mouth. Sea-urchins have been discovered in the Middle Ordovician Period of Great Britain, at which time they were covered with small, irregular plates arranged haphazardly. In the Carboniferous Period these plates were organized into radiating spindles whose number fell to ten from the Triassic Period onwards, but are a structure that has persisted down to modern times.

Graptolites

As we have seen, the stratigraphy of the Cambrian Period has been established chiefly on the basis of successive trilobite remains. In the Ordovician and Silurian Periods, however, this fauna was superseded

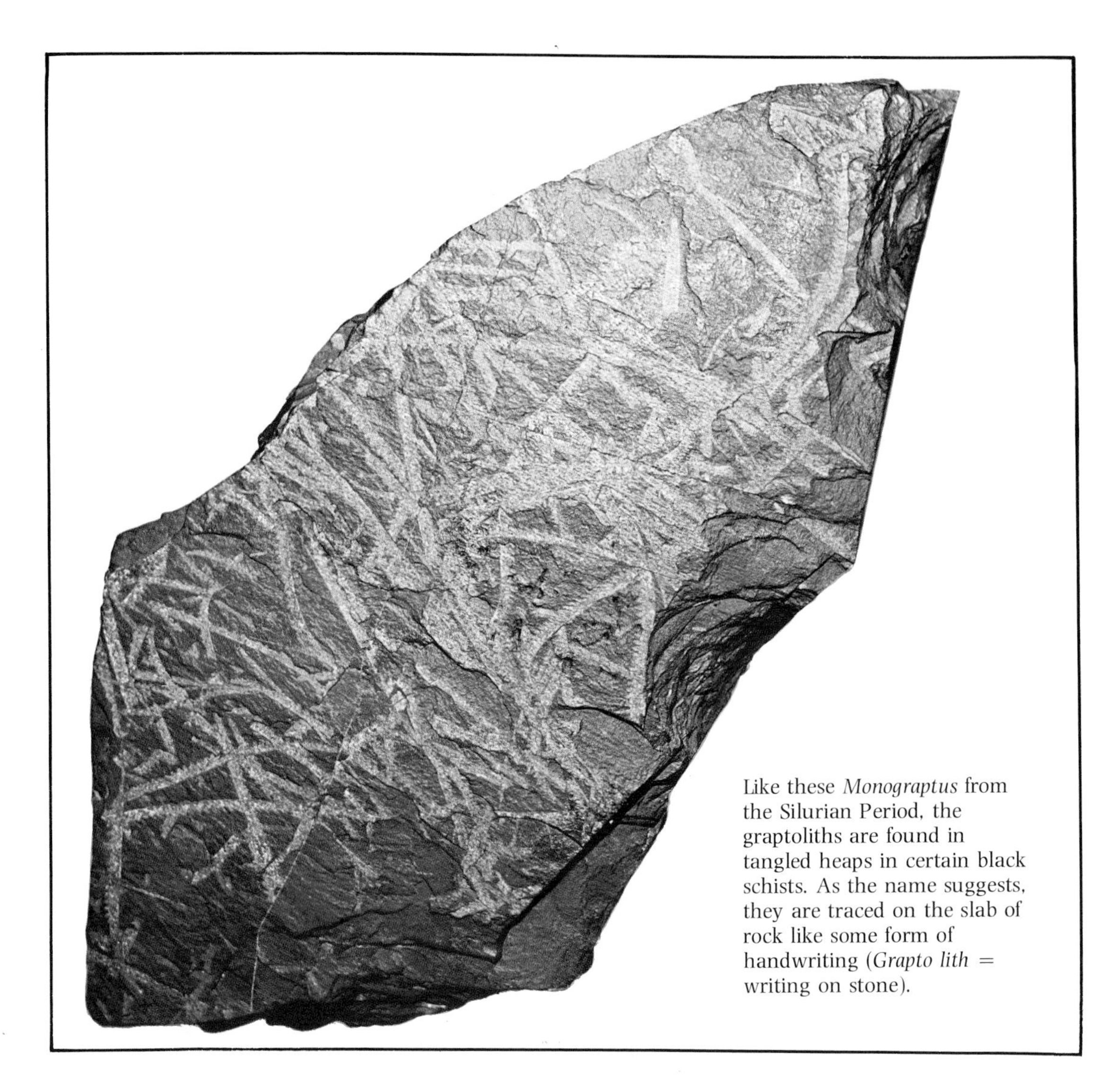

Like these *Monograptus* from the Silurian Period, the graptoliths are found in tangled heaps in certain black schists. As the name suggests, they are traced on the slab of rock like some form of handwriting (*Grapto lith* = writing on stone).

by that of the graptolites, creatures that have been used to identify the 39 zones of these Periods. While continuing to remain colonial in life-style, they were subdivided into two groups. The first, appearing in the Cambrian Period, comprises the essentially stationary, encrusting and erect forms. The major subgroup of these is that of the dendroids, whose conical formation of the genus *Dictyonema* is illustrated on page 101. Starting from an initial bud, a series of chambers arranged in alternating groups of three formed along the creature's branches; the smaller chambers were inhabited by the male individuals and the larger cavities by the females.

Most dendroids were anchored to the seabed, but certain colonies of *Dictyonema* were attached to floating masses of algae or even ended in a small spherical float. This way of life was to predominate in a second group that derived from the first, the graptoloids. Their stratigraphic significance is enormous, since their representatives, carried great distances by the currents of the oceans, have – like *Dictyonema* – attained worldwide distribution of the kind that permits long-distance associations. The graptoloids seem to have been hermaphrodites. Depending on the species, the branches lived separately or grouped together into a 'colony of colonies' beneath a huge float (see page 101).

The graptolites disappeared during the Carboniferous Period. After many transformations, this class became linked to the modern pterobranchs and hence to the Chordata. These very primitive organisms are therefore quite closely related to the very complex vertebrates.

Right: *Dictyonema flabelliforme*, a dendroid typical of the Cambrian–Ordovician Period boundary. Using its adhesive organ, the colony (shown life size) was able to attach itself to a floating object, a fact that explains the practically global distribution of this species. Cross-braces were used to stiffen the longitudinal rows of different-sized chambers. As with modern pterobranchs, this is probably explained by males and females being different sizes. A feature common to both graptoliths and pterobranchs is the very special structure of their chitinous test which is both rigid and elastic. Unlike *Dictyonema*, the chambers of *Climacograptus* from the Ordovician Period, are all equal in size. Graptoliths had both male and female sex organs.
Sometimes very well preserved colonies are found like those shown below. Each branch ended in a central chitinous capsule which was itself supported by a 'pneumatophore', a large vesicle that guaranteed the flotation of the whole colony, while a crown of spherical reproductive organs ringed the pneumatophore.

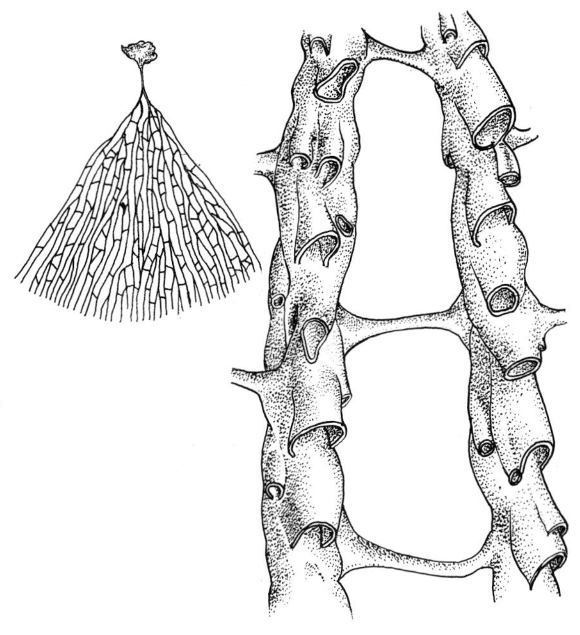

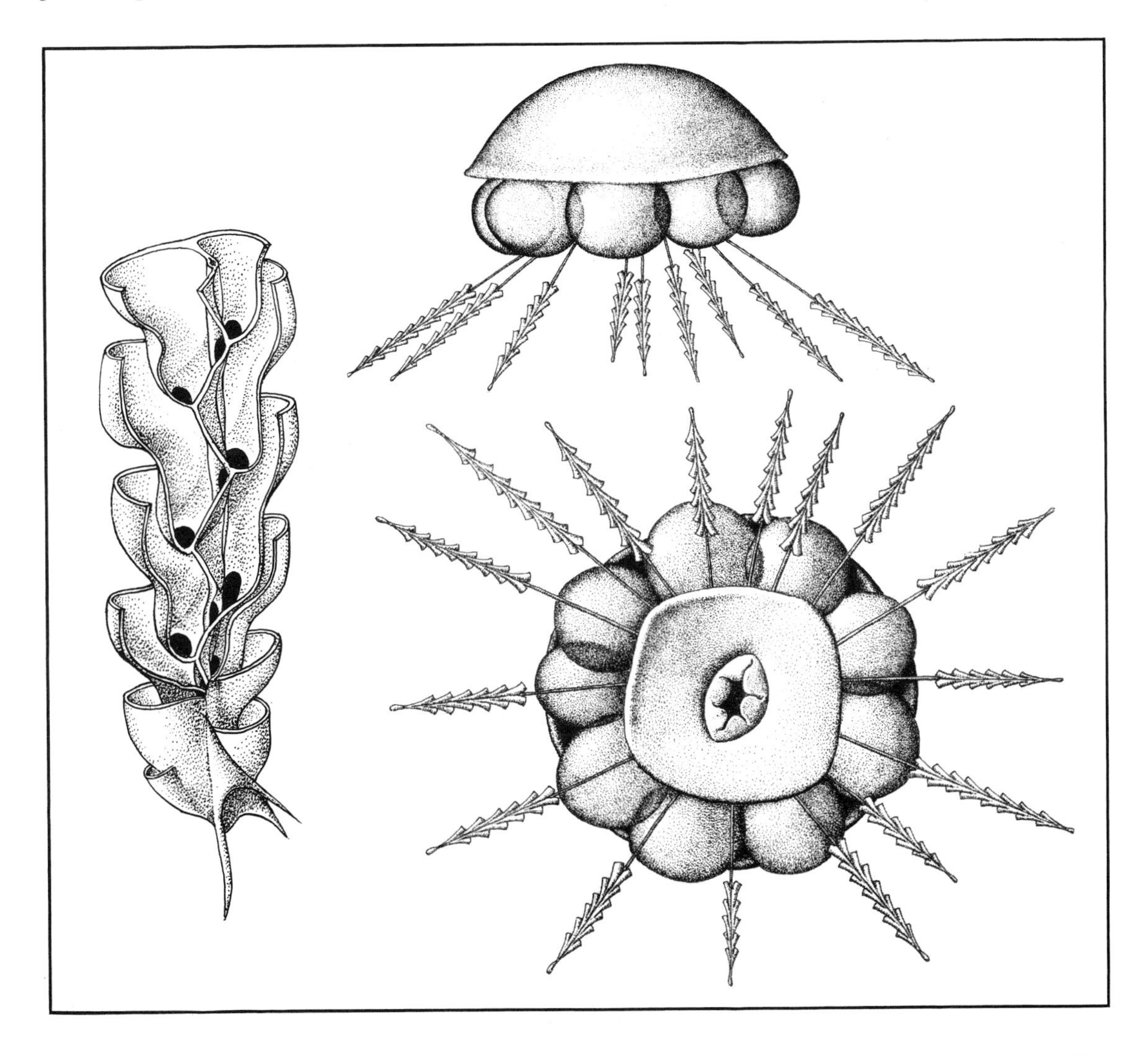

Chapter Four

The World of Water: Fishes

The world of water: Fishes

The origin of the vertebrates

The science of classification, which attempts as far as possible to account for the relationships between animal groups, places together the echinoderms, the Stomochordata and the Chordata in the Deuterostoma. Since the phylum of the vertebrates represents almost the whole of the Chordata, the non-vertebrate Chordata obviously occupy a less significant place in living nature, and include the tunicates and *Amphioxus* (see below). They did, however, undergo considerable development in the remote past, and the carpoids, which were sometimes called 'Calcichordata', are the result of this according to the British palaeontologist Jefferies, a view about which there is much argument, however.

The Chordata derive their name from the notochord or flexible column which supports their body. This is composed of a fibrous sheath comprising numerous cells, and which has been retained in the adult *Amphioxus*, but exists only in the larva or embryonic stages of the Tunicata and higher vertebrates. Above the notochord there is a neural tube that forms a basic component of the creature's nervous system, with the brain being a part of it. Another important feature of the Chordata is that, behind the mouth, the ventral digestive tube forms a 'basket' opening to the outside by a series of lateral slits. These slits, later to become the gills of fishes, originally had a nutritive role, filtering the water taken in through the mouth so as to retain the food particles. This is a device also found among the Stomochordata.

The vertebrates are active animals, and their body structure reflects an adaptation to an active way of life. Their great freedom of movement is underlined in the first instance by an initially elongated body shape of marked bilateral symmetry. The head is clearly distinct and contains the brain, as well as the principal sensory organs; but chiefly it is reinforced by an internal skeleton called the cranium. The rest of the body is supported on articulated components made of cartilage and bone. Bone is found nowhere else in the animal kingdom, but we shall see that not all vertebrates have it. The reason

In the adult state, the ascidians lead a fixed life and the salps allow themselves to be carried passively by the currents, but *Amphioxus*, of which there are 25 known species, swim by means of wriggling. They spend most of their lives with their mouthparts stuck vertically in the sand. Their bodies are divided into a series of similar sections – rather like the simplest of the vertebrates. But despite its 'little fish' look, *Amphioxus* is very much more primitive than any vertebrate.

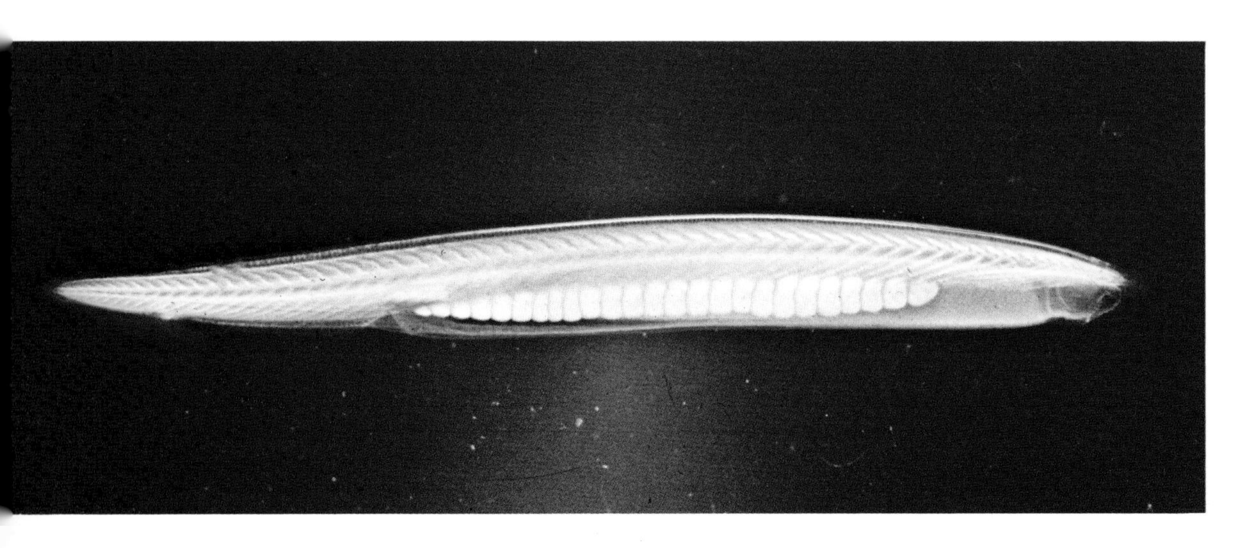

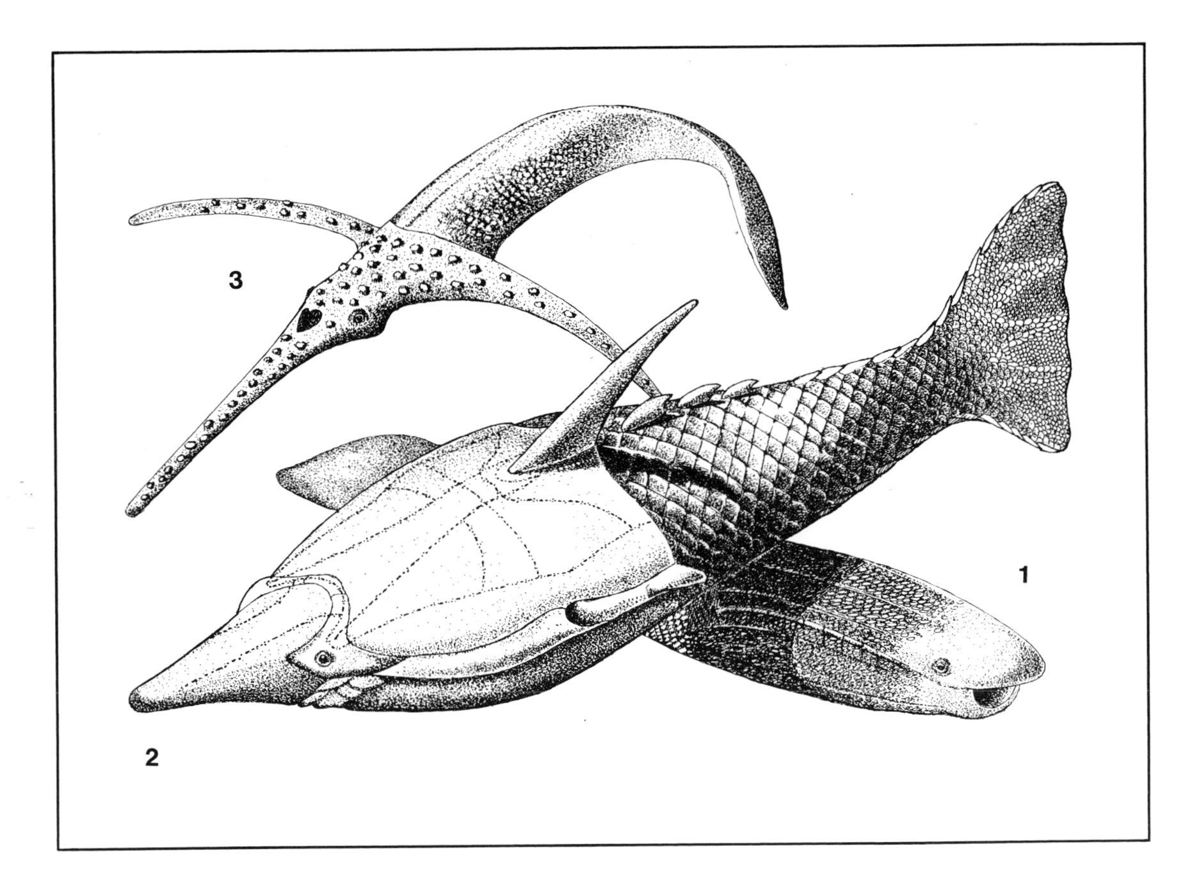

1 *Arandaspis prionolepis* from the Lower Ordovician Period of Australia, the oldest vertebrate we know. It is the most primitive of the Heterostraca.
2 *Pteraspis rostrata* from the Lower Devonian Period of Europe, a typical heterostracan that has been the subject of much study.
3 *Lungmenshanaspis kiangyouensis*, a galeaspid from the Lower Devonian Period. The name of this rather strange creature tells us that it comes from China.

for the existence of bone has been discussed for many years. L.B. Halstead, the British palaeontologist, maintains that it constitutes a reserve of phosphorus, an element particularly essential to muscular activity so vital in vertebrates over a certain size. A primitive skeleton of cartilage would therefore have been joined by a superficial bony skeleton consisting of plates whose secondary function would have been to support the body and to act as protection against predators.

The first fishes

The world is inhabited by some 42,000 different species of vertebrates. This phylum – all the more important because we belong to it – has conquered the continents and the skies, but nevertheless originated in the aquatic environment where the first episodes of its development took place. Half of all modern vertebrates are 'fishes', a convenient but nonetheless vague term since it in fact covers a very mixed group of creatures.

The first traces of the vertebrates go back to the Upper Cambrian Period: minute fragments of bony shell covered in oval tubercules, found in the north-east of Wyoming and identified in 1978 by the name *Anatolepsis*. While the true appearance of this creature still eludes us, the microscopic structure of its bones is very close to more recent forms known as the Heterostraca, the group which provided us with the first complete fossil 'fish'. Shortly before the publication of the finding of *Anatolepsis* put the age of the oldest known vertebrates back to around 500 million years, an Australian researcher found a well-known preserved heterostracod in the Lower Ordovician in the region around Alice Springs, right in the heart of Australia. *Arandaspis* (see above), named after the celebrated aborigine tribe, was discovered in

Cephalaspis lyelli, an osteostracan from the Lower Devonian Period of Britain. This is the classic fossil of the 'old red sandstones', sediments laid down both in the middle and on the margins of the newly emerged lands that made up Northern Europe at that time.

sandstone deposited in a shallow marine basin, the type of environment that seems to have been the vertebrates' original world.

Heterostraca

About 12 centimetres ($4\frac{1}{2}$ inches) long, *Arandaspis* had a streamlined body, the front half of which was enclosed in a bony carapace while the rest was covered in small scales arranged in vertical rows. A longitudinal bone separated two flattened sections on the top of the carapace, while the ventral region was rounded. On the sides were a series of small, lozenge-shaped plates. Finally, and most important of all, *Arandaspis* had no fins or jaws. The first of these two vital features gives it a curious, rather clumsy appearance, and like many Heterostraca it must have been an awkward swimmer, just sculling along with its tail. The absence of any real jaws is a primitive characteristic found in several groups of fossils as well as in the modern myxines (hagfishes) and lampreys.

The present-day myxines are the most primitive of all known vertebrates. For a long time they were classified along with lampreys, but their resemblance to them seems chiefly due to an evolutionary convergence linked by the fact that both groups have rather similar life-styles. The skeleton of the myxines has no vertebrae and consists solely of cartilage, and it is possible, even probable, that their ancestors never had bony skeletons at all. These worm-shaped creatures have no paired fins; their anatomy and physiology contain extremely ancient features which put them a long way from all other vertebrates. Having separated off from the rest of the phylum at a very early stage, probably during the Precambrian Period, the myxines live hidden in the mud on the seabed. They are blind, and come out at night to devour with their horny teeth fishes that are dead or dying, but sometimes living prey as well.

Apart from *Arandaspis*, the Heterostraca are represented in the Ordovician Period only by very fragmentary fossils. *Astraspis*, *Pycnaspis* and *Eriptychius* from the sandstones of the Black Hills of South Dakota and Montana have only yielded pieces of bony armour that have been attributed to the Heterostraca because of their microstructure. Beneath the hollow bone that bore the tubercules, there was a bony foundation devoid of cellular cavities – probably the most primitive type of bone. These jawless vertebrates flourished during the latter third of the Silurian and Devonian Periods only.

The frontal armour of the Heterostraca consisted of a dorsal and a ventral shield and independent plates capable of expanding along their edges, as indicated by a number of concentric striae. These bones were decorated with tubercules, wrinkles or ribs and also contained a network of furrows incorporating the lateral line system, a sensory organ found in living fishes. The jawless mouth was frequently lined with small movable plaques bearing dentine ridges. In typical heterostracods, and probably *Arandaspis* too, once the seven pairs of branchial pouches had been irrigated the water was collected on either side of the head in a channel that opened to the outside through a single opening. The eyes, one on each side, were supplemented by a pineal organ, a central eye covered by a solid plate of tissue. This means that either it was transparent at that point, or the pineal organ had already lost all visual function. Large bony spines, both dorsal and lateral, may have been used to stabilize the creature when swimming, but had no connection with true fins despite their position. Propulsion was carried out by a paddle-shaped tail. The creature's internal anatomy is all but unknown, since although the superficial skeleton was bony, the underlying, wholly cartilagenous skeleton was incapable of being fossilized.

The Heterostraca demonstrate a large diversity of forms, most of which are small in size, although *Tartuosteus* of the Upper Devonian Period attained sizes of up to 1.5 metres (5 feet). While *Pteraspis* represented a sort of average type, certain genera were quite different. For instance, *Eglonaspis* was totally blind, and had a tubular mouth which sucked in minute particles of food from the mud. These bottom-dwelling habits were shared by numerous Heterostraca: *Drepanaspis* for example, with its dorsal mouth and completely flattened profile, was perfectly adapted to life on the sea-floor, while other types provided with dorsal and lateral extensions were doubtless better swimmers.

Thelodonts

It is quite clear that the frequency of the discovery of a group of primitive fishes in the fossil state, and our understanding of them, will depend chiefly upon the extent to which bony tissue was present in early representatives. We may never discover the slightest trace of the very first vertebrates, for in all probability they were completely cartilagenous, and did not fossilize well. The thelodonts, primitive jawless creatures covered in small scales rather than solid plates of skin, are hardly known at all. Scales with the structure of minute teeth and found among the sediments may indeed be attributed to

Here is another osteostracan: *Aceraspis*, a primitive genus from the Silurian Period of Norway (about life size). Behind its cephalo-thoracic shield we can see that the body was divided up into similar sections (this is called metamerization). The line of small polygonal plates running around the shield may have been sense organs of some kind.

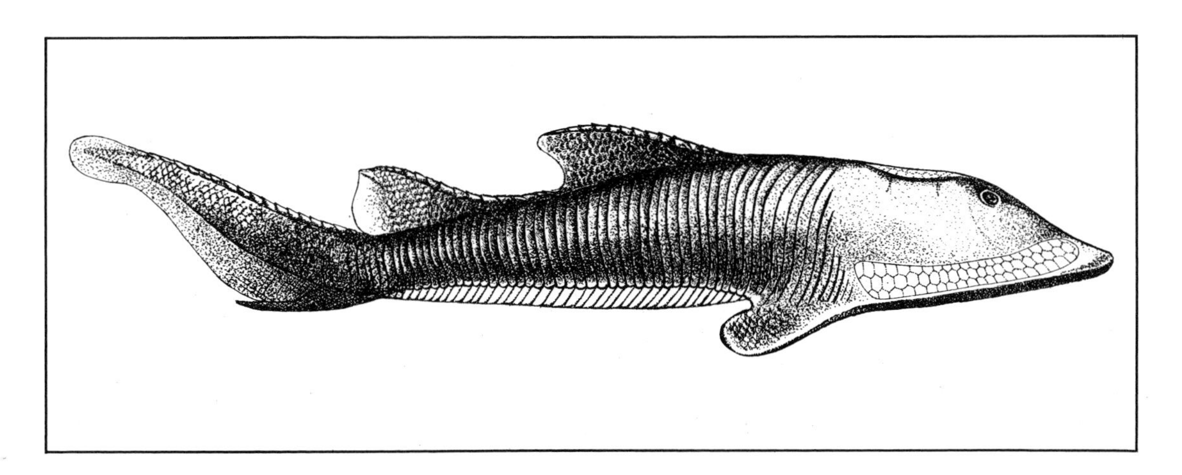

them, but only three species have so far produced relatively complete specimens of this creature.

Although many small-scaled forms which have been discovered have all been called 'thelodonts', this is more because our understanding of the group is still rather poor. As more discoveries are made and our knowledge increases, some of these creatures may need to be re-classified. The best-preserved specimens have a hypocercal tail and two side folds that may have been the beginnings of paired fins, or were perhaps just caused by the flattening of the specimens in the sediments. Beneath each fold can be found a series of approximately eight branchial apertures.

Osteostraca

Unlike the thelodonts, the Osteostraca were extremely bony animals and hence are well preserved. Despite the 350 million years or more that separate them from ourselves, their anatomy is better known than that of many modern groups. Like the Heterostraca, they were protected by a bony armour that covered head and thorax, but this armour was made of a single block (see page 107), and once it was in place, the chances of further growth were slim, which explains the similarity in the size of individuals within the same species. This size never exceeded 50 centimetres (20 inches) and was more often some 10 centimetres (4 inches) in length. A feature unique to these jawless vertebrates was that after the bony parts of the skin were formed, this was followed by a later bone covering of the internal cartilage, which therefore 'moulded' many of the animal's soft anatomical structures.

The shield that covered both head and thorax was frequently semicircular in shape. In the course of their history, which fossils allow us to pursue from the Upper Silurian Period to the end of the Devonian Period, the Osteostraca tended to consolidate and develop their dorsal or lateral body projections, while a vast branchial cavity closed off by a membrane was hollowed out beneath the shield. In front of this was the mouth, and inside the cavity, ten pairs of branchiae were supported on bony 'arms', with each branchial chamber communicating with the outside through a separate aperture.

On the dorsal surface between the two narrowly spaced orbits was a third eye beneath the pineal orifice, and further forward we find a single nostril. The base of this little sac is the origin of the hypophysis, a small gland situated at the base of our own brain. The hormones that it secretes are a constant regulator in our physiology – encouraging growth, making uterine contractions begin during birth, reducing the diameter of blood vessels during a haemorrhage or even regulating the water content of the body.

These decidedly odd creatures had part of their bodies covered with patches of small moveable polygonal plates around the outside of the shield and in the central region (see page 107). These patches – unknown elsewhere in the living world – are a puzzle to palaeontologists. They were armed with nerves, and were perhaps sensitive to variations in the surrounding water presssure. If this is true, they fulfilled a function similar to that of the sensory lateral line of modern fishes. They may also have been electric organs.

The shape of the shield underwent various alterations in different species: a long central guard was seen in *Boreaspis*; this was supplemented by very long lateral points in *Holeaspis*, or even by holes around the outside as in *Sclerodus*. All in all, however, their diversity never reached that of the Heterostraca. Nevertheless, the Osteostraca possessed scaly paired fins that helped them in their movements. The two lobes of the tail were always unequal, the tail being 'epicercal' – in other words, the backbone ran into the upper lobe of the tail.

The Osteostraca have been found in abundance in deposits of old red sandstone, particularly in Britain, Norway (Spitzbergen), Russia and Canada. These sandstones, which owe their colour to the strong presence of iron oxide, were laid down during the Devonian Period in lakes and lagoons. As is suggested by their very flat lower surface, the Osteostraca swam with their mouths pressed to the floor in search of minute particles of food in these expanses of fresh or brackish water.

Galeaspids

A new grouping of jawless fossils has made

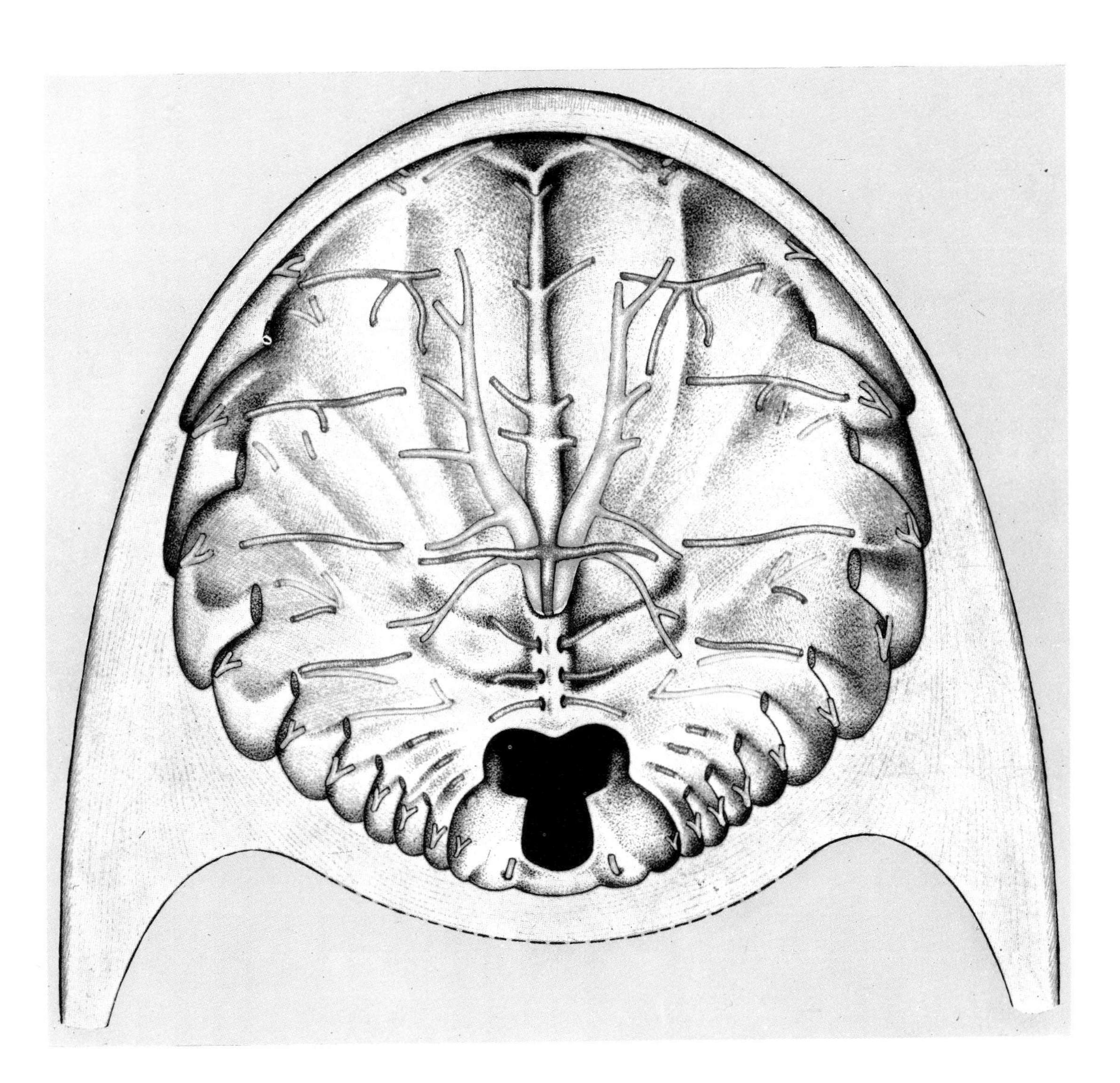

The bones making up the cartilaginous skeleton of the cephalo-thoracic 'shield' that belonged to the osteostracans have given us far more anatomical information about these creatures than is known about the heterostracans. Nowadays these fossils are mostly extracted by using acid, while the technique of series polishing has often been used to explore every detail of such specimens. The shape of the encephalus (part of the brain), the position of the heart and blood vessels, the route of the digestive tract – practically the entire anatomical organization of these animals has been identified, and the above reconstruction is a spectacular example. This is a ventral view of the vast oralo-branchial cavity found within the shield of a *Mimetaspis* from the Lower Devonian Period of Spitzbergen (× 3). The ventral membrane and branchial arches have been removed to show the nerves and blood circulation in the roof of the cavity. The nerves are shown in yellow, the veins in blue and the arteries in red and mauve.

its appearance, so to speak, since 1965, the year in which the galeaspids were first found in the Lower Devonian deposits of Yunnan and Sechuan in China. As the illustration on page 105 shows, their appearance is very reminiscent of the Osteostraca: we find a flattened cephalothoracic (head and body) shield that is sometimes rounded and sometimes ornamented by a central guard and lateral horns. But an extra detail draws our attention immediately – the gaping aperture between their eyes. Oval in shape, it extended transversely; in others it was shaped like a heart or even like the slot in a piggy-bank. The internal anatomy also has some surprises in store: for instance the branchial chambers, whose number in certain genera exceed 20 on each side. These creatures lived on the sea-bed or even buried in mud,

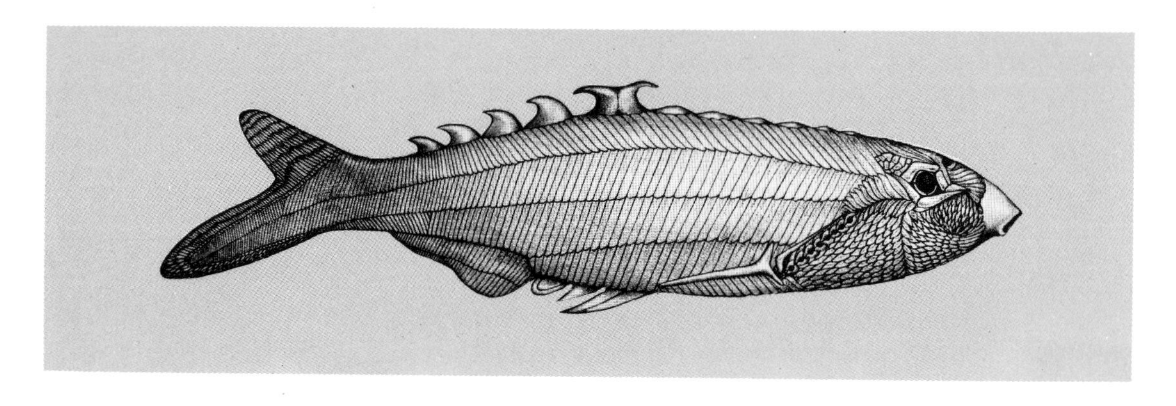

inhaling the water they needed to breathe through their characteristic nasal opening. There is little doubt that this was indeed the function of this duct, which was initially connected to the mouth.

The galeaspids, a large number of whose species have been discovered, had no paired fins. Though rather specialized animals, they nonetheless come lower than the Osteostraca on the evolutionary tree of the vertebrates, and are placed quite close to the Heterostraca.

The anaspids and the origin of the lampreys

The anaspids are clearly related to the Osteostraca; in addition, they showed many of the characteristics of the lampreys of modern times. The anaspids had lost the heavy cephalothoracic shield and were covered in numerous scales which were arranged in herring-bone rows – a pattern which reveals the segmented nature of the body (see above). The body itself was laterally com-

Above: *Birkenia*, an anapsid from the Upper Silurian Period of Scotland. The lateral flattening of the body helped the fish to swim strongly in the open sea but the combination of paired fins and a tail with a large lower lobe (hypocercal) is rare and only found in other agnathans, the ichthyosaurs, Mesozoic marine reptiles and modern flying-fishes. This probably meant that the fish swam rather strangely, moving with their heads downward and their mouths close to the sea-bed where they fed upon small organisms.

Below: Lampreys reproduce in fresh water, but some species reach the sea as adults. Soon after they are born the lampreys bury themselves in the mud on the river-bed and stay there for several years, feeding on the microscopic organisms carried along by the currents. Once adult they are parasites and use their funnel-shaped mouth surrounded with horned teeth to attach themselves to fishes (see illustration opposite). Of course these two ways of feeding, microphagous and parasitic, are in fact the only possibilities open to an animal without any jaws.

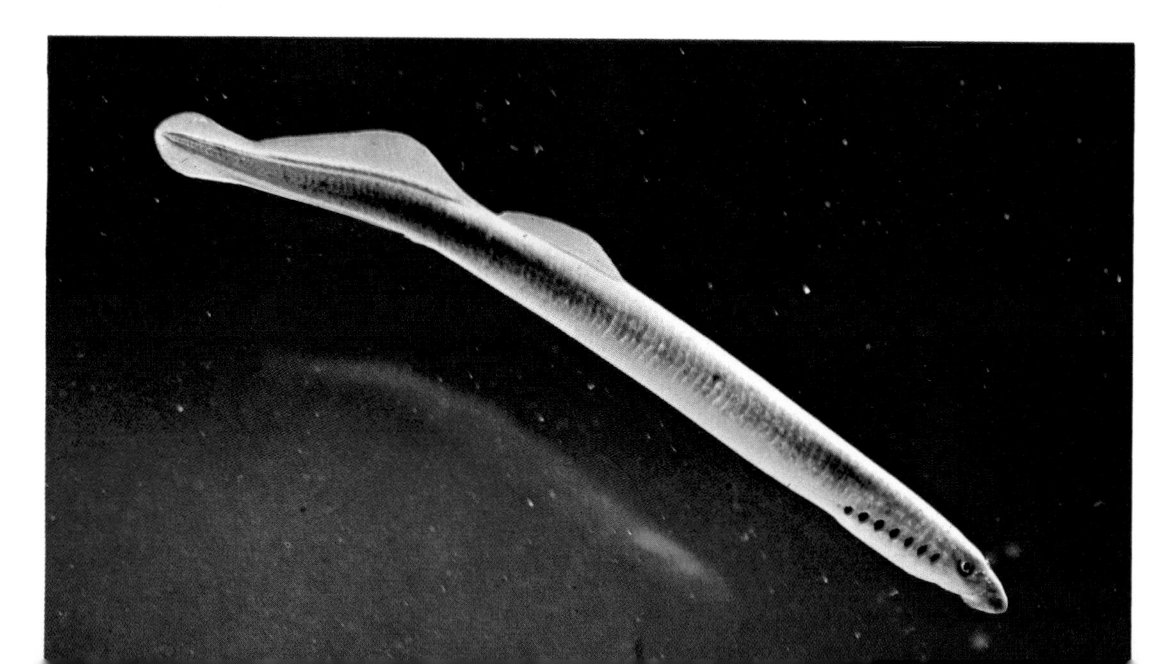

pressed and rarely longer than 30 centimetres (12 inches), while the tail was always 'hypocercal', in other words the backbone of the creature's body was continued into the lower lobe of the tail, a fairly rare feature. In addition to one anal fin there was a dorsal fin, too, but in most instances this was replaced by simple crest scales, as in *Birkenia* (see page 110). The paired fins consisted of bony spines extended rearwards by a lateral fin. The head, covered with small scales, had the same openings as those found in the Osteostraca; flanked by large eyes, it ended in a snout formed by soft tissue and a round mouth. Twelve gill openings were situated well to the rear.

Like the Osteostraca, the anaspids have been found in the old red sandstone of the Devonian Period; they have also turned up in Silurian marine formations as well. They were animals built for swimming and, like most Palaeozoic Era jawless vertebrates, were microphagous (in other words they lived off fine particles). Certain shells of creatures known as the Gigantostraca have been found with circular perforations whose diameter corresponds to that of the mouth of an anaspid. No one has yet attempted to state that the anaspids lived their lives in a similar way to that of the lampreys, some of which attach themselves to large fishes in order to gnaw their flesh and suck their blood, but it is possible that the Gigantostraca suffered the same fate at the hands of the anaspids. Still, a milestone on the road leading to the lampreys is represented by *Jamoytius kerwoodi* from the Upper Silurian Period of Scotland, which is derived from primitive anaspids, and which still possesses paired fins, a feature that disappeared in the lampreys. The first true lampreys were dug from the Carboniferous deposits of Illinois; present-day lampreys have hardly changed at all, however, in the intervening 260 million years.

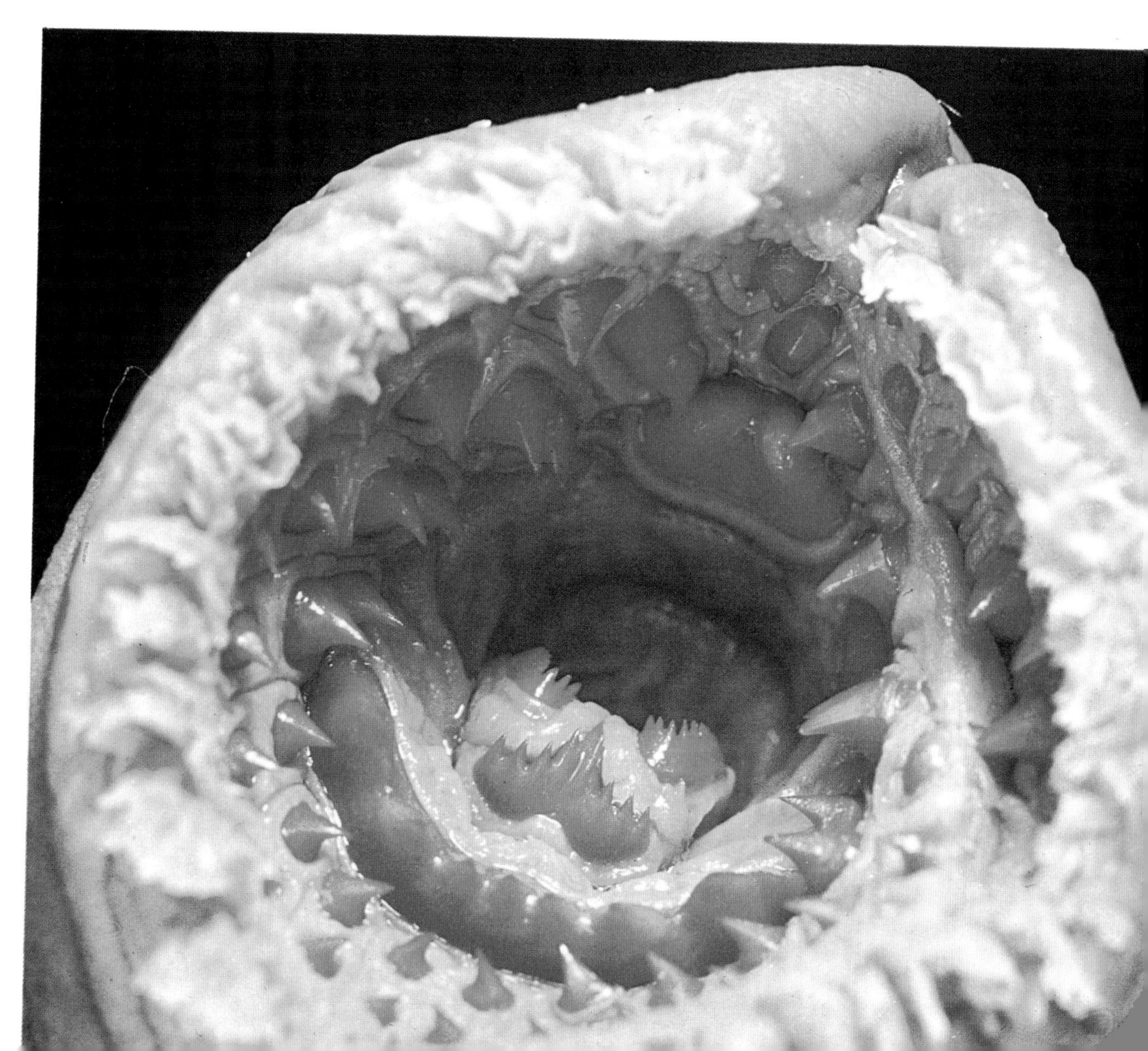

The first jawed fishes

In the evolution of the vertebrates the development of the first species with true jaws was an event of far-reaching consequences. Jaws enabled fishes to eat larger items of food, especially other fishes. Until this happened, vertebrates had been limited to eating small food particles or to living parasitic lifestyles, which restricted their chances to evolve. The advent of jaws was to bring about a diversification in the vertebrates, with the birth of new groups exploring and utilizing the new ecological possibilities that were now open to them.

How did this transition take place? There are two main factors that can help to explain it: the detailed anatomical studies carried out on the fossils of jawless vertebrates, and the analysis of the embryonic development of the jawed vertebrates, which to a certain extent enables us to retrace their history. The results of these two lines of enquiry led to the conclusion that there had been a basic change in the part of the branchial (gill) apparatus.

In fishes, the gill system is supported by a part of the skeleton arranged in a series of arches beneath the brain box. In the jawed vertebrates the first two arches, which were originally cartilaginous, have completely changed, the first of them becoming much larger and jointed in two places to form the jaws. The upper part was to remain more or less free moving in some species, or to become fused to the skull, as in Man. The second arch, on the other hand, was destined to become the hyoid arch, of which the bone located above our own larynx is a vestige. The other arches retained their breathing function.

The ancestor of the jawed animals is undoubtedly one of the jawless ones, but which one? The question is still a matter for debate, and for many years the Heterostraca were the strongest contenders for the title. In the amphiaspids in particular, the first branchial slit seemed to have been modified into a spiracle, a small aperture which we find in sharks especially, located between the mandibular arch and the hyoid arch. However, because of their highly developed tail flippers and paired fins, the Osteostraca and anaspids are now frequently regarded as the cousins of the jawed vertebrates.

Left: A reconstruction of the genus *Dunkleostus* from the Upper Devonian Period, the largest of all the known arthrodires.

Spiny sharks and armour-plated fishes

The oldest known and probably the most primitive jawed vertebrates are the acanthodians. Their hyoid arch was still hardly different from the other branchial arches. They appeared in the Upper Ordovician Period, but flourished chiefly during the Lower Devonian Period. Certain authors like to refer to them as 'spiny sharks', a justifiable name even if they have no direct links with those other sea creatures.

The acanthodians were fairly small; 30 centimetres (12 inches) maximum and often much shorter, with a body completely covered in very thick, small scales placed side by side. Their fins consisted of a triangular membrane held in front by a strong spine, and paired fins were thus arranged in series on either side of the body. A pectoral fin dominated at the front, with a pelvic at the rear. The blunt head bore the large eyes of a hunter. They were also good swimmers, stalking their prey in the rivers and lakes of the Devonian Period continents. Then, in the second half of this Period and during the Carboniferous Period, the acanthodians disappeared beneath the weight of competition from more sophisticated forms. The last of them, *Acanthodes* of the Lower Permian Period, was a very elongated creature whose body was only partly covered in scales.

In the Devonian Period, the acanthodians lived alongside the placoderms or 'armour-plated fishes' – other primitive jawed vertebrates which frequented fresh waters but which also inhabited the oceans in numerous species. As their name indicates, their body was protected at the front by a 'breastplate' of bone. They should not, however, be confused with the jawless vertebrates which had the same feature, since the placoderms possessed jaws, and what jaws! Their carapace nearly always pivoted in two sections, one covering the head and the other the thorax. Behind the armour-plating, the body, which ended in an

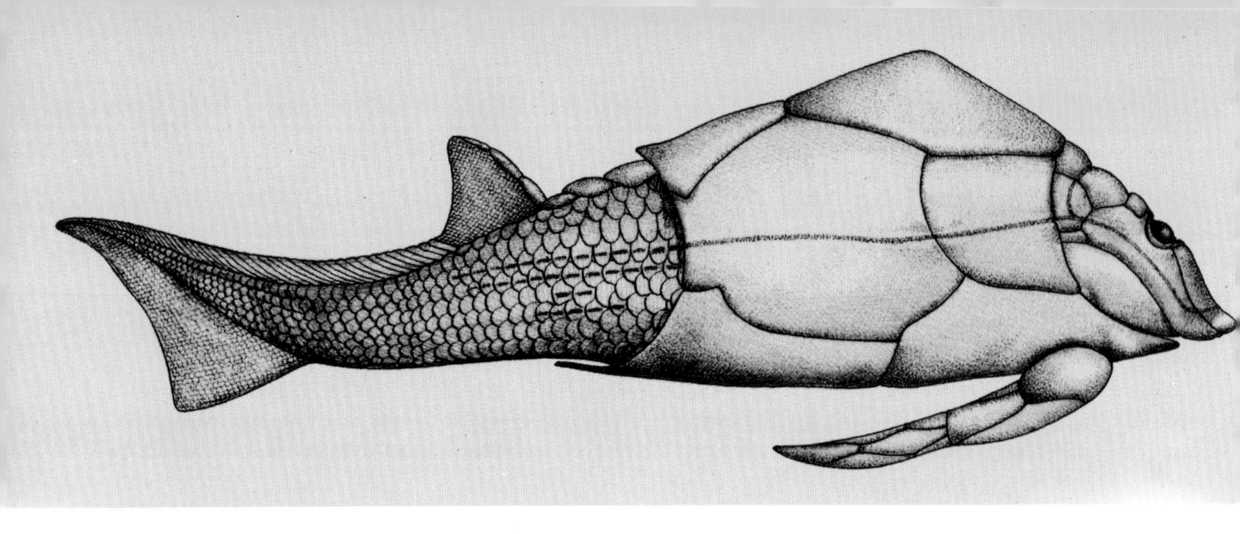

High tides washed up antiarchs on to the many beaches of the 'Old Red Sandstone' Devonian continent, and these creatures were often fossilized in large numbers. They could haul themselves over short distances on land, using their large front fins. Above is *Pterichthyodes milleri* of the Middle Devonian Period, Scotland.

epicercal tail, was either covered by scales, especially in the primitive forms, or bare. The thick plates covering head and trunk were decorated with tubercules, wrinkles and cup-like depressions, criss-crossed by furrows containing the sensory organs of the lateral line system.

Arthrodires

The placoderms are subdivided into several units of which the arthrodires are by far the most numerous and diversified. Our reconstruction of *Dunkleosteus* (see page 112) gives a fairly accurate impression of what these large carnivorous arthrodires looked like. Instead of actual teeth their jaws were equipped with bony picks in the shape of cruel shears, and right at the front were sharp 'fangs', top and bottom. During the Devonian Period they increased in overall size and their thoracic carapace became smaller. While the very first arthrodires never exceeded 50 centimetres (20 inches) in length, the Upper Devonian Period produced gigantic specimens. *Dunkleosteus* – whose armour only has survived – must have been a massive 7 metres (23 feet) or so long, with 'teeth' attaining some 58 centimetres (23 inches) in length. There is no living creature like it. The homostiids, rather less frightening, actually ambushed their prey, half burying their extremely flat, 2 metre ($6\frac{1}{2}$ feet) long bodies in the sand and waiting patiently for any animal unfortunate enough to pass within reach of their mouths.

Not all the arthrodires were such ferocious meat-eaters, however. *Titanichthys*, of which the largest specimens attained a length of 5 metres ($16\frac{1}{2}$ feet), had no cutting plates in their jaws and fed primarily upon algae or molluscs.

In the course of their evolution the placoderms colonized very different ecological niches. *Gemuendina* of the Lower Devonian Period of the Eifel region of Germany looked very similar to the rays, which made their own appearance much later. Among animals with no common descendancy the similarity of appearance can be explained by an identical way of life. In the case of the ptyctodonts, however – creatures astonishingly similar to the living chimaera (see page 119) – a true evolutionary relationship can be seen, in which case the latter would be the sole survivors of the armour-plated fishes that died out 340 million years ago.

Antiarchs

The antiarchs were very odd fishes indeed; if we were to choose a nickname for them it would surely be 'crustacean-fishes'. Preferring to live in fresh water, they were never very large, but multiplied rapidly. Their general organization was similar to that of the arthrodires, and the discovery of *Yunnanolapis*, falling half-way between the arthrodires and the antiarchs, justified – if any justification was needed – the grouping of the latter group with the placoderms. In the antiarchs the thoracic shield was far more extensive than the head shield. As in the arthropods, whose external skeleton is a major feature, the antiarchs' muscles were attached beneath their armour-plated outer skeleton, while the centre of the head incorporated the two eyes, set close together in a single aperture, together with the pineal organ and the nostrils.

Many of the antiarchs, like *Bothriolepis*, were flattened, moving over the sea-bed by means of their strange bony arms, looking for food in the sand and mud with their weak jaws. To a certain extent, the hard bony box which encased their bodies must have protected them against the arthrodires whose prey they often were.

The end of the Devonian Period sounded the death knell for the huge group of placoderms, and by the beginning of the Carboniferous Period there was nothing left of the multitude of species that had not long since inhabited the lakes, rivers and oceans. As usual in these cases, the experts are tempted to find some kind of explanation for this general extinction, and it was recently suggested that the extensive retreat of the seas which marked this particular moment in the Earth's history played an important part at least in the destiny of the giant arthrodires which, while in no danger of competition from other large predators, saw their favourite hunting grounds reduced by the seas becoming smaller. There can be no doubt that these ecological upheavals had a major effect upon Nature's equilibrium, but it is never easy to assess their precise consequences. It should be remembered, however, that in talking about extinctions at the end of the Mesozoic Era, events that we see as being abrupt and simultaneous are only made to appear so by the many millions of years that have since elapsed.

Sharks and rays

The selachians (sharks and rays), which are so clearly related to the placoderms, only appear in the fossil state from the Lower Devonian Period onwards. Their age must be assessed carefully, however, since these were creatures which were hardly suited to fossilization. The selachians, like the chimaera, were 'cartilaginous fishes'. In the placoderms the internal skeleton consisted chiefly

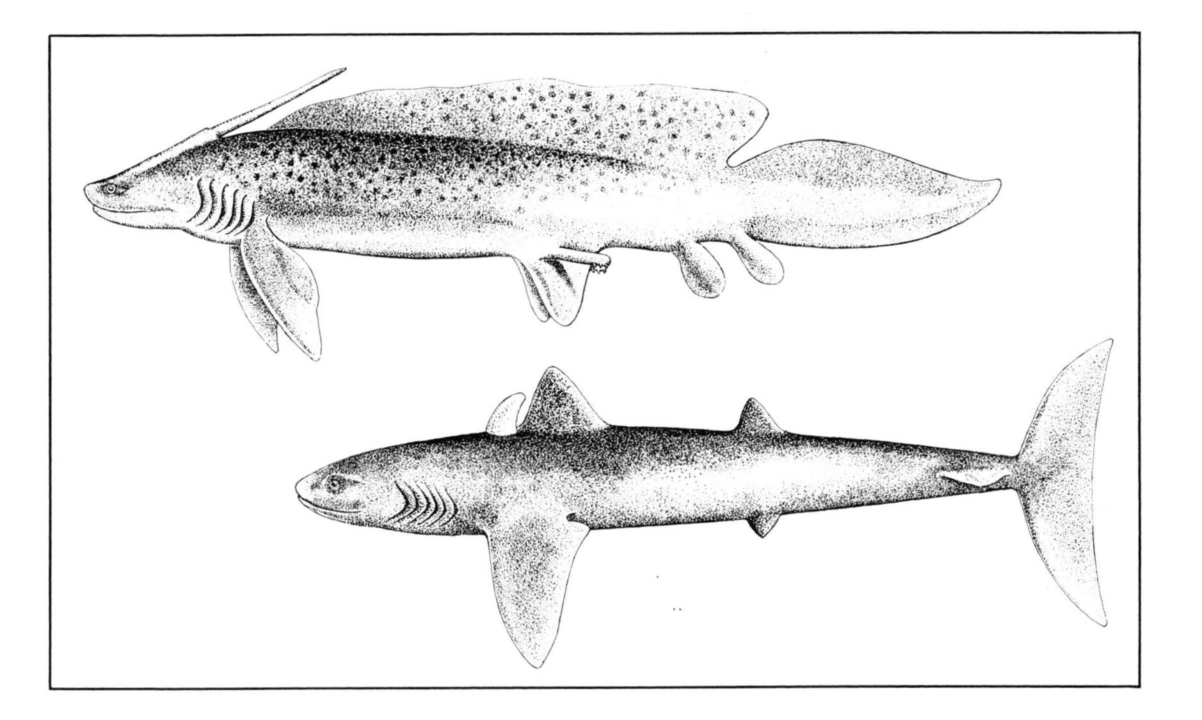

Below: *Cladoselache* of the Upper Devonian Period. Bottom: The chiefly Carboniferous *Xenacanthus*. The sharks have always been predators, and *Cladoselache* was even a cannibal, hunting the young of its own species (these have been found in the stomachs of fossils).

of cartilage while the dermic, or outer body skeleton was, as we have seen, very bony. This was not so with the selachians, and hence their preservation was most difficult – for a long time the first sharks in palaeontological collections were only represented by odd teeth and spines (this still applies to many fossil species). Fortunately for us, however, in some species the cartilaginous skeleton was occasionally calcified, resulting not in bones but in a very characteristic, solid formation of prisms. We have thus been able to find vertebrae, skulls and other parts of certain species. In addition, exceptionally well preserved specimens have provided us with information on the evolution of their soft parts. Nevertheless, the history of the selachians is full of gaps.

Cladoselache (see page 115) of the Upper Devonian Period of Ohio in the U.S.A. is one of the more miraculously preserved genera. We have been able to identify its skeleton, which attained as much as 2 metres ($6\frac{1}{2}$ feet) in length, as well as its skin, muscles and even its kidneys, all of which left their imprint in the Cleveland schists. *Cladoselache* differed from modern sharks by the bony ring surrounding its eyes, by its terminal, rather then ventral, mouth, and by its rigid paired fins. Its tail end was stabilized by a small horizontal 'aileron' on either side of the body. Some specimens featured large, blunt horns situated forward of their dorsal fins, and these were probably males. The latter did not, however, possess the reproductive organ which protruded separately from the pelvic fin on all the chondrichthyans, and which was present on another genus living at the same time, *Diademodus*. This organ is also found on *Xenacanthus*, a small Carboniferous Period freshwater shark characterized by a symmetrical tail and the long spine which extended backwards from its skull (see page 115). The mouth of this more highly evolved form was in the ventral position and its paired fins were much more moveable.

Sharks changed little in the course of their evolution; apart from certain variants, they remained true to the type of large marine predators perfectly adapted to their way of life, making much use of their strong sense of smell to detect prey, and rarely encountering any serious rivals. Their teeth have been found in abundance in Mesozoic and Cainozoic Era deposits, and some of these are of stratigraphic interest. In our own times, the sharks – whose biological success is undeniable – inhabit the waters of the whole World. The largest among them, the basking sharks and the whale sharks which can be up to 15 metres (49 feet) long, are not, in fact, ferocious carnivores, but peaceful eaters of plankton. During the Miocene Period, however, *Procarcharodon megalodon* lived. About the size of the basking sharks, it was a monster whose jaws were armed with serrated teeth as big as a man's hand. The jaws could gape to a width of 2 metres ($6\frac{1}{2}$ feet)! Like the great white shark of today, this super-predator must have swallowed everything unlucky enough to attract its attention, and has left the trace of its terrible teeth on the bones of the great cetaceans (whales) also found in the same deposits.

The origin of the rays is also rather uncertain. These cartilaginous fishes, with their pectoral fins extended to form 'wings', have often been associated with the flat sharks such as angelfishes, but this relationship is far from being proven. The first rays come to our attention during the Lower Jurassic Period, but we have to wait until the Upper Cretaceous Period to see the specialized forms still alive today. Rays also produce bad fossils, and only their ribbed teeth have generally been discovered.

Their particular shape has brought about a whole string of anatomical modifications: the mouth is extended transversely, the very close eyes are located on top of the cranium and, except for the spiracle, the branchial slits have moved into a ventral position. The whole of the rear portion of the body is much reduced in size and ends in a whiplash tail. Like the whale sharks, the giants of the

The teeth of the sharks, often all that are preserved as fossils, are extremely varied. Even in the same individual they can be completely different according to the age, sex and location on the jaw. Very different species, however, may have almost identical types of teeth. A shark's jaw bears several rows of teeth of which only one is used at any one time. When a tooth falls out it is replaced by the one growing in the next row. On the right is a tooth of *Procarcharodon megalodon* of the Miocene Period (× 2 approx.).

group, *Mobula* and *Manta*, have become primarily plankton eaters, swimming along with their mouths wide open, gathering in plankton and small fishes by means of two kinds of flexible horns. Despite their peaceful nature, the sheer size of these animals (maximum wing span 8 metres/26 feet) can make them look fearful and gives them their nickname of 'devil fish', while their habit of leaping out of the water had made them even more terrifying to sailors and fishermen.

Bradyodonts and chimaera

However it happened, the disappearance of the Devonian Period placoderms led to the sudden expansion of a rather mysterious group, the bradyodonts, which occupied some of the ecological niches left vacant by the extinct placoderms. For a long time these animals, which proliferated from the beginning of the Carboniferous Period, were known solely by their teeth and fragments of jawbone resembling the remains of sharks. Recently, however, the discovery of a deposit in the Lower Carboniferous Period limestones of Bear Gulch (Montana, U.S.A.) has given us a far greater insight into the lives of the bradyodonts, some of which, like the edestids that featured spirals of sharp teeth at the front of their jaws, were predators and bore a resemblance to the selachians. Others, probably related to the Holocephala, diversified into a large variety of forms. The cochiolodonts crushed molluscs between their flattened teeth, while the petalodonts gnawed coral just like the modern 'pudding-wife' fishes of the genus *Scarus* do with their beaks. Finally, certain species of bradyodonts from Bear Gulch had a long bony appendage above the head which gave them the name of 'unicorn-fishes'.

The bradyodonts disappeared at the end of the Permian Period, around 225 million years ago, and the chimaera proper first emerged during the Triassic Period. From the Lower Jurassic Period onwards, those found in the fossil state are hardly any different from the modern Holocephala (see page

Above: The complete skeleton of an Eocene Period ray: *Promyliobatis gazolae*. Only in very good deposits like those of Monte Bolca (see page 126) – where this specimen comes from – can we find fossils like these. Apart from the giant pelagic forms (mobulids) most rays are fishes that live on sandy sea floors, as witnessed by their very flattened body shapes. In certain species the long thin tail had a needle at the end which could give an enemy a nasty sting. The rays generally feed on molluscs and echinoderms which they crush with their dental plates like the one shown on the right, belonging to *Ptychodus* of the Cretaceous Period. These teeth are sometimes thought to belong to mollusc-eating sharks.

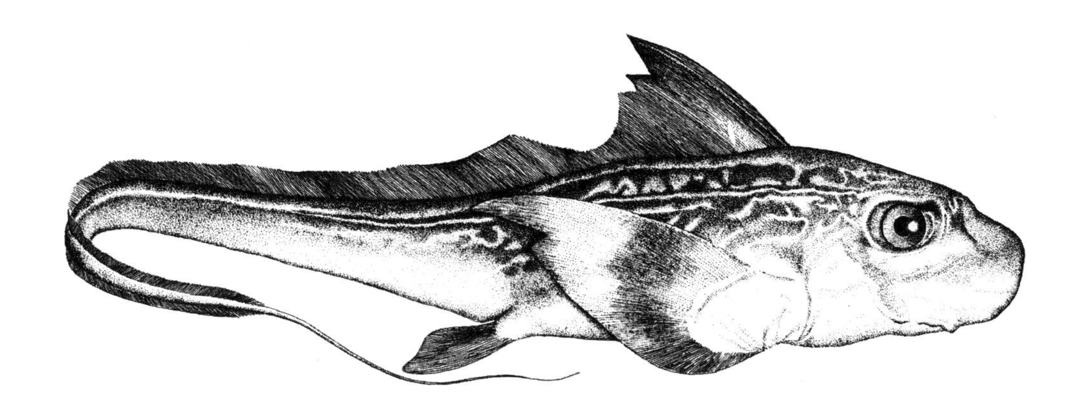

Chimaera – this holocephalian is very primitive, being in some ways similar to the placoderms of the Devonian Period. They are 'living fossils' which must have lived in the safety of the deep oceans for millions of years.

119). Today, these cartilaginous fishes inhabit depths between 200 and 1000 metres (650 and 3280 feet), while the Rhinochimeridae, with their snouts elongated in the shape of a knife, descend as far as 2600 metres (8530 feet). The body of the chimaera narrows down to a wire-like tail which has long since lost its propulsive function. The pectoral fins, on the other hand, have substantially grown in surface area, while the large, stubby snout is often extended by a guard.

The chimaera, whose jaws are equipped with pick-like teeth, catch small animals on the sea-bed; the males possess a clasper, a sort of claw enabling them to hold on to the female during mating. This very special device was found on certain bradyodonts in Bear Gulch, and it is also a feature of the Mesozoic Era chimaera. Initially, this organ was very powerful and common to both sexes, probably as a defensive weapon. From the Middle Jurassic Period, however, it diminished in size and became found only on the males.

Bony fishes

Among the huge numbers of aquatic vertebrates which appeared in the Palaeozoic Era, the class Osteichthyes made its appearance at the end of the Silurian Period. These were the 'bony fishes' with an internal skeleton of bone, unlike the 'cartilaginous fishes' and the placoderms. The acanthodians (see page 113) are sometimes regarded as the most ancient osteichthyans, but their many primitive characteristics make their classification very difficult, and that is why they are dealt with separately.

The osteichthyans were typified by the complex nature of their skeletons, both in the cranium and jaws, as well as the region of the gills and fins; these fins were from then on supported by scales that were modified into thin rods called lepidotrichs. The skin-forming part of the skeleton was well developed in the early forms which were covered in thick scales. So far as the anatomy of the soft parts was concerned, part of the digestive tract was separated off in the bony fishes to become a lung or swim bladder. The function of the former was to breathe air; the latter was a pouch – frequently divided by constrictions into successive chambers – that contained a mixture of gases rich in oxygen. This hydrostatic device enables a fish to balance its own weight and to hold its position in water. The sharks and the large, wandering rays which do not possess this organ have to spend their entire lives swimming or they sink.

The class Osteichthyes is split into two main branches, both of huge importance in the story of evolution. The first is the actinopterygians which were very successful and cover practically all modern fishes – some 20,000 species. The second – the sarcopterygians – finally managed to produce the first vertebrates to be able to survive on land. This was achieved at the end of the Devonian Period.

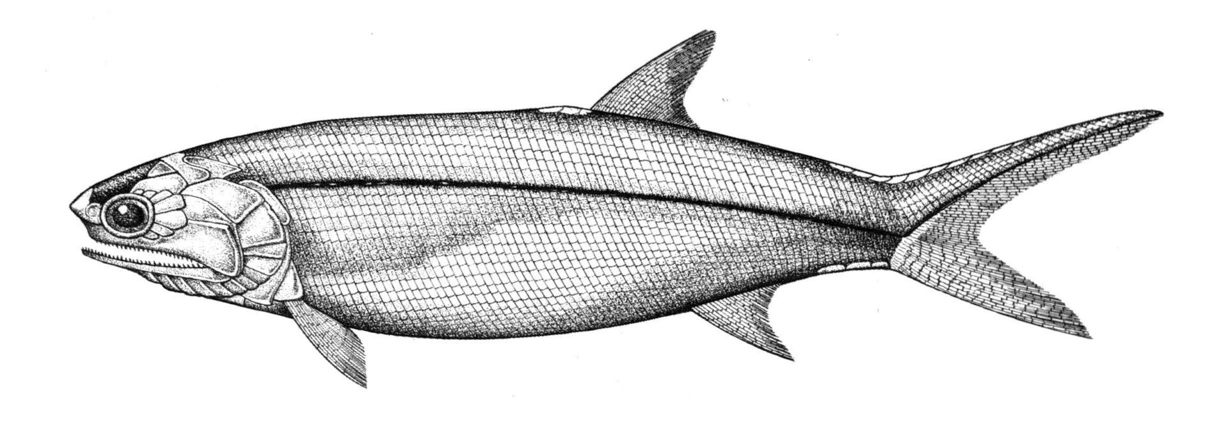

It is the genus *Palaeoniscus*, from the Permian Period, that gives its name to the most primitive actinopterigians: the palaeonisciforms, whose fin lepidotrichs were parallel at first but as time went on spread out.

A living fossil: the coelacanth

The sarcopterygians owe their name, which means 'fleshy fins', to the presence at the base of both their paired and single fins of a fleshy lobe containing pieces of skeleton (see page 121). These animals, which occupy a key position in the evolution of the vertebrates, will be examined in detail in the next chapter. They included the coelacanthiforms – sometimes known as actinistians – which frequented oceanic environments and led a life which is somewhat difficult to understand.

The discovery of a living coelacanth at the end of the 1930s was a piece of news that exploded like a bomb upon the scientific world. The experts had supposed them to have died out at the end of the Cretaceous Period, and yet here were natives of the Comores Islands catching them on the end of their fishing lines just a few hundred metres beneath the surface of the Mozambique Channel. The natives did not share the excitement of the western scientists, since the low-quality flesh of the coelacanth had been dried, salted and sold on the markets for generations. Its rough scales were even used to clean the inner tubes of bicycles before a repair patch was applied to a puncture! It was on 22 December, 1938 that Miss Latimer, in whose honour *Latimeria* was later named, noticed an odd-looking fish some 1.5 metres (5 feet) long on the fish tip in the port of East London, South Africa. She had it taken to the museum where she was curator and sent a drawing of it to Professor J.L.B. Smith who, as J. Dugan reports, was stupefied: 'If I'd met a dinosaur in the street I wouldn't have been more astonished.' From 1952 onwards, the rewards offered to fishermen along with the eagerness of researchers led to the study and collection of a number of fresh specimens. The outcome was quite reassuring for the palaeontologists since, apart from a few details, the reconstructions that had been made on the basis of fossils were in the main justified by the living creature.

Latimeria chalumnae (see page 121) still remains a surprising animal; it dwells in cool waters – 15–16°C (59–61°F) – at depths between 400 and 1000 metres (1312 and 3280 feet), living mainly off fishes and crustaceans. The females are larger than the males and lay eggs that are few in number but large in size – as big as apples. The lobed fins and symmetrical tail astonished its first observers, but when it was dissected many other oddities were revealed as well. The coelacanth has no vertebral column, its body being supported by a fibro-elastic tube filled with a thick fluid. This tube or cord ended in a cranium that was jointed in two parts. Its brain was minute, with most of the cerebral cavity being filled with fat.

Coelacanth-like fossils are known from the Upper Devonian Period to the Cretaceous Period and flourished chiefly during the Permian and Triassic Periods. There were no major developments in their structure; in fact as time went on their skeletons became less complex and they lost the maxillary bone in the upper jaw. Their lung is sometimes bony in fossils, and is fatty in *Latimeria* for which it was a reserve organ. Contrary to what was popularly reported in the Press at the time of the coelacanth's discovery, this group was never at the origin of the land vertebrates (some enthusiastic journalists even presented it as a direct ancestor of modern Man). In fact *Latimeria* was the last of a line which evolved very little and which became more and more dependent on a marine existence.

Below: One of the first fossil coelacanths to be found: *Undina* of the Jurassic Period. Underneath it is the modern *Latimeria*. Both animals show the characteristic actinistian features. We can see the 'diphycercus' tail (arranged symmetrically either side of the line of the body which is extended by a fleshy lobe). A small mysterious organ close to the nasal capsules which had been noticed in a Devonian actinistian called *Nesides* was discovered in *Latimeria*. Its precise function is not known and it is not found in any other living vertebrate.

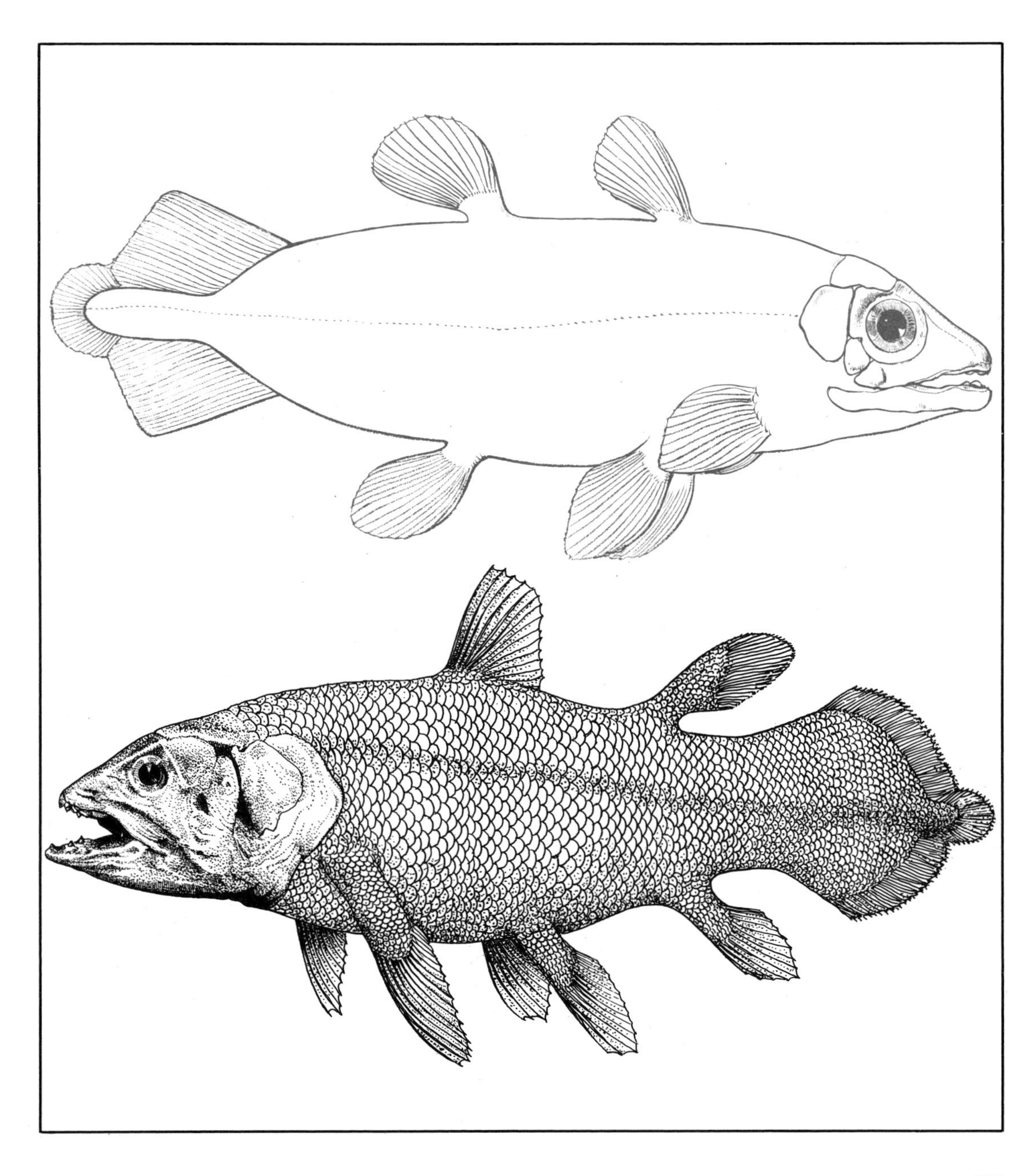

The mysterious polypterus

The Devonian Period was a time of real growth and expansion for many groups of fishes, and subsidiary branches were grafted on to the two main ones of actinopterygians and sarcopterygians, representing genera which are difficult to classify. The struniiforms, which existed chiefly during the Upper Devonian Period, are examples of a group which is hard to define, and their fossils, although still largely unknown, show a confusing blend of characteristics. Their double dorsal fin and symmetrical tail made them appear closely related to certain sarcopterygians, but their truncated snout and large eyes – indicating that sight was more important than smell – their maxillary structure and in particular the shape of their fins, are distinctly actinopterygian traits.

Polypterus of the Nile (see above right) is another puzzling creature, and was identified for the first time in 1802 by the celebrated French naturalist Etienne Geoffroy Saint-Hilaire who, like so many learned men of the day, had accompanied Napoleon Bonaparte during the Egyptian campaign. *Polypterus* and its only related genus, *Calamoichthys*, are uniquely African, the meagre fossil remains of *Polypterus* which we have, just a few scales and vertebrae, all coming from that continent. The vestiges do not go back further than the Cretaceous Period, and yet these very singular creatures must have appeared earlier, during the Palaeozoic Era. Like the coelacanth, they are regarded as 'living fossils.'

Some features of *Polypterus* are reminiscent of the most primitive of the actinopterygians, others of the sarcopterygians; many, however, are quite original. The base of the paired fins consists of a fleshy lobe, but its skeletal organization is very special, with the dorsal fin being fragmented into a series of flag-like structures, each extended behind a small 'mast'. The mandible reminds one of that of *Latimeria*.

Polypterus was a predator with a keen sense of smell, capable of living in murky waters low in oxygen. Beneath the intestine it possessed two primitive, asymmetric lungs and was thus able to breathe oxygen from the air. *Calamoichthys*, which frequented the coastal rivers of the Gulf of Guinea and sometimes ventured into brackish waters, was a sort of eel-shaped *Polypterus* which had lost its pelvic fins.

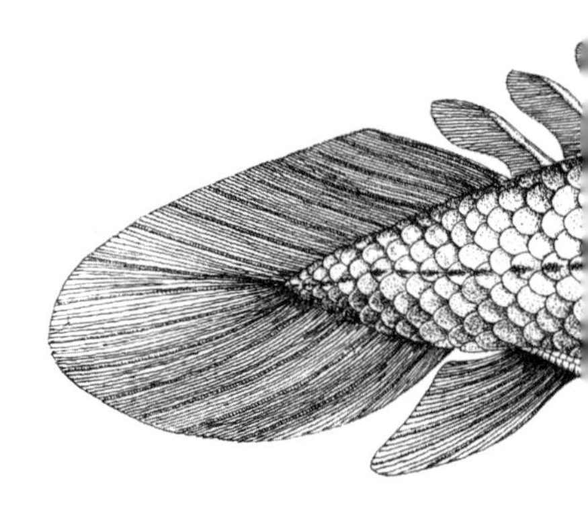

In the genus *Polypterus* the largest fishes can be 1.2 metres (4 feet) long, and there are nine different species. At the beginning of the 19th century, Etienne Geoffroy Saint-Hilaire said of the genus that it 'refused to come under any of the known divisions.'

Actinopterygians

Apart from the bony fishes known as sarcopterygians whose fins were supported by a fleshy lobe, there were also the 'ray-finned' specimens known as actinopterygians, in which the skeletal pieces at the base of the fins were within the body and the lepidotrichs were fan-shaped, not rod-like. This type of fin, supple and very mobile, together with a swim bladder, made the actinopterygians very 'at home' in the aquatic environment.

The origins of these fishes go back a very long way indeed. *Andreolepis* of the Upper Silurian Period was probably an early actinopterygian, and its existence is clearly proven from the Lower Devonian Period onwards by the scales of *Dialipina* collected at Spitzbergen and in Nova Zembla in the U.S.S.R. *Cheirolepis* from the Middle Devonian Period is the first complete specimen we have; its scales, similar in shape to those of the acanthodians, remind us of the possible relationship which exists between these two groups. The continued growth in the importance of these ray-finned fishes was reflected during the second half of the Mesozoic Era by a huge explosion of forms, culminating in their total supremacy in both

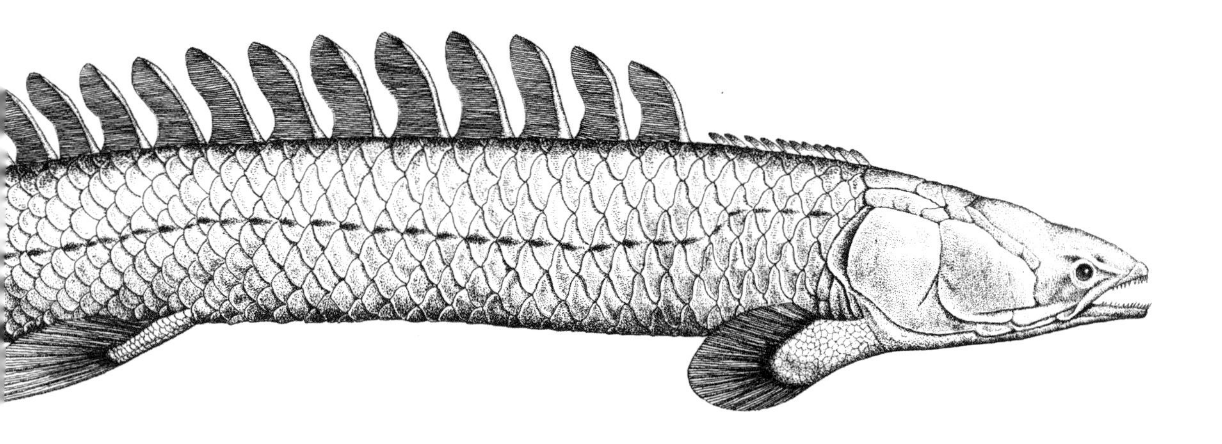

fresh water and sea water in all latitudes and at all depths.

Zoological studies have always split the actinopterygians into three units: the 'chondrosteans', now represented only by the sturgeons and paddlefishes, the 'holosteans', of which *Amia* and *Lepisosteus* of the American fresh waters are the only survivors, and finally the 'teleosteans' which included all the other living actinopterygians. However, palaeontology has shown that at least the first two units are not natural, but are 'grades', that is successive morphological types through which different strains passed, guided by the same evolutionary tendencies.

The palaeonisciforms are placed in the first of these grades – these are the most primitive and most ancient of the actinopterygians, although they are found as far back as the Cretaceous Period (see page 120). They are easily distinguished from other bony fishes of the Devonian Period, since they are the only type with just one dorsal fin. The tail was epicercal, the head large, with a short snout and large eyes. The speckled scales that covered its body were very thick, and generally speaking the skin skeleton was well developed, although the vertebral column was hardly ossified. From this point of view,

Colobodus of the Triassic Period of Besano (Lombardy, Italy). This still primitive actinopterygian had very thick scales.

the modern sturgeons are not representatives of fossil chondrosteans since, not only is their internal skeleton very cartilaginous, but their skin has only very few scales. With their wide mouths, these ancient forms were on the whole rather monotonous in outline, and only a few genera stand out, like *Tarrasius* of the Lower Carboniferous Period all of whose single fins have joined together, or *Cleithrolepis* of the Triassic Period which acquired a raised profile.

The second grade, the holosteans, is primarily of the Mesozoic Era (see below and opposite). While the vertebral column advanced to its full potential, the rest of the skeleton regressed, an evolutionary phenomenon that we have already seen on more than one occasion. The scales became thinner; the tail tended to become symmetrical with the vertebral axis of the upper lobe becoming smaller and curving upward, while the lepidotrichs were rearranged (see pages 120 and 125). The forms diversified: *Microdon* and *Macromesodon* of the Upper Jurassic and Lower Cretaceous Periods became disc-shaped, while *Aspidorhyncus*, of approximately the same age, was very elongated and its snout was extended forward by a long pointed guard.

All of these tendencies were accentuated even further in the teleosteans, which flourished mainly towards the end of the Mesozoic Era, with an increasing lightness of skeleton giving them a growing degree of agility. Of the four layers of scales in the palaeonisciforms, only a thin bony sheet of very varied structure remains; the vertebral column, however, contained solid vertebrae which, along with teeth and otoliths (small pieces of the inner ear), are frequently found in the fossil state. The pelvic fins were often displaced, being positioned beneath the pectoral. From the Devonian Period onwards the mouth of the actinopterygians became

Like the fossil shown on the preceding page, *Paralepidotus ornatus* may be thought of as a holostean. This particular example comes from the Upper Triassic of Val Imagna (Bergamo, Italy).

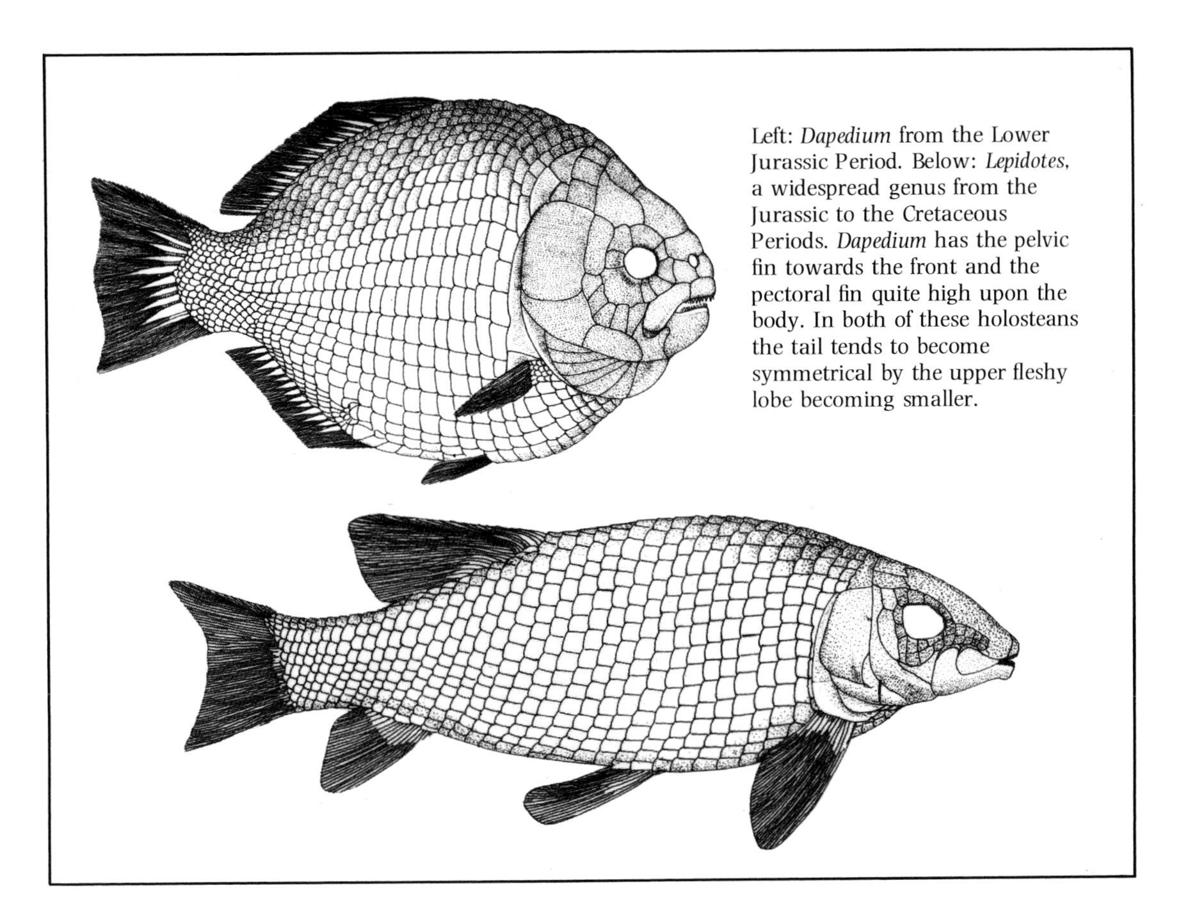

Left: *Dapedium* from the Lower Jurassic Period. Below: *Lepidotes*, a widespread genus from the Jurassic to the Cretaceous Periods. *Dapedium* has the pelvic fin towards the front and the pectoral fin quite high upon the body. In both of these holosteans the tail tends to become symmetrical by the upper fleshy lobe becoming smaller.

shorter and shorter; in the teleosteans the jawbones are very free-moving to the extent of being protractile, that is capable under some circumstances of dislocating to engulf enormous prey. An extreme example of this development can be seen in *Chiasmodon*, a deep-sea fish that was capable of swallowing other creatures two to three times its own size. These creatures from the ocean's depths illustrate well the degree of specialization which the teleosteans attained in order to conquer new ecological niches. Swimming along with their mouths wide open, they attracted their victims with the aid of luminous 'bait'. In a number of species the minute male spent his entire life attached to a female who might have had several such companions, all fed from her own blood circulation, though the males possessed their own independent respiratory system.

This group produced the strangest of the vertebrates and was very rich in the variety of forms. The layman would scarcely guess that *Hippocampus* (the sea-horse) found around our own coasts are fishes: they are in fact teleosteans, just like the butterfly-fishes (chaetodontines) and angelfishes (pomacanthines), those colourful specimens that dwell in tropical reefs; other teleosteans include marlins, sailfishes and swordfishes, roaming animals known from the Oligocene Period. Hunting for smaller species among the countless banks close to the surface, they are able to swim faster than any other living aquatic creatures, with top speeds exceeding 80 kilometres per hour (50 miles per hour).

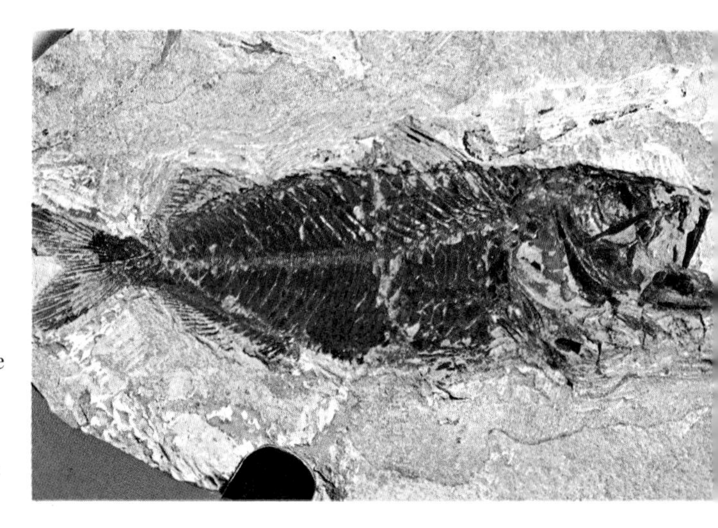

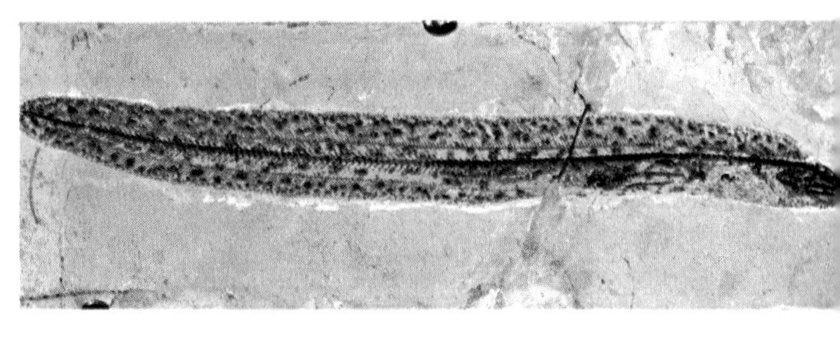

The marine deposits of Monte Bolca near Verona, Italy, were discovered in the 15th century and are very famous. Over 100,000 splendid specimens have been found there and are the pride and joy of public museums and private collectors the World over. During his Italian campaign, Napoleon and his aides took such an interest in the collection belonging to the Count Gazola that the Count offered to give a large part of it to them 'of his own free will'. The Monte Bolca deposits were created under exceptional conditions at the beginning of the Middle Eocene Period. The whole area was occupied by a sheltered lake along a very broken coastline which now and then was shaken by violent volcanic activity. Burning carbon dioxide gas welling up from the lake bottom killed the animals by suffocation and 'cooking'. For a while nothing lived in the lake and the corpses were buried in calcaro-magnesian sediment without being attacked by necrophagous organisms (those feeding on dead material). This process was repeated several times and led to the fossilization of a rich fauna consisting mainly of teleosts (bony fishes).

Chapter Five

The Conquest of Dry Land

The conquest of dry land

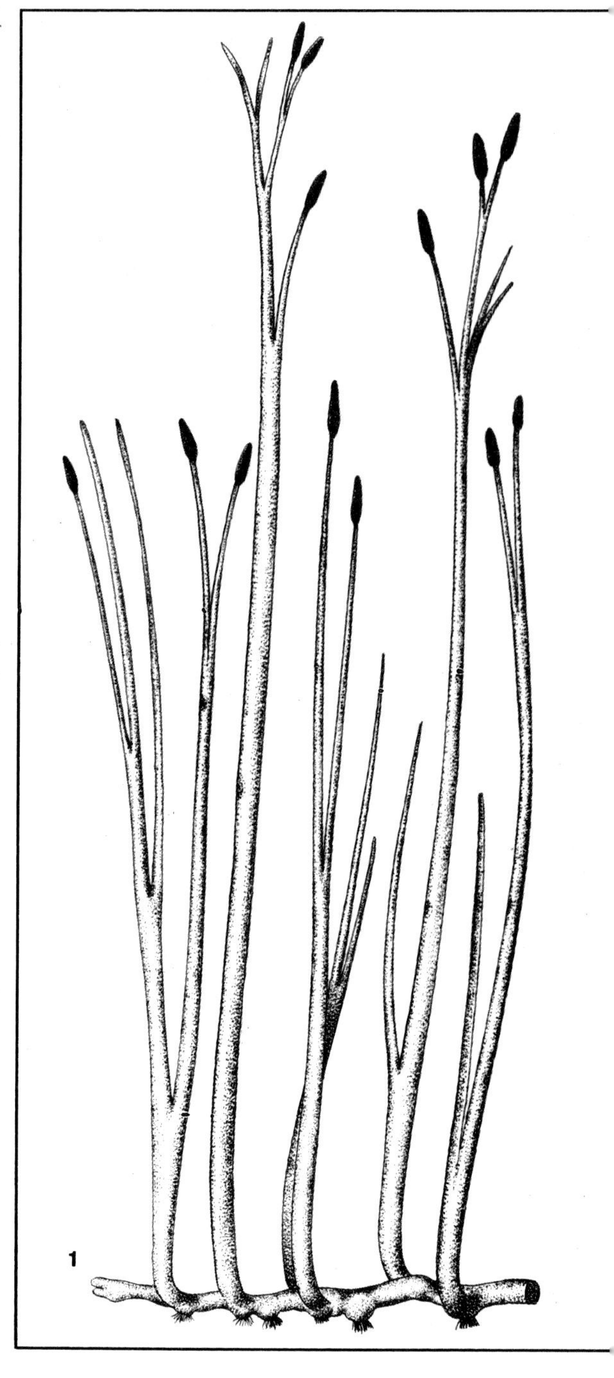

The continent of the 'Old Red Sandstones'

At the end of the Silurian Period and during the Devonian Period, the vast continent that was centred on modern North America and which covered most of northern Europe presented a desolate landscape. The Caledonian mountain range that had been thrown up during the first half of the Palaeozoic Era was subjected to intense erosion that wore rocks smooth and resulted in huge masses of sediment. In Britain, for example, these deposits attained a depth of 4000 metres (13,120 feet), consisting of schist, conglomerates and, above all, those red sandstones that have given this zone the name of the 'Old Red Sandstone' continent. One has to imagine vast stretches of desert where the erosion produces a red sand that is gradually drawn towards the sea. This brick colour was due to an accumulation of iron oxide which the trickling waters were unable to dissolve. The rocks were sculpted by a wind that rounded down each grain of sand and cut polished facets on small stones. One might think that, to begin with at least, this arid scene was chiefly characterized by the rarity of land plants which might favour the wettest regions, rather than to any climatic cause; but it was in fact this hostile landscape that was to see plant life for the very first time.

Having invaded the world of water, animal life was set to colonize dry land as well. Still, on land as in water, a rich fauna was unable to exist without the plants that form the first links in the food chain, and hence the development of even primitive plants was an absolute necessity if animals were to be able to live entirely away from the aquatic environment. As we shall see, this liberation was a laborious process, and for a long time many creatures remained greatly dependent upon their original watery world.

The transition from water to air posed a number of problems for plants, before they beset the animals as well. In particular there were problems over protection and support. The plant organism first had to overcome the problem of drying out, and so it developed an outer layer called a cuticle to preserve its internal humidity. Secondly, while seaweed is supported by the water that surrounds it, on land the plant stalk had to be developed to support the plant's own weight in order to

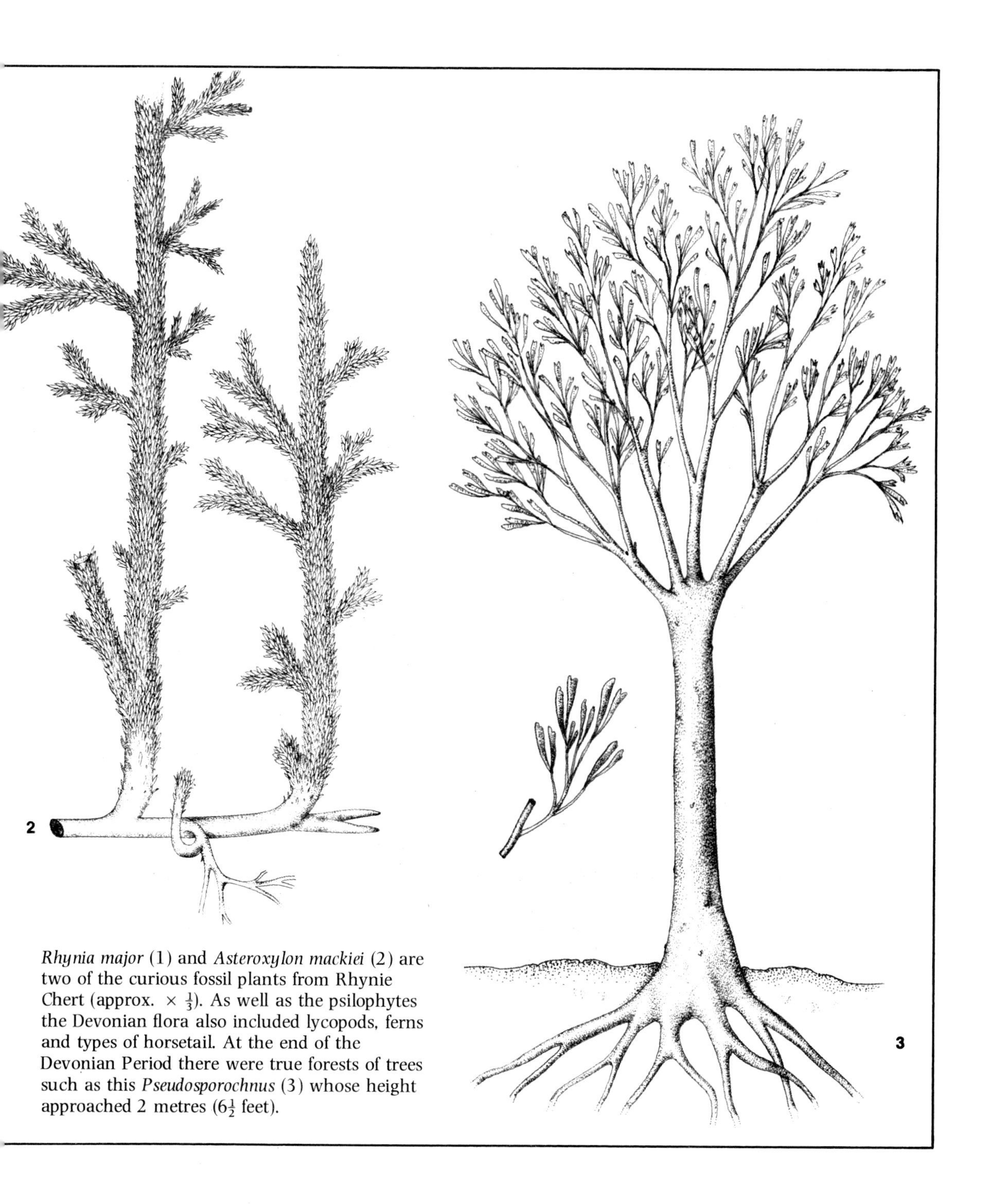

Rhynia major (1) and *Asteroxylon mackiei* (2) are two of the curious fossil plants from Rhynie Chert (approx. $\times \frac{1}{3}$). As well as the psilophytes the Devonian flora also included lycopods, ferns and types of horsetail. At the end of the Devonian Period there were true forests of trees such as this *Pseudosporochnus* (3) whose height approached 2 metres ($6\frac{1}{2}$ feet).

remain erect. This second problem was not fully solved until the advent of very strong vessels which allow the plant to draw the water it requires from the soil, and which also serve as its 'skeleton'. Plants that are equipped with canals of these essential vessels are called 'vascular plants'.

The first vascular plants go back at least to the Upper Silurian Period, their most ancient traces having been discovered in Libya and Australia, although the best known, *Cooksonia*, was found in Wales. Although these plants were never very tall, vegetation became much more varied in other ways during the Lower Devonian Period. Our knowledge of these primitive plants has been greatly advanced by the study of the deposits at Rhynie near Aberdeen, Scotland. Here, a hot water spring had been surrounded by silicified plants whose finest structures had been preserved in a bank of chert (see page 131) about 2 metres ($6\frac{1}{2}$ feet) thick. The plants of Rhynie belong to a group that is represented by only two very localized genera today, *Psilotum* and *Tmesipteris*.

Growing around the margins of swamps, *Rhynia* had no leaves and its stalks grew to about 50 centimetres (20 inches) above ground level. In the course of reproduction two equally important phases alternated in the life cycle: when the spores contained in the end capsules of the stalks landed in a suitable place they produced a branched, vascularized gamete-bearing structure (the gametophyte) which itself provided the gametes. Then after fertilization of these gametes and the growth of a new plant, the cycle was complete. Among the genera similar to *Rhynia*, *Asteroxylon* represents a more evolved type, decorated with non-vascular leaf-organs that did not as yet, however, constitute true leaves (see page 131). The Rhynie flora is not unique, however, and other sediments, like those in Gaspé, Quebec, demonstrate the variety of vegetable life from the Lower Devonian Period onwards. Plant life was to evolve very rapidly, concentrating first and foremost on the acquisition of leaves and roots – more efficient soil stabilizers than rhizomes – and then on the development of reproductive mechanisms, to allow plants to colonize drier environments. The next period, the Carboniferous, was to see vast and luxuriant forests.

Among the animals, it was chiefly the vertebrates and the arthropods which were to set about conquering the dry continents, and within these two great divisions, several groups attempted this transition independently and with varying degrees of success.

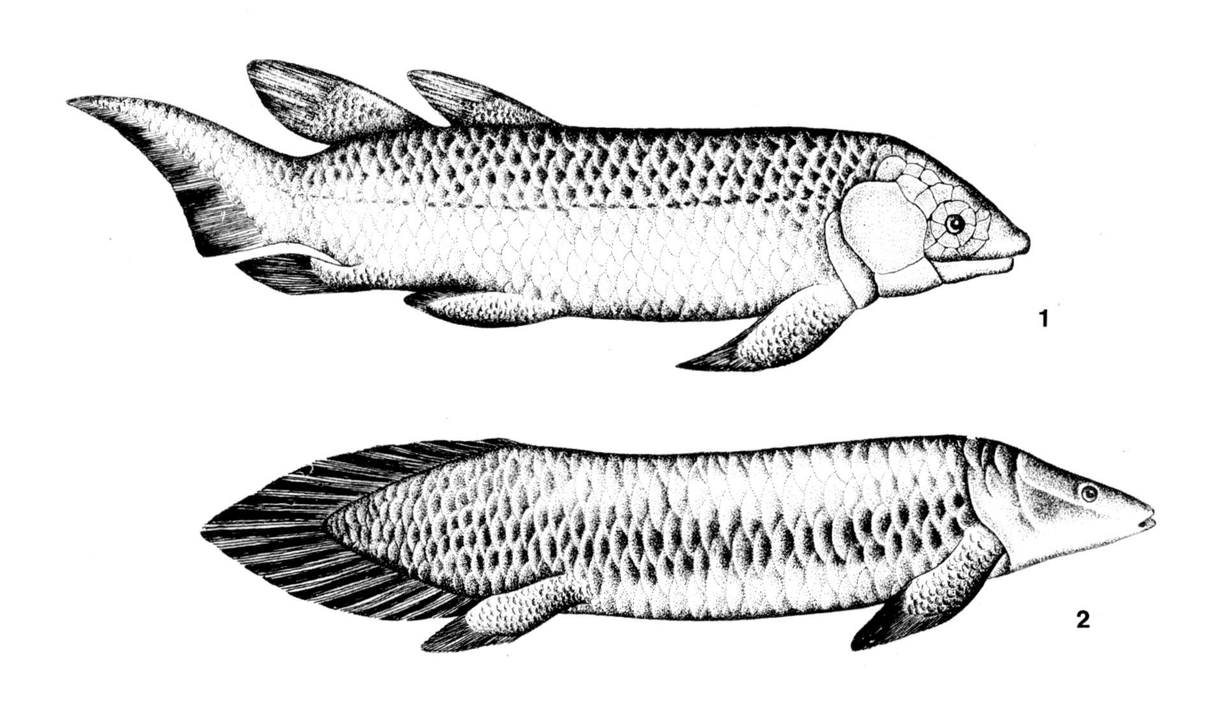

1 *Dipterus* ($\times \frac{1}{4}$) is a dipneust of the Middle Devonian Period that still shows the primitive fin layout of the sarcopterygians. All the fins had a fleshy base – there were two dorsal fins, one heterocercal caudal fin and one anal fin.
2 *Neoceradotus* ($\times \frac{1}{10}$) the modern Australian dipneust, has a much smaller set of fins. The animal now has just a single non-paired diphycercus fin.

Dipneusti

Of all the sarcopterygians, the lobe-finned bony fishes, the Dipneusti, for a while at least, passed for the possible, if not probable, ancestors of the land vertebrates. In the last century, these creatures were also associated with the amphibians, owing to their superficial similarity to certain salamanders. Furthermore it is true that the way of life and some of the anatomical features of species alive today, seem to mark them as beings adapted to both aquatic and terrestrial life.

Modern Dipneusti are divided into three genera: *Lepidosiren* from South America,

Above: *Neoceradotus*, which the Australian aborigines call 'Barramunda', is quite different from its African and South American dipneust cousins. It may grow up to 1.8 metres (6 feet) and weigh up to 50 kilograms (110 pounds). It is only found in a few rivers of Queensland where it hunts small fishes, molluscs and crustaceans which it crushes between its dental plates. It also feeds on marine vegetation, especially when young.

Below: The unique and roughly eel-shaped species of *Lepidosiren* (1) and the four species of *Protopterus* (2) belong to the family of lepidosirenids. As we can see, they are actually very closely related. Their adaptations enable them to withstand the periodical extreme droughts that would be fatal to *Neoceradotus*.

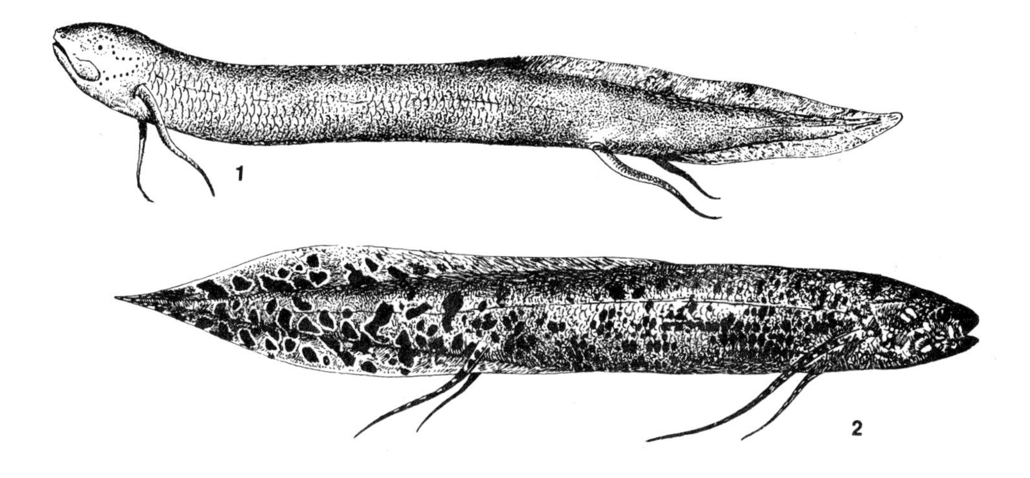

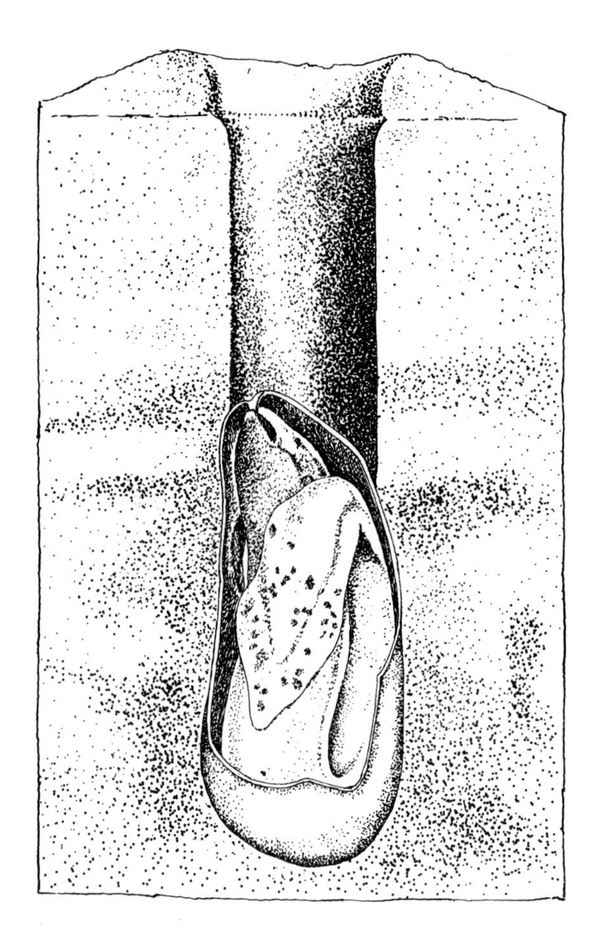

A section through a *Protopterus* burrow which is usually closed by a plug of cracked mud which allows air to pass through. Surrounded by its cocoon of mucus, this creature spends the dry season folded up with its tail covering its head. The dipneust is able to breathe oxygen from the air through a small 'funnel' opening to the outside world.

Protopterus from Africa and *Neoceradotus* from Australia (see page 133). These creatures inhabit fresh water only. As their name suggests, they possess lungs – indeed they are known as lungfishes. There are two in the African and American species, but only one, the right lung, in *Neoceradotus*. In the former, the body is eel-shaped and the pectoral and pelvic fins very narrow and wiry. All modern Dipneusti have in addition a single odd fin that ends in a symmetrical tail.

The Dipneusti are endowed with a dual respiratory system: thanks to their lungs they can breathe atmospheric oxygen, and with their gills they are able to take in oxygen in water. But these two facilities are not used equally: in *Neoceradotus* for instance (the species with one lung) the gills are very important, while the lung only becomes vital during the dry periods when the decomposition of organic matter in evaporating pools and ponds considerably reduces the water's oxygen content and causes the death of fishes possessing only gills. Still, *Neoceradotus* is not safe from the danger of drying out completely. In *Lepidosiren* and *Protopterus* on the other hand, whose gills are far smaller, the lungs do most of the work of respiration, which means that *Protopterus* is quite capable of leaving the water and moving over dry land. In the dry season when the more or less temporary expanses of water which they inhabit dry up, *Lepidosiren* and *Protopterus* take refuge in a burrow, and the latter even buries itself in the mud in a mucus cocoon (see illustration left). Inhaling the air's oxygen, they lead a slow-motion life for several months until the next rainy season starts. As an experiment, the 'hibernation' of a *Protopterus* has even been artificially prolonged for four years.

Fossil Dipneusti have been discovered from the Lower Devonian Period, and they are represented among the sediments, above all, by the very special dental plates (see page 135) that lined their mandible and palate. Unlike modern forms, the old Dipneusti lived in salt water. Their profile featured a heterocercal tail quite clearly separate from the one anal and two dorsal fins (see page 132). The primitive scales were thicker than on modern Dipneusti, and the body with its cartilaginous vertebrae bore a strong resemblance to that of sharks.

Despite their lobe-fins, that seem to bear the rough outline of a leg, and despite their respiration by lung and the ability of some of them to leave the water temporarily, the Dipneusti which we know cannot be taken to be the direct ancestors of the tetrapods – that is the purely terrestrial vertebrates – but they are certainly related to them. The structure of their cranium follows an organizational scheme that cannot be seen as a precursor of that of the first amphibians. Their maxillary bone has disappeared and their vertebrae are very unusual. In addition, the perforations linking the nasal sacs to the digestive tube and to the lungs, the 'choana', are unlike those of the tetrapods. In a fish, water penetrates the nasal sacs through a forward positioned nostril and is then expelled through the posterior nostril. The 'false choana' of the Dipneusti, however, would appear to be only posterior nostrils which have moved and not additional perforations. Finally, the skeleton of the paired members is nothing like the limbs of the first tetrapods.

To sum up, therefore, the Dipneusti do not constitute a real attempt at colonizing dry land; their characteristics correspond with an adaptation to temporarily drying out which was acquired relatively quickly. In the United States, from the Upper Carboniferous and Permian Periods onwards, fossil burrows of these creatures have been found, some even still with the original occupant. The Dipneusti evolved slowly and, as often happens, tended to lose some of the advances they had made as they became more bony: they have left only a few relics in living nature.

Rhipidistians

While the Dipneusti and the actinistians (see page 120) made only predictable and unexciting advances, a third group of 'lobe-finned fishes' proved to be more dynamic, and some of its members started along a road which led to the conquest of dry land by the vertebrates. These were the rhipidistians, which appeared at the beginning of the Devonian Period.

Unlike the actinistians and the Dipneusti, the rhipidistians are known only in the fossil state and never survived the end of the Palaeozoic Era. They were long-bodied predators covered in thick scales; they possessed two dorsal fins and their tail was both

All that remains of many fossil dipneusts are their dental plates. The photograph shows some Mesozoic specimens.

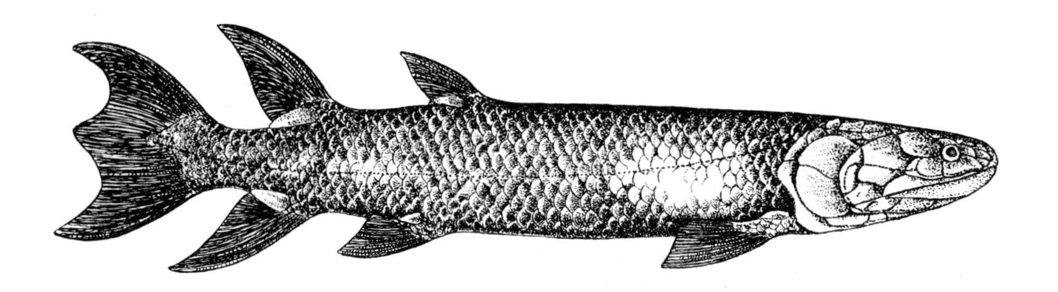

Eusthenopteron foordi (× 4 approx.) is one of the better-known osteolepiform rhipidistians; the first tetrapods descended from similar forms, but were already living when *Eusthenopteron* was swimming in the Upper Devonian waters over 350 million years ago.

symmetrical and heterocercal. Their eyes were small and surrounded by bony plates, and their elongated snout was characterized by a wide mouth. A further feature is the ability of the bones of the cranium to move slightly against one another, probably in order to absorb shocks when the animal closed its mouth around some prey.

To all appearances these are the only sarcopterygians to possess any true choana: the bones of their cranial roof, with an opening at the top that probably contained a third eye, were organized in a manner similar to that of the tetrapods. Finally, although like the Dipneusti and actinistians some possessed paired-fins consisting of bones from which symmetrical rays diverged, others had a structure from which it is not difficult to imagine the development of tetrapod limbs (see page 137), an association which is further supported by other similar characteristics like the structure of the vertebrae and teeth.

We therefore site the roots of the tetrapods within the framework of the rhipidistians, near to the 'osteolepiforms', and it was to this group that *Eusthenopteron* belonged. Because of its age, however, this creature cannot possibly represent the ancestor of terrestrial vertebrates. Even further distant are the 'porolepiforms' which can be classified on the basis of certain characteristics but which are less well known. Their head was shorter, their pectoral fins more pointed, and large areas of their body were devoid of scales. By reference to certain features of their cranium and especially the structure of their snout, the Swedish palaeontologist E. Jarvik claimed to see the origins of the urodele, that is tailed, amphibians such as salamanders and newts in these porolepiforms, while the osteolepiforms had been at the roots of the other tetrapods. According to his theory that the tetrapod vertebrates arose from two separate lines of evolution, the class Amphibia, which today includes the Urodela and Anura (that is both tailed and tailless amphibians, the latter including frogs and toads) is a totally artifical group since the Urodela are foreign to all other land vertebrates. These views have been strongly criticized owing to the numerous specializations that are common to both Urodela and Anura, and even putting the osteolepiforms and porolepiforms into the same 'rhipidistian' unit is frequently rejected. The question remains therefore as to how lineages that were different – at least from osteolepiforms – were separately able to acquire their four legs.

Transition to terrestrial life

There are various theories given to explain the even partial abandonment of the aquatic for the terrestrial environment by the rhipidistians, and the different views usually flow from the different reconstructions that have been made of the Devonian world which witnessed this major event in the history of life.

The alternation of rain with dry periods may have had disastrous consequences for the river and lake fauna of the 'Old Red Sandstone' continent, and the point of departure for the transition to dry land was initially situated here: it was claimed that, to escape asphyxia and starvation, the rhipidistians underwent a series of adaptations that enabled them to survive in waters low in oxygen content and to move across land in order to find new lakes when drought dried up the old ones.

Much criticism has been levelled at this hypothesis, and regular doubts have been expressed as to whether the Devonian climate was actually arid. In spite of such criticism, however, the existence of deposits of salts and layers of cracked mud would seem to favour the theory of a periodical drying-up of the land and to contradict the idea of permanent humidity. It remains to be definitely proved whether the ancestors of the tetrapods actually inhabited fresh water; like the Dipneusti, the rhipidistians evolved in the marine waters near the coast, and some modern researchers are tempted to regard these waters as the original environment of the land vertebrates. They claim that, thanks to their various adaptations, the rhipidistians would have been able to survive on immense sandy beaches uncovered by the tides, and could even have prospered in the specific surroundings of shallow lakes, coastal swamps and estuaries. They may even have sought freshwater pools from time to time in order to mate and find shelter for their spawn. Their fleshy-lobed fins seem to have been ideal for moving through very shallow waters where air-breathing was indispensable, rather than made for crawling over dry land itself. The shallow margins of rivers, lakes and pools doubtless offered protection against the threat of more powerful predators, especially for the young. Finally, the arthropods of these regions provided these creatures with fresh sources of food that could be enjoyed without competition from lungless fishes. Still, it is hard to imagine how vertebrates which had not yet learned to wriggle across land could regularly have caught spiders, millipedes and scorpions, animals that were already perfectly adapted to terrestrial locomotion. In any case the teeth of rhipidistians were typically those of fish eaters.

Whatever the theory it seems that the very first tetrapod characteristics were acquired

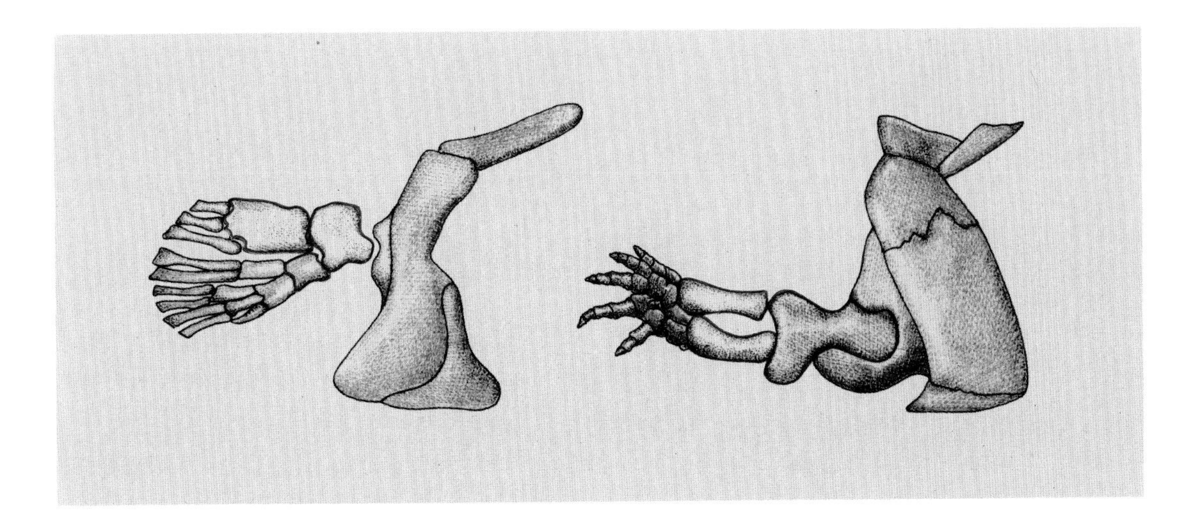

The skeleton of the fleshy lobes that carry the paired fins of an osteolepiform did not have to change a great deal to become a tetrapod limb. Both possess the same basic branched layout. Once the lepidotrichs had disappeared, bones in the fins became smaller and formed a hand which, to start with, had seven fingers. The elbow joint was also formed.

to improve the chances of survival in an unreliable aquatic environment rather than to conquer a new domain, and that this response, different from that made by the Dipneusti to a similar problem, 'pre-adapted' the rhipidistians to the revolutionary transition that was to give birth to the most elaborate forms of the vertebrate phylum.

Ichthyostega

It was in the beds of the 'Old Red Sandstones' of the Upper Devonian Period in Greenland that the oldest-known tetrapod was discovered. With a length of around 1 metre ($3\frac{1}{4}$ ft), the essential characteristic of *Ichthyostega* was that it had four legs whose five-fingered skeleton (plus two very small fingers or phalanxes) had lost the bony rays that supported the fins of the rhipidistians. The phalanxes had become distinct and an elbow had appeared, while the new legs were able to move on solid bone girdles. In front, the pectoral girdle was separate from the skull, and the head was connected to the trunk by a very short neck (see pages 139 and 140). At the rear, the pelvic girdle was attached to the spinal column by the formation of a sacrum to permit a 'walking' motion. The vertebrae had modified and, although forming a vertebral cord, this could now move sideways to a considerable extent. The creatures's very robust ribs overlapped to form a relatively well-rounded thoracic cage that housed completely functional lungs. However, the body ended in a tail which – like that of a fish – still bore lepidotrichs and thus suggests still very marked aquatic habits, as confirmed by the existence on the skull of sensory canals similar to those of the rhipidistians. This differentiation in the 'lateral line' system was to be replaced by simple grooves in the amphibians. The internal structure of the cranium was still in two blocks. It is likely, too, that the gills, at least in the adult, have disappeared. And there are other features which rule out *Ichthyostega* as being a direct ancestor of modern land vertebrates. Nevertheless, from the Upper Devonian Period onward other tetrapods became more and more varied, as shown by *Acanthostega*, another genus less well known than *Ichthyostega*, as well as by trace-fossils and some meagre remains in rather older deposits in Australia.

There is no doubt that with its flat tail, flexible spinal column and lateral line nervous system, *Ichthyostega* was a good swimmer. It must have spent most of its time in water, feeding like its ancestors upon small fishes caught between its conical teeth. Its life also began in water, for it was to be a long time before vertebrate eggs acquired the type of shell necessary for their preservation on dry land. The young *Ichthyostega* must thus have been born equipped with gills and grown up in water.

Life out of water

Animals as well as plants had to have acquired a number of mechanisms before being able to survive in the less hospitable environment of dry land. The osteolepiforms have shown us that some of these mechanisms were already developed, or at least outlined, prior to the transition from water. Such was the case with the lung, and we should not forget that an answer to the problem of air breathing was not unique to the vertebrates, the teleosteans having perfected several devices other than lungs which made it possible to breathe atmospheric oxygen. The same applied in part to paired limbs whose basic components were laid down at a very early stage. *Ichthyostega* elaborated and perfected respiration by lung by developing a thoracic cage which was later supplemented by a diaphragm, and its whole skeleton was reinforced in order to meet two new requirements, that of the support of the body in air rather than water, and that of walking.

Other adaptations were also necessary for life on land. Organisms first had to overcome the problem of drying out or desiccation in order to conserve their internal moisture, and the tetrapods solved the problem by means of a skin covered in dead cells rich in keratin, a covering that has the advantage

A reconstruction of *Ichthyostega* from the Upper Devonian Period of eastern Greenland. This very primitive tetrapod (4-legged animal) inhabited the shorelines of freshwater lakes and rivers. It fed on fishes and was unable to allow its skin to dry so it never moved far away from its lakeland home.

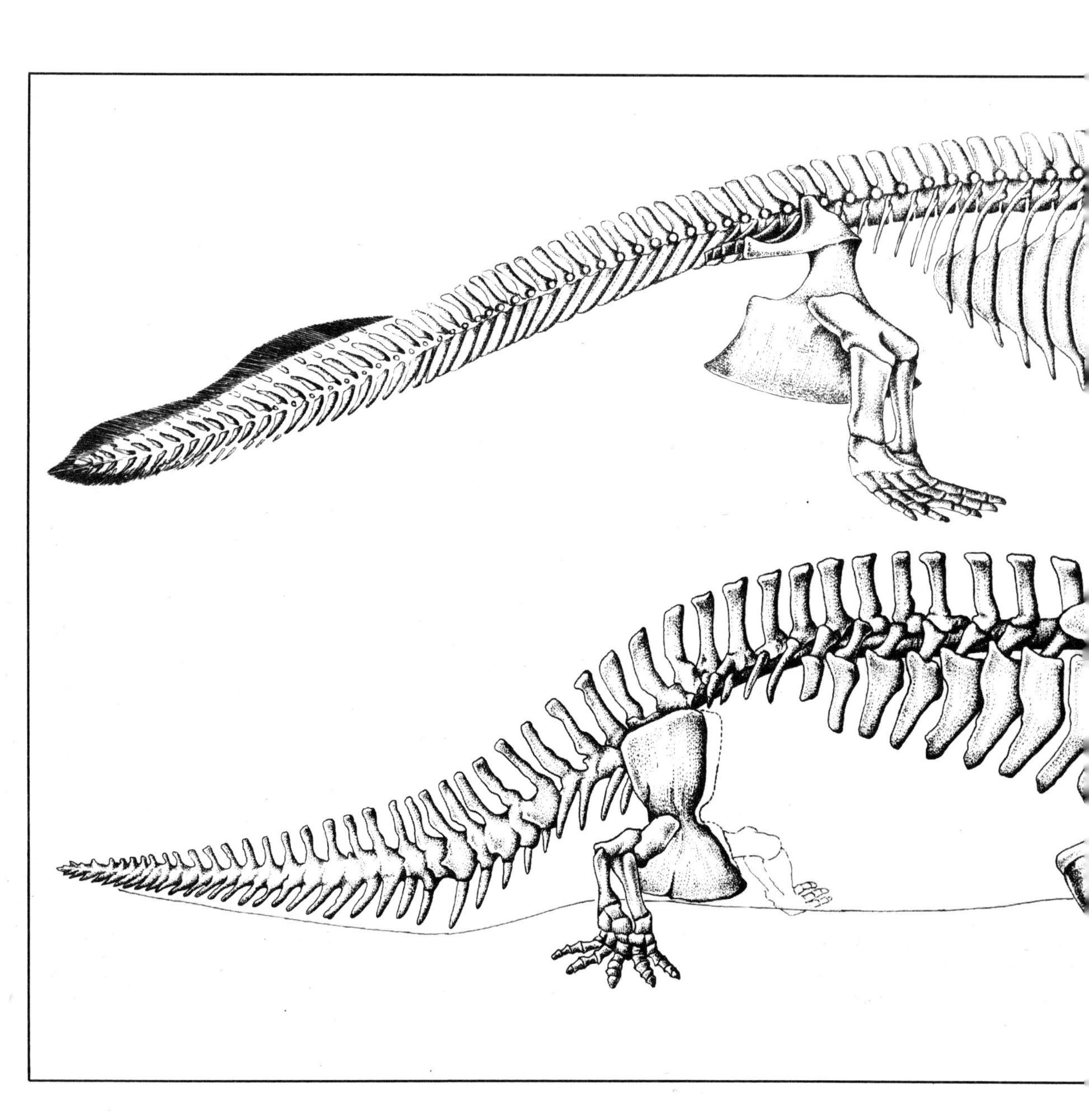

over the cuticles and shells grown by other living creatures of being highly flexible. However, the very first amphibians must have been compelled to keep their skin moist, and so were forced to stick close to water. In modern amphibians the layer of keratin is thin and at least partially, though varyingly, permeable to water: when covered in mucus the skin retards the drying process of the body in open air, but allows water to penetrate when the animal is immersed. The ability to breathe through the skin is vitally important to these creatures.

The sensory organs, too, had to change in order to respond to the new environment. The eye transformed to be able to operate in air which does not refract in the same way as water. Most important of all, however, the eye was now equipped with a tear gland to keep the cornea moist. At the same time, the cornea was kept clean by the periodical wiping motion of a protective eyelid. A true tongue made its appearance, and with it, taste. This organ is still lacking in modern tetrapods of the highly aquatic kind, such as the Pipidae frogs which retain the lateral line nervous system in the adult. This device, which is able to sense changes in water pressure, gradually regressed during the transition from water to dry land. In the immediate successors of *Ichthyostega* it was still apparent from the hollow furrows in the

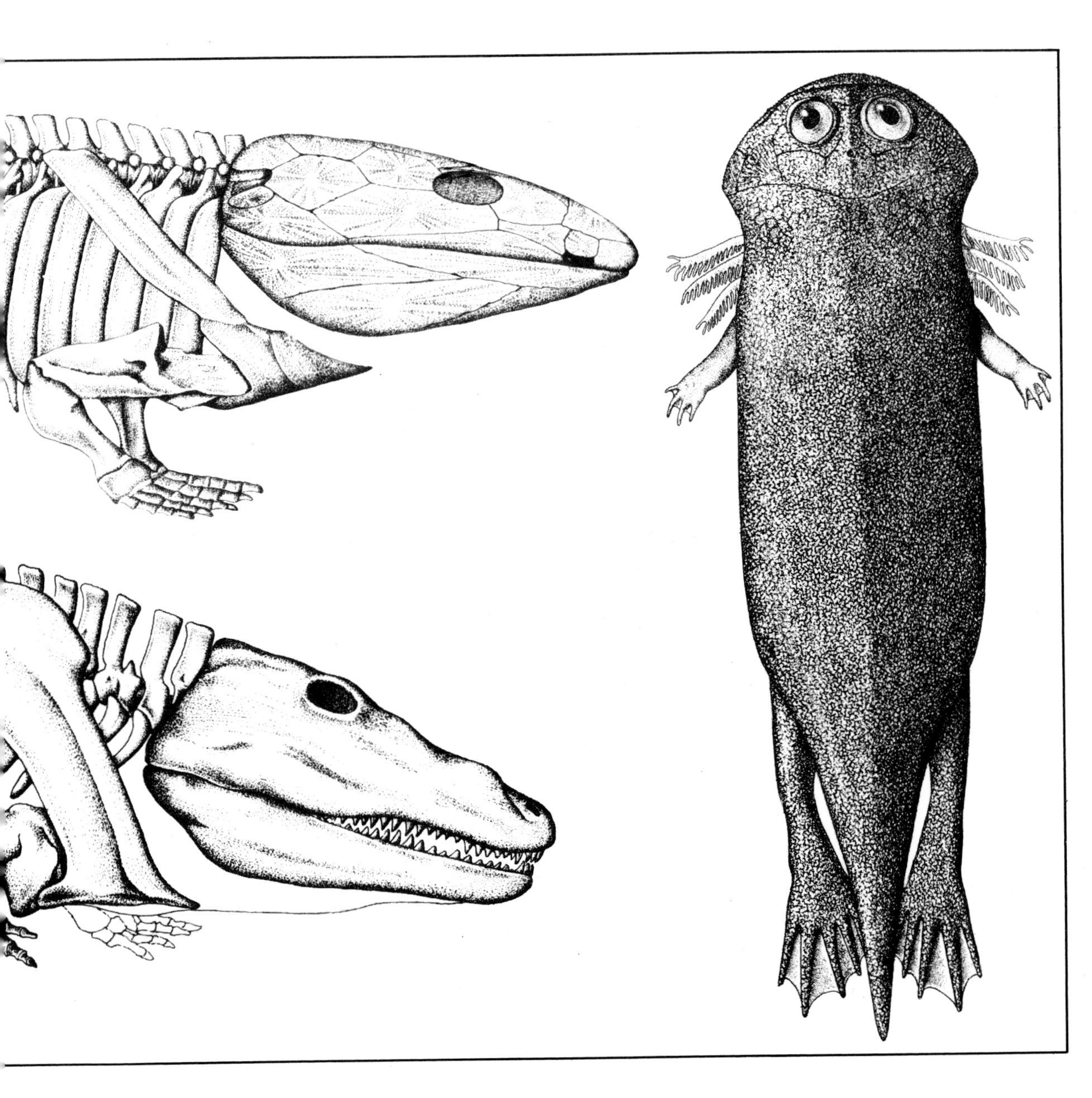

surface of the cranial bones, but today in amphibian spawn and in those adult forms that have retained it, the furrows are marked in the skin only. In the new environment, the perception of vibrations was dependent on a totally new system, the middle ear. The inner ear and its balancing function were retained, but the first tetrapods used existing structures to perfect a system of sound amplification. The branchial chamber which corresponded with the hyoid arch (see page 113) has now become a sort of resonating box, while the spiracle which opened to the outside is now nothing more than a simple aperture across which the membrane of the eardrum is stretched. Between this eardrum

Left: The skeletons of *Ichthyostega* and *Eryops*. All the bones of *Ichthyostega* have not yet been found but it looks like a large rhipidistian that has lost its gills and acquired legs. It has some strange features like the lepidotrichs on the tail and nostrils on the rim of the mouth (see page 139).
Eryops, which dates from the Permian Period was a more recent creature, about twice as big as *Ichthyostega*. It had massive girdles and limbs better adapted to walking, while its pelvis consisted of three bones on each side (pubis, ischium and ilium). *Ichthyostega* had a single pair only.

Above: Along with the fossil spawn of primitive amphibians we also find neotenic forms such as *Gerrothorax* from the end of the Triassic Period of Scania (southern Sweden, approx. × $\frac{1}{6}$). These solely aquatic creatures kept their gills in the adult stage.

and the inner ear, the vibrations are transmitted by means of the stirrup-bone, a development of the hyomandibular which was the chief component in the hyoid arch and hitherto used to suspend the jaws. Later on, the mammals added other small bones to the middle ear mechanism.

Lepospondyls

The very favourable environmental conditions of the Carboniferous Period allowed the tetrapods to enjoy a period of expansion and diversification. In low-lying regions, a warm climate with frequent rainfall provided the amphibians with a world of swampy forests (see page 148) where they flourished in a multitude of different forms which bear very little resemblance to their present-day descendants. There was a fresh turning-point, too: it was around this time that some tetrapods began to find most of their food on dry land.

In addition to often gigantic species (see pages 141 and 143) that were related to *Ichthyostega* and known by the name of 'labyrinthodonts', less conspicuous but highly original creatures also abounded – these were the lepospondyls. Of these, the most curious were without doubt the aistopods. The aistopods had already lost their recently acquired legs, moving for all the world like snakes, but being in no way associated with them. Of the mandible, they retained only two bones on either side of the head. The presence of aistopods from the early Carboniferous Period would seem to indicate that the Devonian types of tetrapods persisted for perhaps longer than is generally supposed.

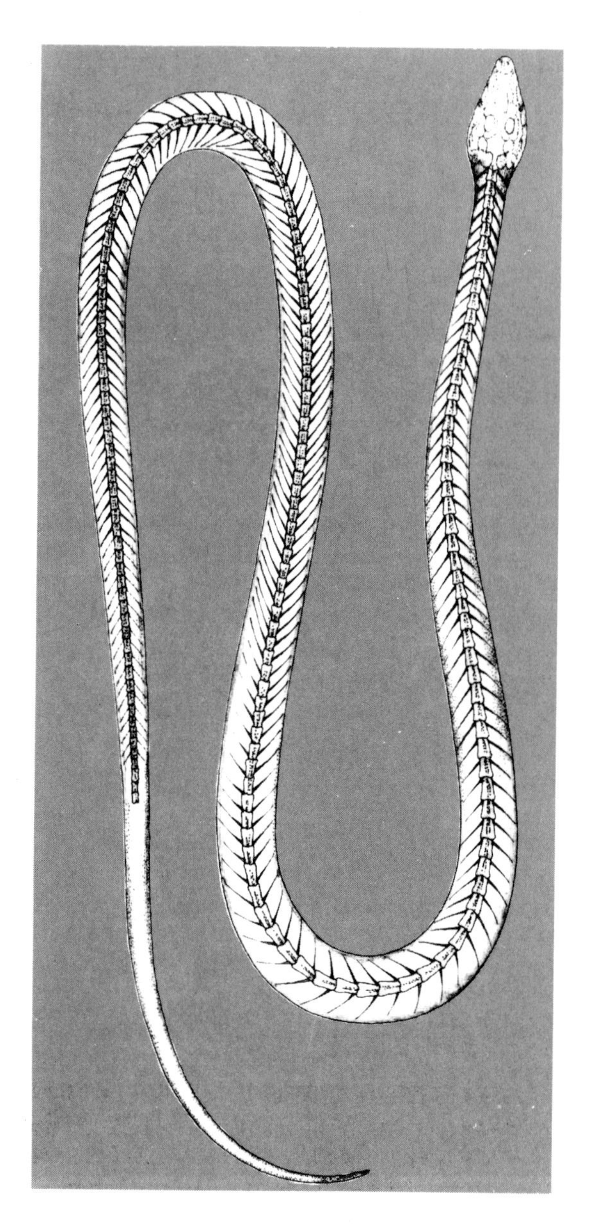

Dolichosoma of the Bohemian Carboniferous Period ($\times \frac{1}{2}$). Many people thought that the aistopods with their 200 vertebrae must have lived in the water but in fact they lived on dry land, crawling amid the rotting debris of the Carboniferous Period flora.

The nectridians, on the other hand, resembled large salamanders with delta-shaped heads, sometimes very long or flattened to form two 'horns' as was the case with *Diplocaulus* from the Lower Permian Period of the U.S.A. The head was also out of proportion to the body. With their small limbs, elongated bodies and long, laterally flattened tails, the nectridians were for the most part swimmers.

Finally we have the third lepospondyl unit, the microsaurs, which despite their name were indeed amphibians which looked like lizards. Although usually around 20 centimetres (8 inches) in length, certain specimens attained 1 metre ($3\frac{1}{4}$ feet) in length. Their bodies were covered in small scales and possessed several advanced features, especially a smaller number of phalanges. The microsaurs have even been claimed as possible ancestors of the reptiles, but certain of their characteristics exclude this possibility.

With their rather simply structured cylin-

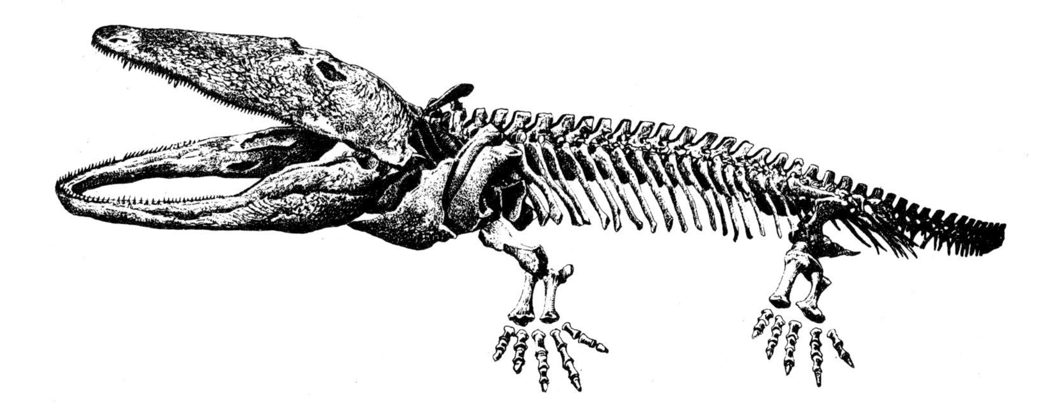

The skeleton of *Cyclotosaurus*, an enormous temnospondyl of the Upper Triassic Period. This animal was a close relation of the giant of the group, *Mastodonsaurus*, whose skull alone measured as much as 1.2 metres (4 feet) long. *Mastodonsaurus*, frequently found in the Triassic Period of Germany, was the first labyrinthodont to be identified.

drical vertebrae, the lepospondyls are not represented by any surviving animals, and did not outlast the Permian Period. Still, the possible origin of the Urodela is frequently seen in the nectridians, while the aistopods and microsaurs are often taken to be the ancestors of the Apoda, those curious worm-like, legless animals which today inhabit tropical and sub-tropical regions.

Labyrinthodonts

The expression 'Labyrinthodont' is a reference to the singular structure of the teeth which these creatures inherited from the osteolepiforms. In the latter, pairs of fangs were regularly arranged along the rows of small teeth, and in each pair one had remained functional while the other had fallen into disuse and been lost. A cross-section through their teeth reveals a very characteristic 'maze-like' pattern (see page

In section, the teeth of the stegocephalians reveal numerous folds of dentine, and it was this particular feature which gave their owners the name of 'labyrinthodont'.

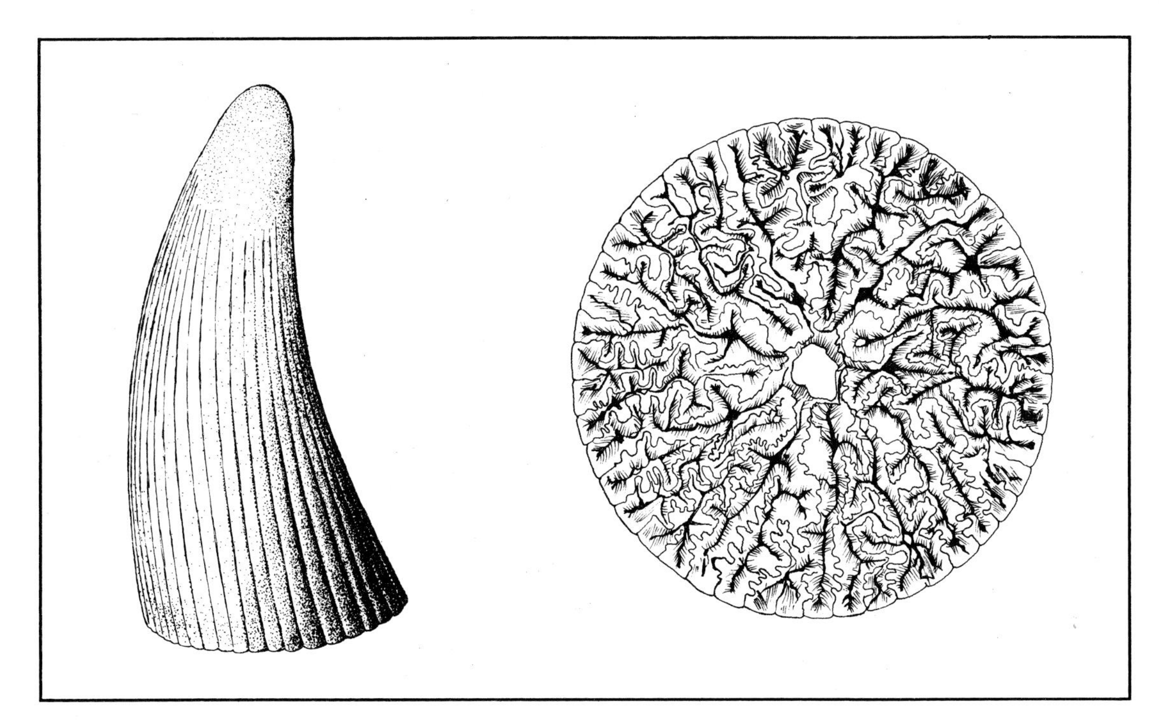

143) that is also found in the labyrinthodonts. These amphibians are also called 'stegocephalians' because of the flat roof shape of their very bony skull. Except for the exclusively Devonian Period forms mentioned above, the labyrinthodonts are divided into two groups: the vast unit of the temnospondyls which could probably be classified into more groups and that of the batrachosaurs which can be more clearly defined.

Eryops of the Permian (see pages 140 and 141) is a fairly typical temnospondyl, being a heavy, marsh-dwelling fish eater. Its skeleton is unusual in that the head is massive, the limbs short and the girdles powerful. These heavy quadrupeds must have moved around chiefly by creeping and sliding, while a related genus like *Cacops* was more obviously terrestrial with a hardening of the skin on its back. In this particular strain, the evolution of the cranium was primarily marked by the development of huge 'windows' in the palate, similar to those observed in the skulls of Anura. It is through these openings that the eyeballs of a frog are activated by retractive muscles to 'push' their prey, once caught, down their throats – it is common knowledge that a frog closes its eyes when swallowing large prey.

The most extraordinary of all the temnospondyls were without doubt the genera which had reverted to a purely aquatic way of life, retaining into adulthood immature features, especially the gills (see page 141). It is, however, not always easy to recognize the fossil young of the large species of animals which retain their immature features into adulthood.

The drying and cooling of the Permian Period climate compelled certain temnospondyls to become increasingly terrestrial, while other genera turned their backs for ever on dry land to return to the water, doubtless encouraged in this by competition from reptiles who were flourishing at that time. Thus it is that – during the Triassic Period – forms persisted that were basically adapted to a marine or lake-dwelling existence. In the oceans, the trematosaurs spread throughout the globe: their flat heads often ended in a very long snout while their skulls bore very marked sensory furrows. Swimming was made easy by long, streamlined bodies and powerful tails. They are the only purely oceanic amphibians we know, although they may well have returned to fresh waters to reproduce, unless they were viviparous (that is live bearers), a theory based on the fact that no amphibian spawn has ever been discovered in marine deposits. The freshwater forms like *Metoposaurus* of the Upper Triassic Period were totally incapable of surviving on dry land. Huge accumulations of their skeletons have been found, especially in Morocco where, crowded together in the drying lakes, they perished, unable to survive out of the water.

The origins of the Anura are without doubt to be found among the temnospondyls – they possess the same arrangement of the cranial roof, the same 'windows' in the palate, and the acquisition of features which allowed the skull to move on the vertebral column (a characteristic that appeared independently in mammals). Unlike the temnospondyls, however, whose bone structure became harder as they adapted to a more and more terrestrial existence, the Anura with their more aquatic ways underwent a marked lightening of the skeleton.

At first sight the batrachosaurs of the Carboniferous and Permian Periods were not all that different from the temnospondyls, being carnivores (we know of no fossil amphibians that were obviously herbivorous). They were less varied, however. Aquatic at the outset, they did not become truly terrestrial until late on in their lineage. Their more raised cranium was characterized by a compact roof and their vertebrae, upon which a major portion of the classification of all these primitive amphibians is based, are also very easily identifiable. They have attracted much attention because they are at the origin of all the non-amphibian

The seymouriamorphs were small animals and hardly ever grew to more than 1 metre, ($3\frac{1}{4}$ feet). *Seymouria baylorensis* from the Lower Permian Period of Texas is one of the better-known species. It possessed the robust limbs of a terrestrial animal and its cheekbones were firmly attached to the roof of the skull, unlike the more primitive batrachosaurs. It evolved too late in geological time to be a direct ancestor of the first reptiles (which appeared in the Middle Carboniferous Period).

tetrapods. The group with the largest number of reptilian characteristics is the seymouriamorphs (see page 145). Nevertheless, while they show us what the ancestors of the reptiles might have looked like, these animals are in fact just a side branch of amphibians that died out without any descendants. Their young, equipped with gills, always started life in water. More primitive

Palaeophonus was a scorpion of the Upper Silurian Period of Gotland and Scotland. It is shown here about twice life size; it is often regarded as the oldest terrestrial arthropod, but has no 'stigmata', those tracheal openings typical of the air-breathing forms. Terrestrial scorpions made their first appearance in the Middle Carboniferous Period.

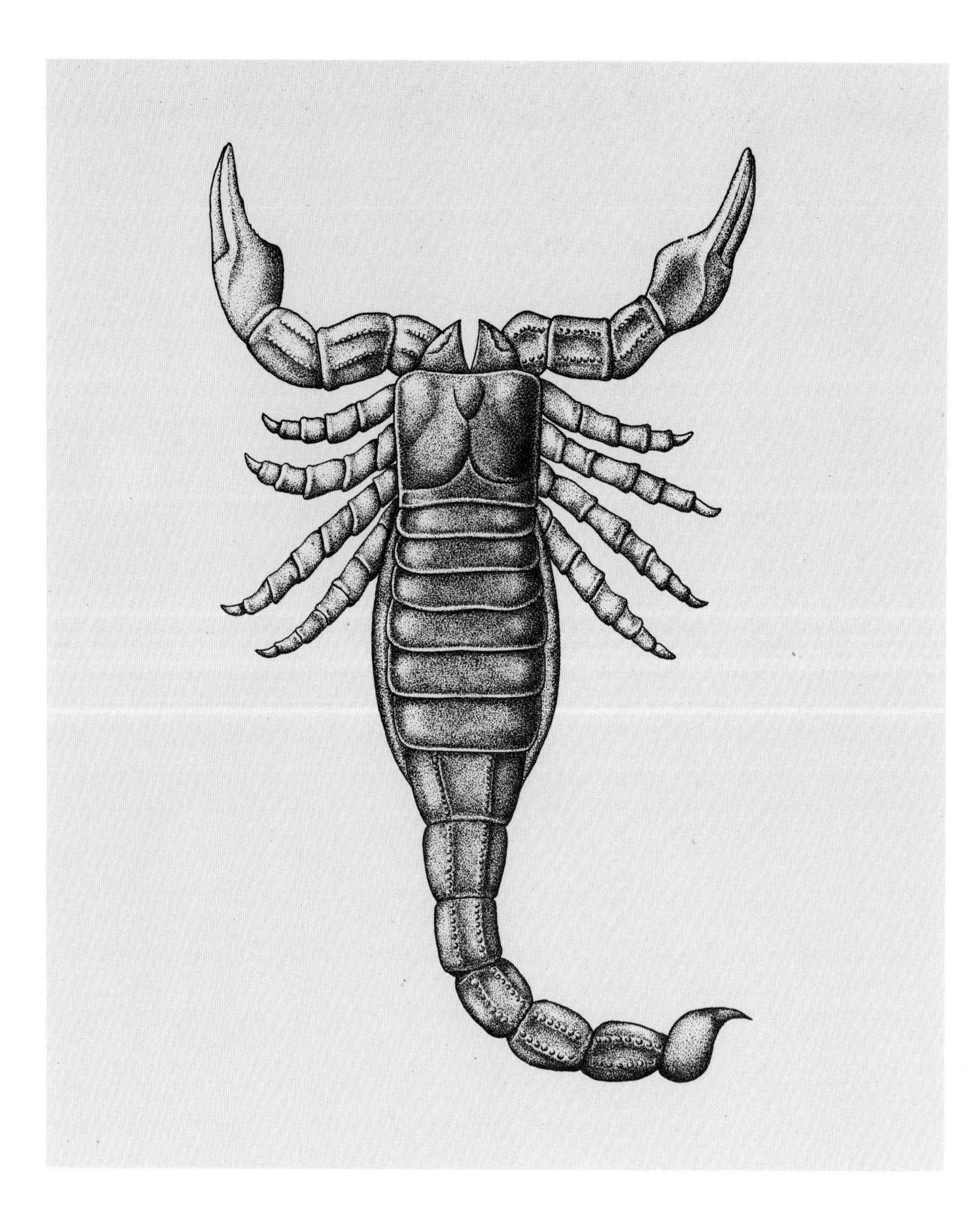

genera like *Gephyrostegus* from the Upper Carboniferous Period do not have the specializations of the seymouriamorphs and are surely much closer to the origins of the reptiles.

Arthropods conquer the land

Running parallel to, and sometimes ahead of, the vertebrates, the arthropods also had an attempt at colonizing dry land. It is likely that the very first land arthropods derived from several aquatic groups, but their departure from the waters could not be accomplished until the rate of ultraviolet radiation through the Earth's atmosphere had fallen relatively low (see page 52), until which time only night time excursions along the shorelines would have been possible. The second requirement was the adequate development of vegetation and the resulting humus to provide shelter and food. Most authors agree on the probable marine origin of these creatures, saying that the estuaries of rivers and tidal swamplands with their rich plant remains were the scenes of the first attempts at penetrating the land.

The arthropods needed only a few modifications to be able to survive on land. Their chitinous exoskeleton supported the body while giving protection against drying up. They had to acquire a suitable respiratory system that was frequently composed of a long maze of tubes opening out at the surface of the skin. These 'tracheae' with their various air sacs formed a very dense network that supplied oxygen to all parts of the body as far as the tissue layer. Their role was similar to that played by the bronchi (air tubes leading to the lungs) of vertebrates, but they are not concentrated into localized organs such as the lungs.

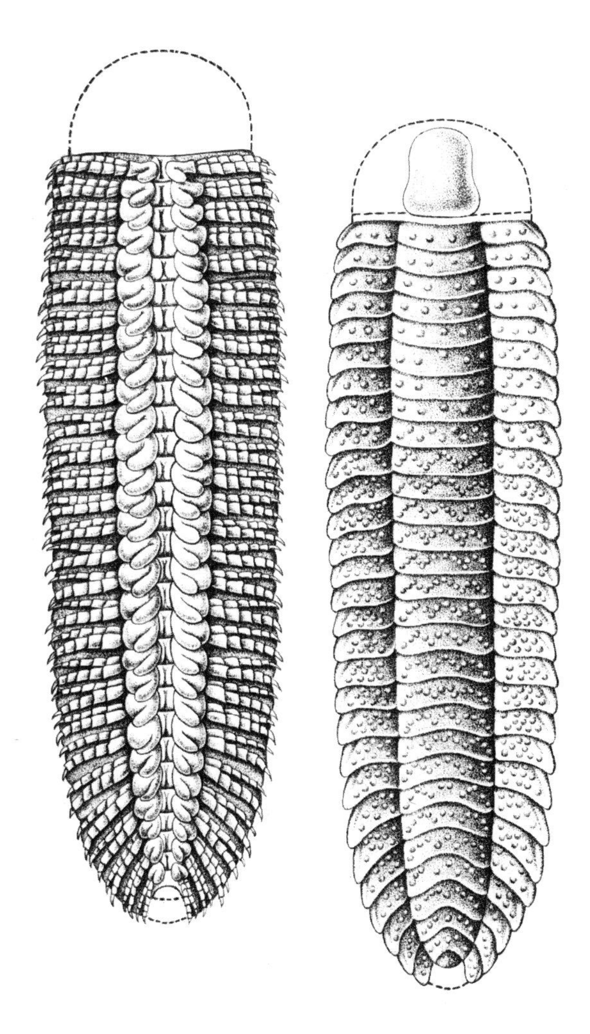

Arthropleura armata ($\times \frac{1}{7}$ approx.) from the Middle Carboniferous Period coal beds of the Saar. These animals, looking like very long trilobites (see page 89) grew very large, and are now classified either with the myriapods or as a separate but very similar group.

The first to have succeeded in the attempt seem to have been the myriapods, a group whose most familiar modern representatives are the millipedes and the iulus. Some very questionable vestiges have been discovered from the Silurian Period of Britain, and it is only really among plant remains in the Old Red Sandstones of the Lower Devonian Period in Scotland and England that their presence has been definitely proved.

Once again, the extraordinary deposits at Rhynie, Scotland, have provided us with most information on Devonian Period arthropod fauna, and Collembola (wingless insects), Acarida (mites and ticks) and spiders have been found ideally preserved in the silicified peat. Like their present-day descendants, minute creatures which dwell in the ground or in rotting timber, the very first (of the genus *Rhyniella*) must have fed upon spores and micro-organisms. Unlike numerous modern forms such as ticks, however, the Acarida of Rhynie could not have been parasites of vertebrates. *Protocarus* probably sucked the sap of *Rhynia* (see page 130) whose fossils often show perforations along the stalks. These micro-arthropods were also the prey of spiders that attained a size of some 1.5 centimetres ($\frac{3}{5}$ inch) and which hunted them in the spore cases of psilophytes that had fallen to the ground. The disjointed debris of their victims has been found in large amounts, and at Alken an der Mosel in Germany a Devonian Period lagoon has yielded up millipedes and spiders which, like those at Rhynie, must have stalked and hunted their prey.

Inside carboniferous forests

The most familiar geological deposit (and the most sought-after) from the Carboniferous Period is coal, the substance from which the period above all derives its name. This precious fuel originated in swampy areas covered in luxuriant forests (see below). The ferns and horsetails which grew there attained great heights, and their stalks – often as thick as treetrunks – were not much shorter than those of the *Sigillaria* and *Lepidodendron*, lycopods which grew to tens of metres in height. The *Cordaites*, ancestral to the true conifers, also added their many-branched trunks and huge oval leaves.

During the second half of the Carboniferous Period enormous plant masses were thus accumulated in swamp waters whose low oxygen content ensured that they would rot slowly. The layers of peat formed in this way were covered from time to time by coats of clay and sand brought by flood waters or the rise in sea level. Then fresh vegetation took root (we have found fossilized traces in these

banks) and a whole new wealth of rotting leaves and trunks was formed. This cycle, repeated over millions of years, buried the layers of peat deeper and deeper. These layers, now covered by more recent sediment, then dried and subjected to every-increasing pressure, slowly transformed into lignite and later on into coal.

In this environment the arthropods underwent a tremendous burst of evolution. The abundance of living and dead plants offered practically unlimited food supplies to the vegetation eaters who began to multiply in profusion – as did their predators, of course. Vertebrates starting to look for food out of the water also benefited by this proliferation, and under this double pressure the plant-eaters evolved rapidly.

The hot, wet climate of the Carboniferous speeded up the conquest of land by the animals. Arthropods became common among the luxuriant flora, offering fresh sources of food for the vertebrates. A typical view of this landscape shows a tree-living fern and the bare trunk of a *Sigillaria* in the foreground, while in the distance we can make out huge specimens of *Lepidodendron*. *Calamites*, a kind of giant horsetail, took root at the edge of the swamps, and at the extreme right of the picture appear the huge leaves of *Cordaites*.

These creatures prospered for hundreds of millions of years. True insects first made their entrance from the middle of the Carboniferous Period, multiplying quickly from then on. Many of them possessed specialized organs designed to pierce the bark of trees and other plants and suck the sap, while using their wings to escape from their natural predators, the spiders, who were still running after their prey though some spiders responded by spinning webs in the air to catch flying prey. By this time the spiders had all but completed their evolution; nevertheless, the species capable of weaving the most efficient of these aerial traps were not present until the Cainozoic Era. In addition, the absence of flying vertebrates meant the insects had few enemies: cockroaches, mayflies and dragonflies were the chief representatives of this winged fauna.

The insect's best means of defence against the unwelcome attentions of the vertebrates has always been to make its flesh unpleasant to taste or even poisonous, warning off potential aggressors with characteristic colours. There can be little doubt that the eye-like spots borne by *Protodiamphipnoa* of the Upper Carboniferous Period – similar to certain modern butterflies – were intended to drive away, maybe actually to frighten, potential predators.

Towards the end of the Carboniferous Period some insects took on enormous proportions. *Meganeura*, (see pages 148 and 150) a huge dragonfly with a wingspan of 65 centimetres (26 inches), must have had little to fear from spiders' webs. This development of giant forms had also started to affect other arthropods, and certain Devonian Period scorpions, probably amphibious by nature, were not much under a metre (39 inches) in length. Nor did the millipedes lag behind, with *Arthropleura* attaining a length of 1.8 metres (6 feet) (see page 147) and leaving behind some impressive tracks on the surface of certain sandbanks. The multiplication of land vertebrates, however, reduced these groups to smaller proportions, and although the spiders were able to hold their own, the scorpions regressed considerably. Only the insects in fact continued to be successful.

The boundary between the Carboniferous and Permian Periods is marked by changes in insect fauna that correspond with far-

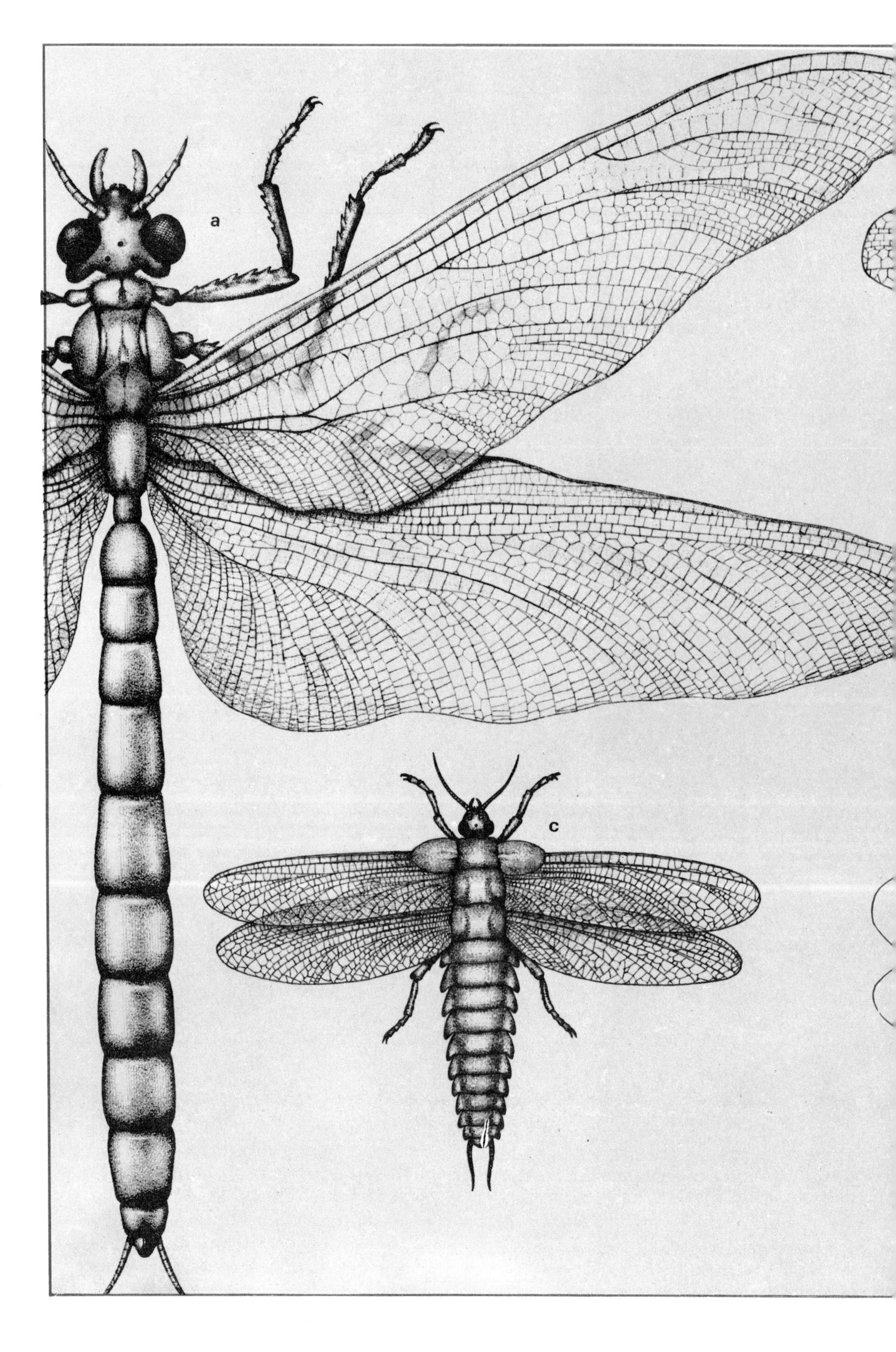
a
c

Some flying insects from the Carboniferous swamplands:
(a) *Meganeura*, with a wingspan of 65 centimetres ($25\frac{1}{2}$ inches), the largest insect ever to have lived.
(b) A form similar to the Orthoptera (modern-day grasshoppers).
(c) *Stenodictya lobata* belonging to the Palaeodictyoptera, a group only found in the Palaeozoic Era resembling the most primitive of all winged insects. They have three equal thoracic segments, a fixed pair of horizontal balancers and a very primitive arrangement of wing veins.
(d) *Mylacris*, one of the many cockroaches which lived at the time.

reaching changes in climatic conditions. As the environment dried out and the beginnings of a seasonal cycle appeared, so did the very first cases of metamorphosis. Although the descendants of *Meganeura* still existed, travelling through the air, the general tendency was to decrease in size. During the Mesozoic Era the ancient strains declined, to be supplanted by the modern orders. Evolution of the insects during the Cretaceous Period is little known, but one thing at least is certain: the advent of the flowering plants gave the insects a whole new source of food to exploit. Finally, during the Tertiary Period, the development of a social life-style among certain groups was completed, and gave them the appearance we know today.

Insects rarely fossilize owing to their frailty and flying abilities, and so truly exceptional conditions are required for fossil preservation. Apart from the fauna of Carboniferous forests, other specimens have been extracted from more recent deposits such as those at the end of the Jurassic Period at Solnhofen, Bavaria. But the very best fossil insects are undoubtedly those found in the Oligocene Period ambers from the shores of the Baltic Sea. Like the Coleoptera shown above, they have usually retained their fine structures in the coniferous resin that began to form some 30 million years ago.

Fossil insects found in deposits just a few tens of millions of years old are often little different from their present-day relatives. But some, like the Lepisma (above) and the Ephemera (below), have hardly evolved at all since the Carboniferous Period.

Chapter Six

The Reign of the Reptiles

The reign of the reptiles

The amniotic egg

As we have seen, the very first land-living vertebrates were still largely dependent upon the aquatic environment since this remained for a long time their chief source of food. Thus they rarely strayed very far from it. It was the way they reproduced, however, which linked them to the world of water to begin with; their eggs, once laid in swamps, lakes or rivers, gave birth to tadpoles equipped with gills but which were incapable of crawling onto land without undergoing a major change in the structure of their body organs. Having reached adulthood they were then compelled to return to their original environment in order to breed.

Some modern amphibians have succeeded in reducing their dependence upon water. The eggs may be laid on land with the young returning to water after hatching, or the process of metamorphosis may be shortened. But it was only with the advent of the 'amniotic egg' that the vertebrates were really able to free themselves from water altogether; this vital step in the development of the vertebrates was accomplished from the end of the Palaeozoic Era by a group of creatures that it is convenient for us to call 'reptiles'.

The yolk sac containing food reserves is already present in eggs of the lower vertebrates, but in the reptilian egg this is supplemented by new structures, the most important of which is the amnion, a membrane which envelops the embryo, holding it suspended in a bag of liquid. In a way the amnion can be thought of as a small piece of the ocean or river which had, until this development, seen the first stages of growth. This new device was accompanied by a structure called the allantoic sac, which served as a 'dustbin' for organic waste products, thus preventing the embryo from being poisoned by its own waste products. Finally, a tough shell encased and protected the whole. Thus, well sheltered and armed with the essentials of survival, the embryo tetrapod was able to develop completely on dry land.

Reptiles: an artificial group

To the human mind, the word 'reptile' evokes a clearly defined (though not always correct) picture of a cold-blooded, usually crawling animal covered in scales and laying eggs that are surrounded by a shell. Nevertheless this name has little meaning for a palaeontologist since although some fossil 'reptiles' are indeed related to the modern turtles, crocodiles, snakes and lizards, other forms are at the origin of birds and even mammals. So that in terms of evolution, a crocodile is closer to a chicken than to an iguana, as amazing as that may seem. Therefore, rather than speaking of reptiles we should speak of the origin of the Amniota, a group distinguished by the development of the amniotic egg, and divided into several lineages all of which had reptilian features but whose final development was not necessarily into a reptile, since some went beyond this structural stage.

The flourishing of the Amniota group was accompanied by the appearance of new anatomical features, especially bony structures by means of which we are able to identify the first 'reptiles'. The labyrinthodont teeth and the lateral line nervous system have now been lost forever, and the latter did not even reappear in those Amniota which eventually reverted to an aquatic life-style. More positive signs of progress are that the skull took on a new form; more specifically, the notch over which the eardrum was stretched became closed, and the eardrum moved back, while the inner ear turned downwards. To make room for a reinforced chewing muscle, openings appeared on each side of the skull (see page 157). While the more primitive reptiles have none of these openings (and are termed anapsids), the more evolved forms have one or two. The arrangement of these openings is very important since their pattern is the basis for a classification of the various Amniota lineages. The first two neck vertebrae became specialized to form an atlas bone pivoting on one axis. The limbs too became modified: the tarsal bones fused to form an

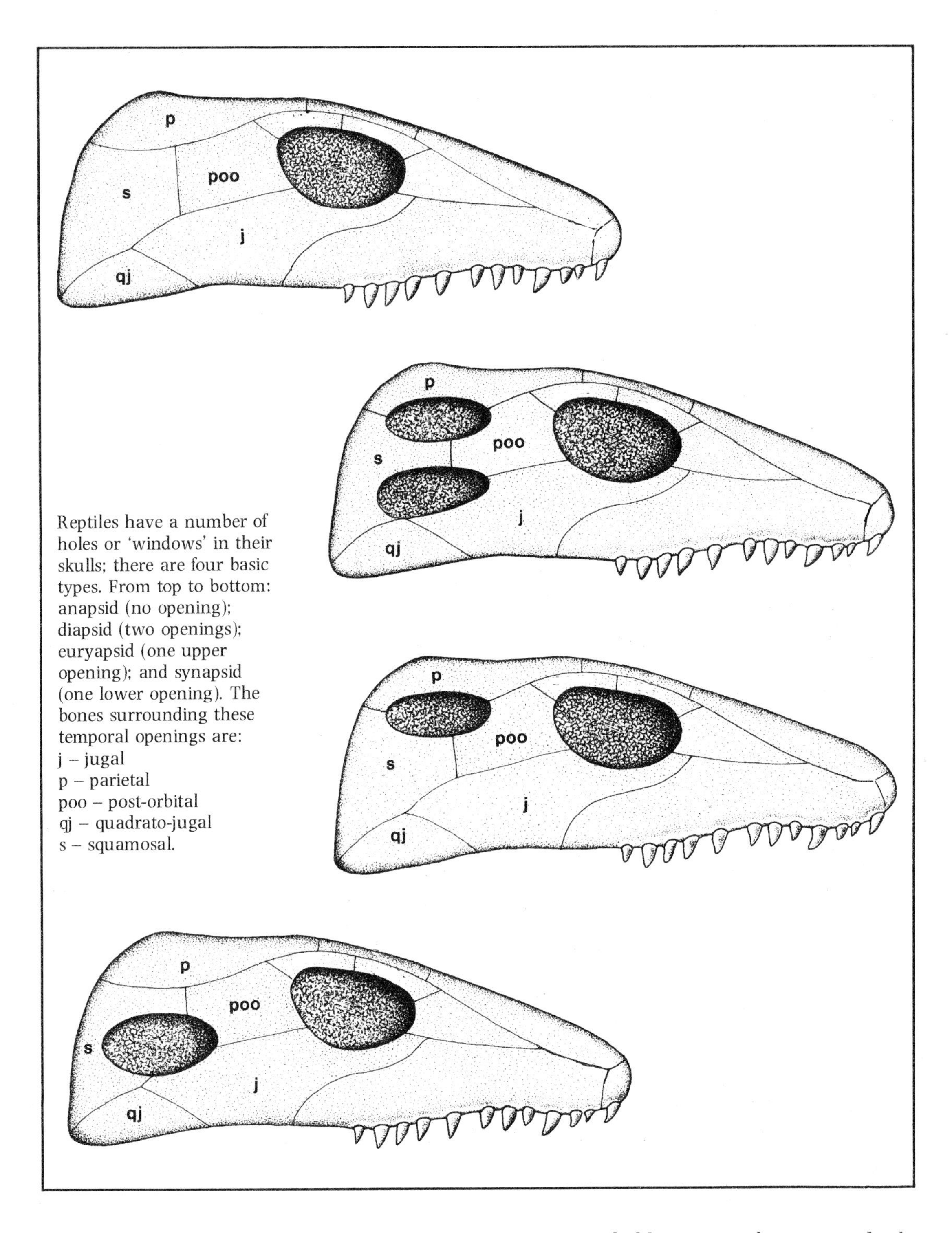

Reptiles have a number of holes or 'windows' in their skulls; there are four basic types. From top to bottom: anapsid (no opening); diapsid (two openings); euryapsid (one upper opening); and synapsid (one lower opening). The bones surrounding these temporal openings are:
j – jugal
p – parietal
poo – post-orbital
qj – quadrato-jugal
s – squamosal.

astragalus (ankle-bone) and a calcaneum (heel-bone), whilst the number of individual bones diminished. Overall, the skeleton tended to become more mobile and better adapted to terrestrial movement than ever before.

Of all the refinements that differentiated the Amniota from the amphibians, one of the most remarkable was without any doubt connected with the circulatory system. Although no fossils have been left behind to show its appearance we are still able to observe that, in modern reptiles whose hearts are divided into partitions, the oxygenated blood which is supplied by the vein from the lungs can only partly blend with the

This complete skeleton of *Labidosaurus* was found in the Lower Permian Period of the United States, and was a representative of the captorhinomorphs – a group of very primitive reptiles characterized mainly by a skull without any temporal openings (anapsid). This creature was about 1 metre ($3\frac{1}{4}$ feet) long.

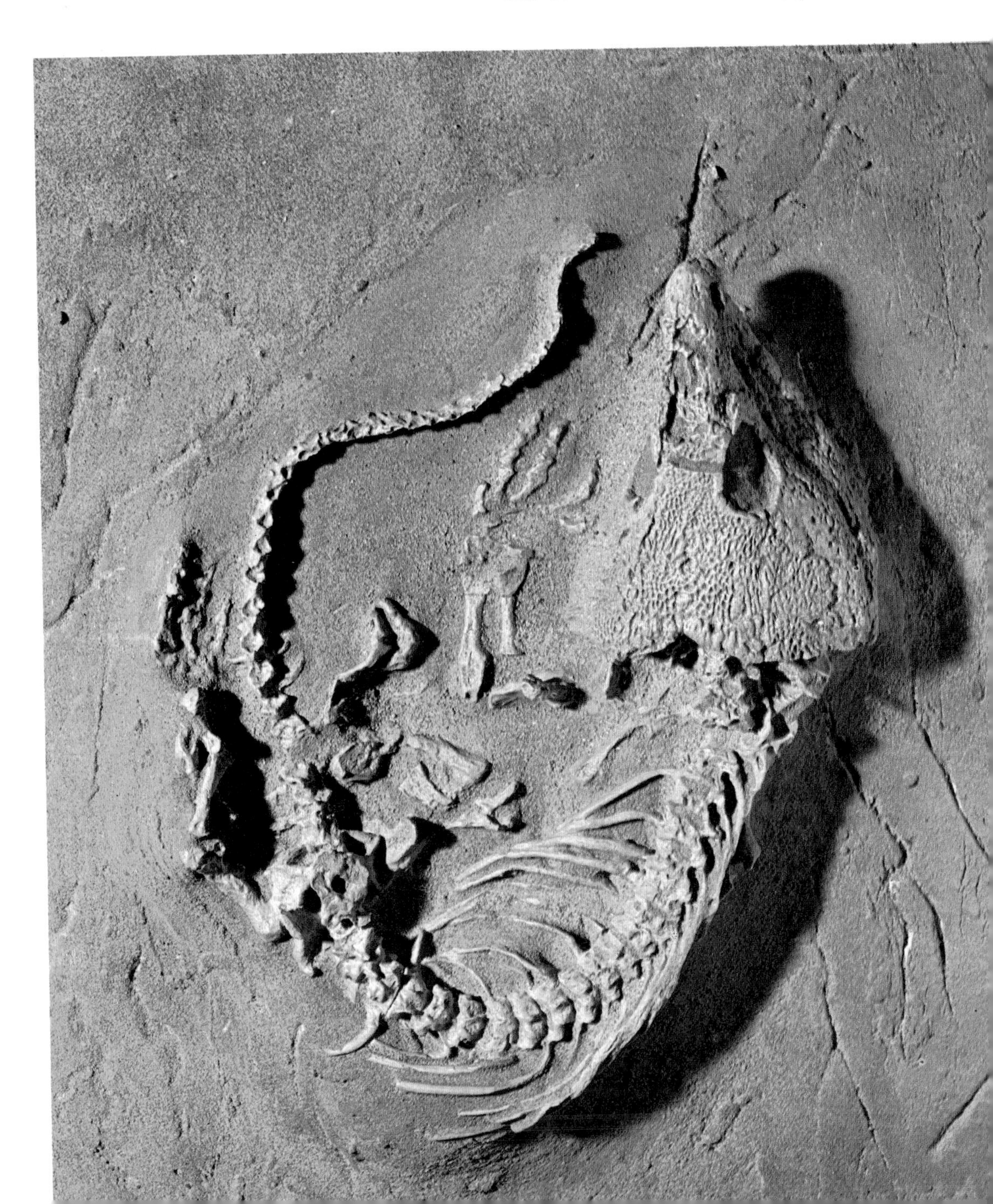

de-oxygenated blood which has already gone round the system. The ventricles of the heart are in fact separated by an incomplete wall that is absent in amphibians. This gradual separation of the heart into two chambers, one right and one left, was not perfected until the mammals and birds evolved, but still it was capable of increasing the performance of the respiratory system. Fossils of reptile skulls show an increase in the size of the brain compared with that of the amphibians, although this was still fairly small. Finally, the reptiles' excretory system was also improved, the kidneys now eliminating the waste products as urea or uric acid and no longer as ammonia, a highly toxic substance that had to be diluted in a large volume of water not to be dangerous. This was yet another modification which helped the Amniota to free themselves from the world of water.

The oldest reptiles

As with nearly all ancestral lineages, the greatest confusion reigns among the very first reptiles. The characteristics we have already looked at did not appear all at the same time and are not automatically connected to the acquisition of the amniotic egg. We saw in the seymouriamorphs how the amphibians may have acquired some, but not all, reptilian traits, hence certain forms hover on the borderline without us being able to identify them positively as either amphibians or reptiles. The basic problem is to know whether several amphibian strains acquired the amniotic egg; this is an idea which is almost impossible to prove, and which has generally been regarded as highly unlikely. Nevertheless the true reptiles were contemporary with genera such as *Diadectes*, a heavy herbivore related to the specialized group of seymouriamorphs but still having most of the reptilian characteristics – at least the ones we can see from fossils of hard features.

It is likely that the reasons which compelled certain four-footed creatures to lay their eggs on land stemmed from the need for extra protection, since the lakes, rivers and swamps were packed with predators: arthropods, fishes, and many other creatures which were no doubt tempted by this source of food. The far more sparsely populated dry land offered much better chances of survival and success. Still, the very first reptiles appeared in the swampy environment of Carboniferous forests, and of those forms that retain certain amphibian traits, *Hylonomus* of the Middle Carboniferous Period is an undisputed reptile, with remains discovered for the first time in 1852 in the hollow stumps of *Sigillaria* (see page 148) from the Joggins deposits of Nova Scotia, Canada. These small creatures owe their fossilization to these natural traps which were filled up by mud. They resembled large lizards some 30 centimetres (1 foot) long, with hands and feet very well developed in comparison with the rest of the leg. They fed on insects and must have had a style of life not unlike that of modern lizards.

Hylonomus belongs to the group of captorhinomorphs whose small anapsid (that is, with no skull openings) forms are known right down to the Middle Permian Period (see page 158). This group also gave rise to other rather primitive families, some of which like the procolophonids were very small in stature, while others like the pareisaurids of the Permian Period were up to 3 metres (10 feet) in length. These were heavy quadrupeds, compact in shape and with a skull usually covered in spikes and outgrowths. Numerous complete skeletons have been found in South Africa; they were rather slow-moving herbivores protected by both their weight and by bony plates covering their backs. Despite their size, however, they fell victim to the representatives of another strain of reptiles – the synapsids – many of which were ferocious carnivores. In parallel with the captorhinomorphs and their descendants, a second group had in fact appeared that dominated the four-footed creatures at the end of the Palaeozoic Era and beginning of the Mesozoic Era. This second strain led – much later on – towards the mammals.

We can see therefore that the Amniota separated into at least two branches at an early stage. The first, the sauropsids, are represented today by the reptiles and birds. They reached their peak during the Mesozoic Era. The second, the theropsids, were a much less diverse group, and they gave rise to the mammals. Apart from their synapsid type of skull, a number of features distinguished the theropsids from the sauropsids, in particular the original arrangement of the aortic arches which leave the heart, and the excretion of urea rather than uric acid. In Chapter 8 we shall see how the theropsids evolved.

Mesosaurs

The first mesosaur was discovered in South Africa but we do not know exactly where it was found because the slab of rock bearing the fossil was once used as a cooking-pot lid by a tribe of Hottentots! Other mesosaurs were found in Brazil towards the end of the last century, and this geographical distribution may appear puzzling at first sight. However, it is understandable when we remember that South America and Africa were closely connected at the end of the Palaeozoic Era (see pages 232–233).

The mesosaurs, which all lived at the beginning of the Permian Period, are the very oldest totally aquatic reptiles. Their body shape was poorly adapted to life in water (see page 161) and their limbs in particular did not form the 'paddles' peculiar to more specialized aquatic forms (see pages 164 and 169). It is interesting to note that the ribs were thickened and this is a characteristic of other groups such as the sirenian mammals. This increase in weight probably evolved due to the extra buoyancy these animals gained from the surrounding water. The mesosaurs also had nostrils placed well back on the snout and this greatly helped them to breath when they were swimming.

The streamlined body of the mesosaurs was propelled through the water by lashing the powerful tail up and down. They hunted small fishes and crustaceans, seizing them between their long jaws armed with many needle-like teeth. These teeth were probably too fragile actually to bite but served as a kind of grille that kept the prey trapped in the jaws.

Turtles

The only modern-day anapsids are the turtles or chelonians which, despite being rather specialized, are related to the most primitive reptiles. Exactly which groups they arose from is still a mystery, but they probably evolved from the captorhinomorphs, or perhaps the procolophonids. As often happens, the fossil record is incomplete; intermediate forms from the Permian Period are unknown, and the first chelonian, *Proganochelys* of the Triassic Period, already had some of the features of a true turtle (see page 163). The most primitive turtle so far found is a creature called *Eunotosaurus* from the Middle Permian Period of South Africa, and this is sometimes seen as an ancestor of the chelonians. It is an incomplete fossil showing only the underneath of the animal. The neck was quite long and the teeth are found on the palate and along the jaws. There are few vertebrae (a characteristic of the turtles), and the dorsal ribs, like enlarged petals, grew outwards from the body to form a kind of shell. According to D. Watson who first described *Eunotosaurus* these enlarged ribs may have supported bony plates of skin (the creature's outer layer of armour). Despite these features, however, *Eunotosaurus* may in fact be merely the result of convergent evolution from ancestors quite separate from those of the chelonians.

Proganochelys, from the upper Triassic Period of Germany, was about 1 metre ($3\frac{1}{4}$ feet) long and all the bony components of the shell can be seen, although the head, tail and limbs could not have been pulled within the carapace and the creature's defence was provided by many sharp spikes and bony knobs. The palate was still armed with teeth but the jaws were strengthened by the characteristic powerful turtle's beak. Enclosed in their bony 'box', covered with its horny plates, the chelonians were superbly protected from enemies and they were able to spread successfully into many different environments. They have hardly changed their basic appearance since Permian times.

The most important feature of turtles after the Triassic Period was the ability to retract the vulnerable parts of their body such as their legs and head into the shell. Two mechanisms were developed to withdraw the head. In the pleurodires, which first appeared in the Cretaceous Period, the neck

This skeleton of a *Mesosaurus* found in the province of Parana (Brazil) was about 75 centimetres (29 inches) long and belonged to a young specimen: the carpals (between hand and forearm) and the tarsal bones (between foot and leg) had not yet been properly turned into bone and could not therefore be preserved by fossilization.

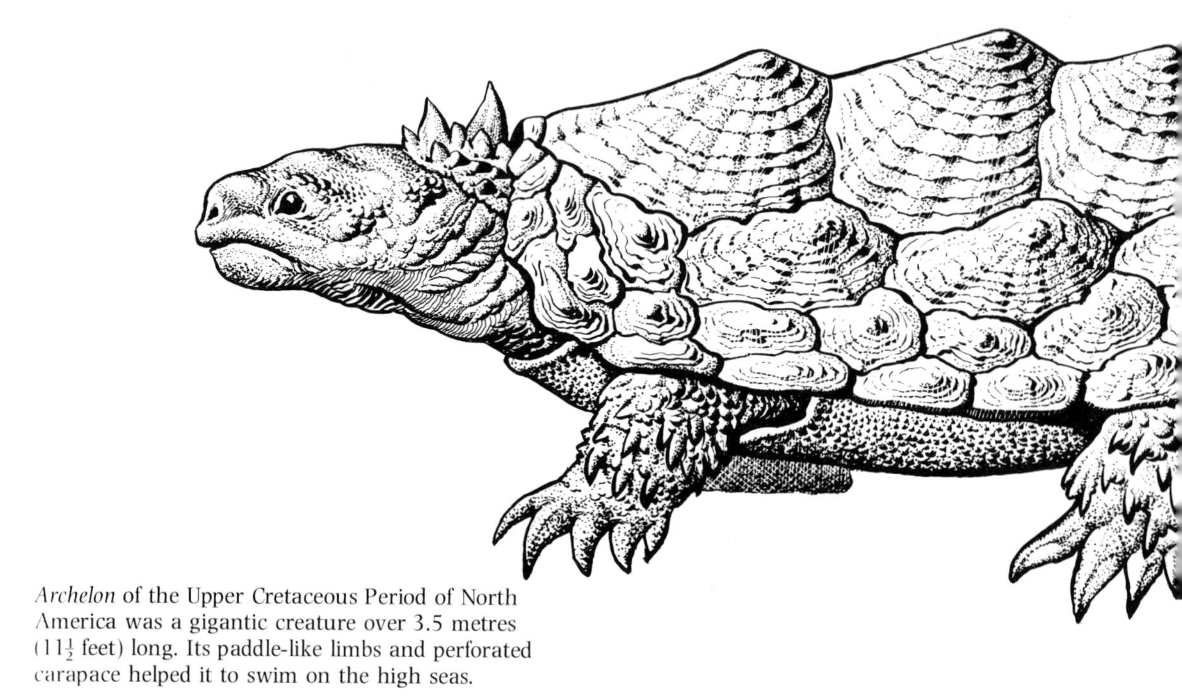

Archelon of the Upper Cretaceous Period of North America was a gigantic creature over 3.5 metres ($11\frac{1}{2}$ feet) long. Its paddle-like limbs and perforated carapace helped it to swim on the high seas.

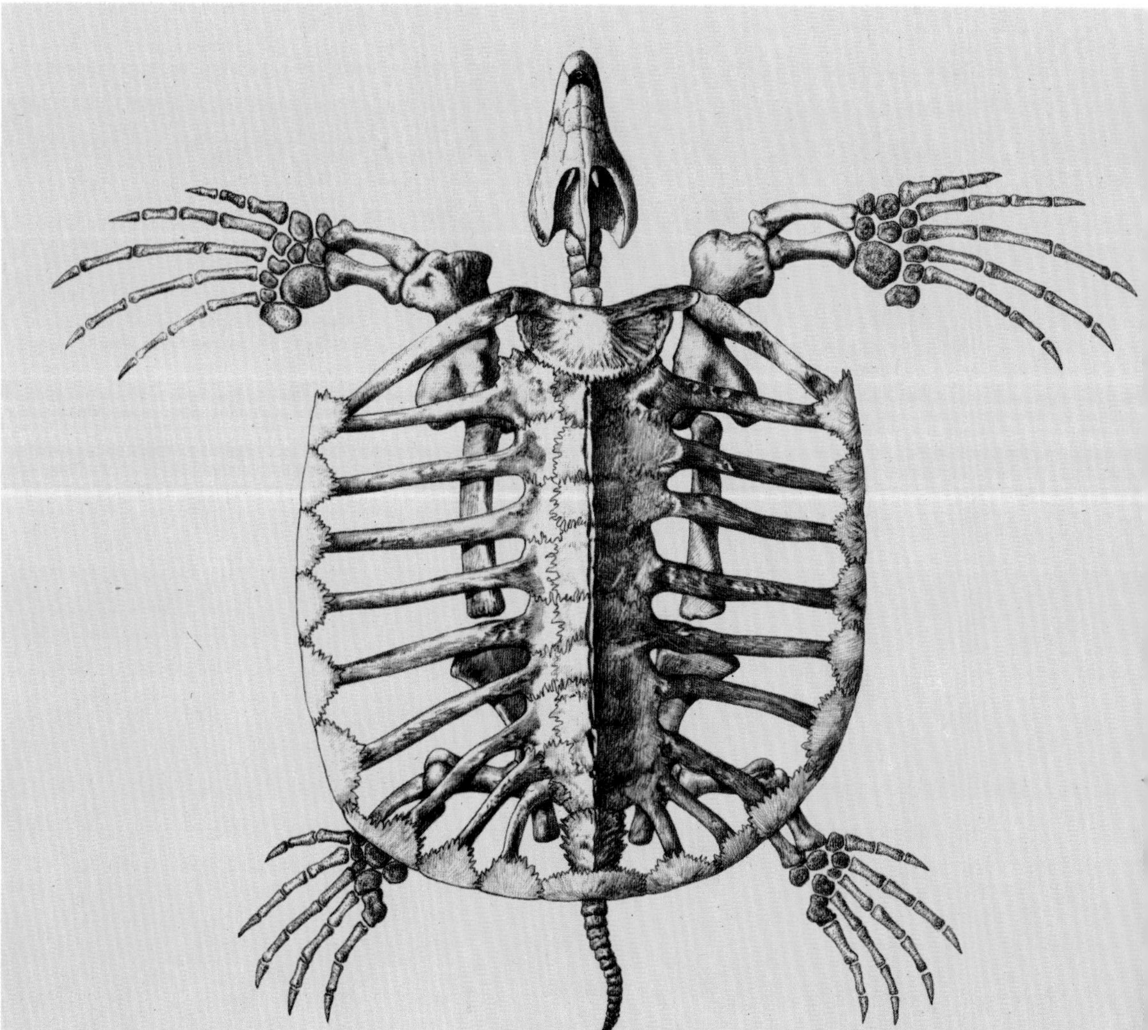

is folded sideways in a Z-shape. Although they were at one time widely distributed, the pleurodires are now found only in the continents of the southern hemisphere. The cryptodires (see below) pulled in their heads by folding their necks into a vertical S-shape, and this group makes up the vast majority of present-day turtles. The group arose at the end of the Jurassic Period, and has given rise to many different types. While the primitive chelonians probably led a semi-aquatic life, some cryptodires lived on land and could survive even in the hot desert zones. Others conquered the oceans by modifying their limbs to form paddles and by reducing the weight of their shells. On land and in the sea the cryptodires often became very large (see page 162). The marine turtles have never been able to free themselves completely from their dependence on dry land for they have to return there to lay their eggs. This habit now threatens the survival of today's marine turtles. The large number of huge eggs they bury along the shorelines may offset the destruction of a large part of

Proganochelys (below) of the Upper Triassic Period of Germany is easily recognized as the oldest of the turtles. The trionychidans were cryptodire turtles highly adapted for life in fresh waters; they were covered by a soft, scaleless carapace, and as this ventral view of the skeleton shows, their bony plastron or 'breastplate' was lightened by fontanelles. Although limited to tropical regions in modern times, this family inhabited Europe during the Cainozoic Era.

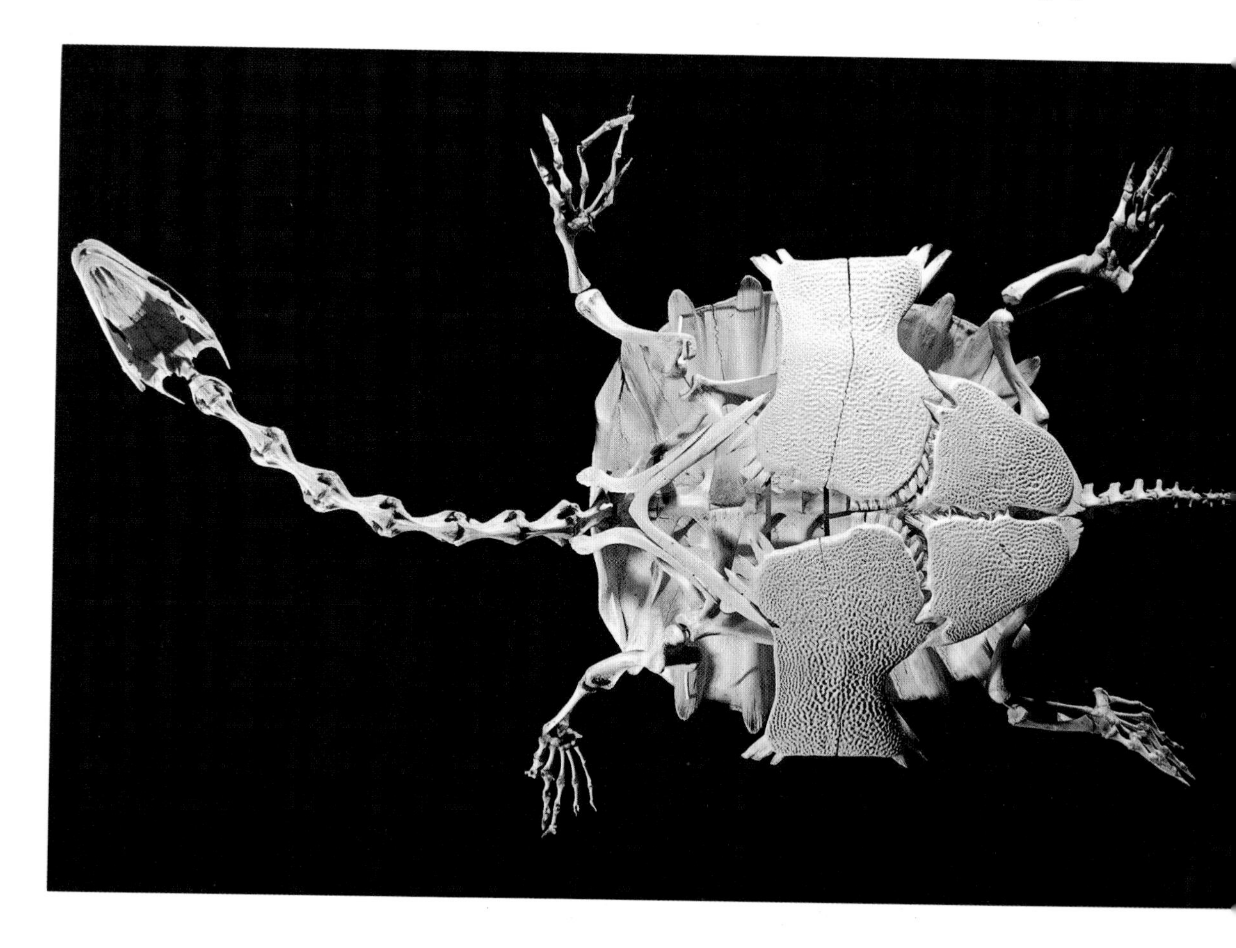

their progeny by their natural predators. However, this is no protection against their most savage enemy: Man.

Ichthyosaurs

The name ichthyosaur literally means 'lizard-fish', and these reptiles which went back to the oceans were very successful. The magnificently preserved specimens found in deposits such as those at Holzmaden (Wurttemberg) in Germany give us an accurate picture of what these animals looked like. Not only have many complete skeletons been unearthed but the soft parts of these animals can also be seen (see illustration above). The 'fish shape' of these reptiles is probably the best known example of evolutionary convergence. The slender body ended in a vertical hypocercal tail. There was no real neck, while a dorsal fold on the back formed a triangular blade looking very much like a shark's fin, and the paired limbs were transformed into flippers. The arms and legs were extremely short but powerful, while the number of bones on the hands and feet – and even of the fingers themselves – had grown. Like the fishes, they had many biconcave vertebrae, and the nostrils were placed closer to the eyes. Finally, the animals' huge eye sockets had a ring of bony plates that may well have protected the eyes from huge pressures when diving very deep.

The origin of the ichthyosaurs is also uncertain. They were once thought to be the possible descendants of several groups, in particular of the mesosaurs or primitive mammals, but these ideas had to be abandoned for various reasons. The oldest-known specimens come from the Middle Triassic Period. They were already very specialized and had only moderately elongated snouts, limbs that were less modified than in Jurassic forms, and a vertebral column that was straight, not bent downwards in the tail. The tail itself was long and the animal had a small upper fin. Typical forms as illustrated above and right come from the Lower Jurassic Period: the front limbs have become stronger and larger and the tail is nearly symmetrical.

The biology of the ichthyosaurs must have been fairly similar to that of modern dolphins. Hunting in packs, they mainly ate belemnites (see pages 20 and 21), whose

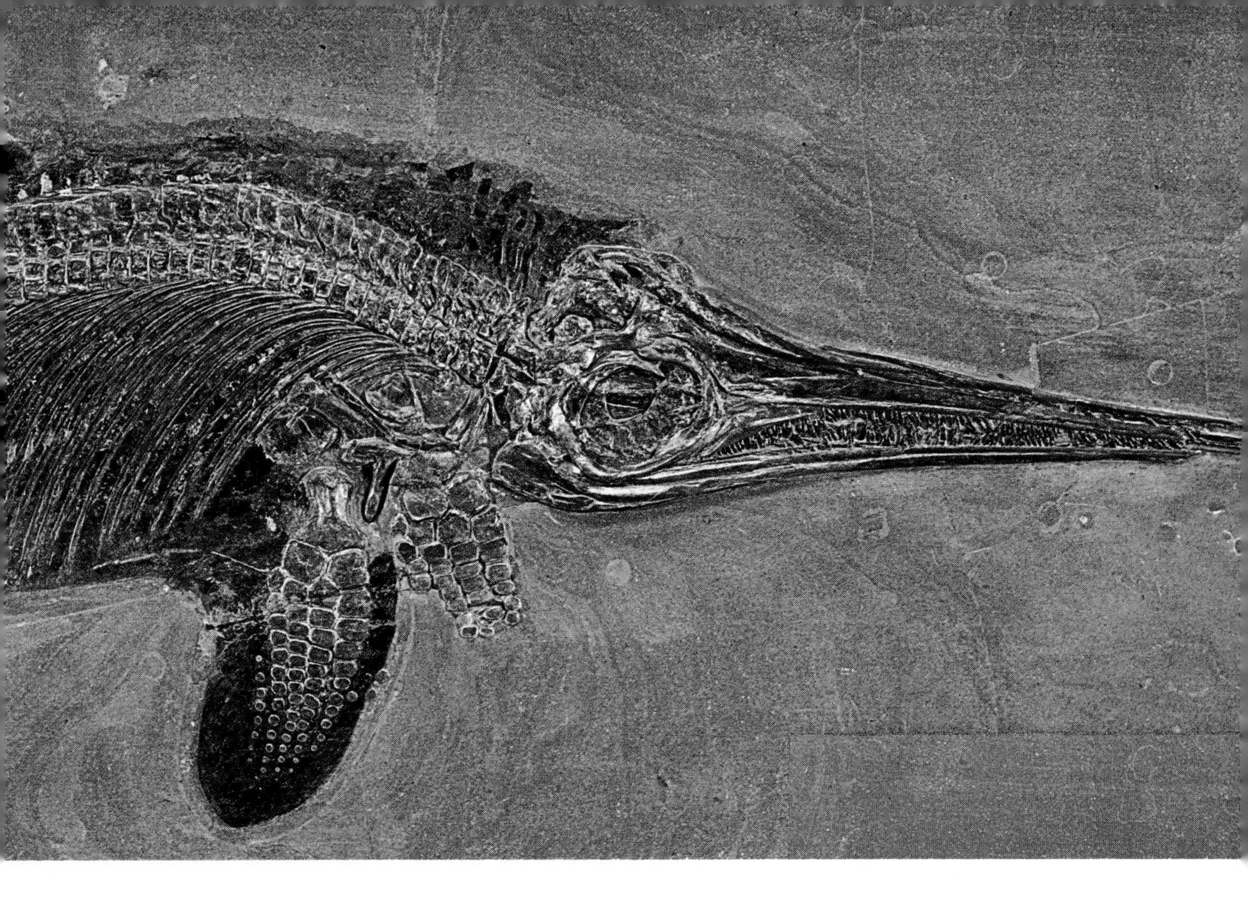

Above: *Stenopterygius acutirostris* from the famous Lower Jurassic formations of Holzmaden, Germany. Of the dozen or so complete specimens found here, only a few reveal the outline of their bodies as traced on the rock by a carbon film.

Below: The conical teeth of the ichthyosaurs were inserted along their jaws in furrows rather than sockets. By the strange phenomenon of 'convergence' they show similar adaptations to the unrelated odontocete cetaceans.

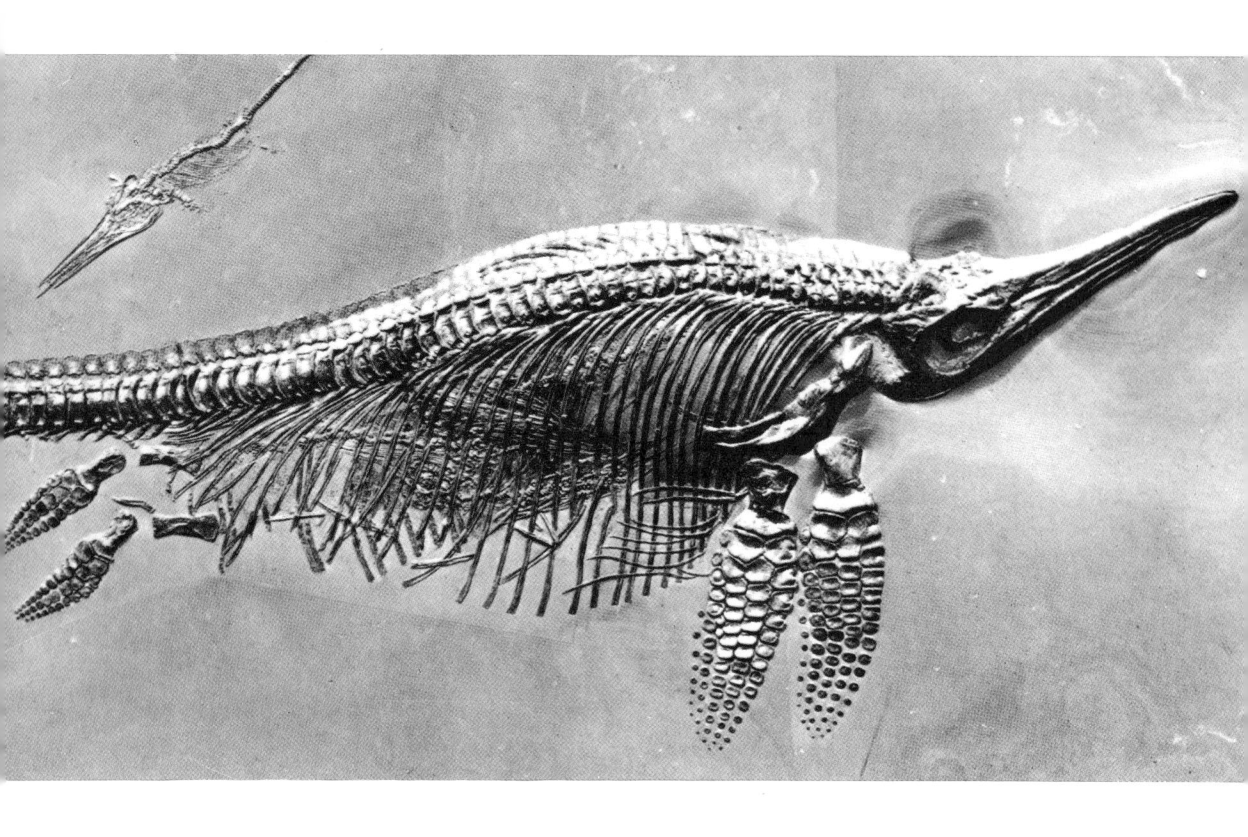

A certain number of fossilized ichthyosaur 'births' have been uncovered, sometimes with the baby still partially inside the pelvis of the mother, which may have died in the process. On the other hand, the embryo may have been expelled following an accidental death.

guards have been found in large numbers inside the fossil stomachs of the specimens discovered at Holzmaden. Amongst the debris contained in these skeletons, young ichthyosaurs have been found, some of whom had probably been eaten by the adults. Some embryos have also been found still within their mothers. The ichthyosaurs, which could not crawl on to dry land, were forced to reproduce in the sea, and so they became ovoviviparous, in other words the eggs were not laid, but kept inside the mother until hatched. It is these young animals, rolled up in a spiral fashion, which can be seen inside certain fossils. The ichthyosaurs were not the only group to show ovoviviparity, it is found in other reptiles too. It should not be confused with the true viviparity of the mammals, since in ovoviviparity the egg is never fed by the mother.

The ichthyosaurs were most abundant during the Upper Triassic Period and some of them became giants – *Shonisaurus* found in the Shonshon Mountains of Nevada, U.S.A. was as much as 15.5 metres (51 feet) in length. Throughout the Jurassic Period they remained common but less diversified. They became less common throughout the Cretaceous Period, and were probably competing with other groups of marine reptiles. They finally died out completely at the end of the Mesozoic Era.

Placodonts

The euryapsids make up a second very important group of marine reptiles (see page 157). If we overlook some terrestrial Permian Period genera whose position in the Euryapsida is doubtful, the owners of this type of skull break down into two groups adapted to an aquatic life: the placodonts and the sauropterygians.

Represented during the Triassic Period only, the placodonts were a group of species adapted to a very special way of life. The best-known placodont belongs to the genus *Placodus* and was a sort of enormous lizard with short legs and neck and a cylindrical trunk. Its most notable feature was its teeth (see page 167) which were those of a mollusc-eater. Propelled by its powerful tail, it dived in coastal waters in search of food on the sea-bed. With its robust front teeth, it tore away the mollusc shells which it then crushed against its palate. The shape of its jawbone tells us that it had very strong muscles for chewing its prey, and the skin on its back was covered with a whole series of bony nodules which were a type of primitive carapace. Some very specialized forms such as *Placochelys* and *Henodus* have this feature quite well developed.

Henodus was encased in a flattened bony 'jacket' very similar to that of the turtles, but with a much larger number of sections. This convergence with the marine chelonians is most astonishing. The tail was shortened and the legs worked like paddles. The jaws had lost most of their teeth, and their forward part was armed with a horny beak. *Helveticosaurus* was a genus of placodont with several interesting features including many sharp teeth which showed that it was a carnivore. It is interesting to note that recently it was suggested that the placodonts were possible ancestors of the ichthyosaurs. Otherwise, the 'lizard-fish' are more often regarded as euryapsids.

Sauropterygians

While the world of the reptiles had its dolphin equivalents in the ichthyosaurs, the Triassic Period nothosaurs were its 'seals'. Most of these animals were about 1 metre ($3\frac{1}{4}$ feet) long, but certain species grew up to four times that long. Like the placodonts, they occasionally returned to land and never went very far from it. They also had a skeleton which was still quite similar to that of a land animal. Their limbs especially – which ended in claws – were adapted for swimming but still permitted walking on dry land. Still, their similarity to the placodonts stops there: with their elongated necks and pointed teeth they must have led a very different kind of life as hunters of fishes and cephalopod molluscs.

The much more specialized plesiosaurs

The cranium of *Placodus* from the Middle Triassic Period ($\times \frac{2}{5}$ approx.) seen from the side and below. The word placodont means 'plate-teeth', and these crushing devices covered a very large palate. Because they were discovered separately, they were first thought to belong to certain types of shark.

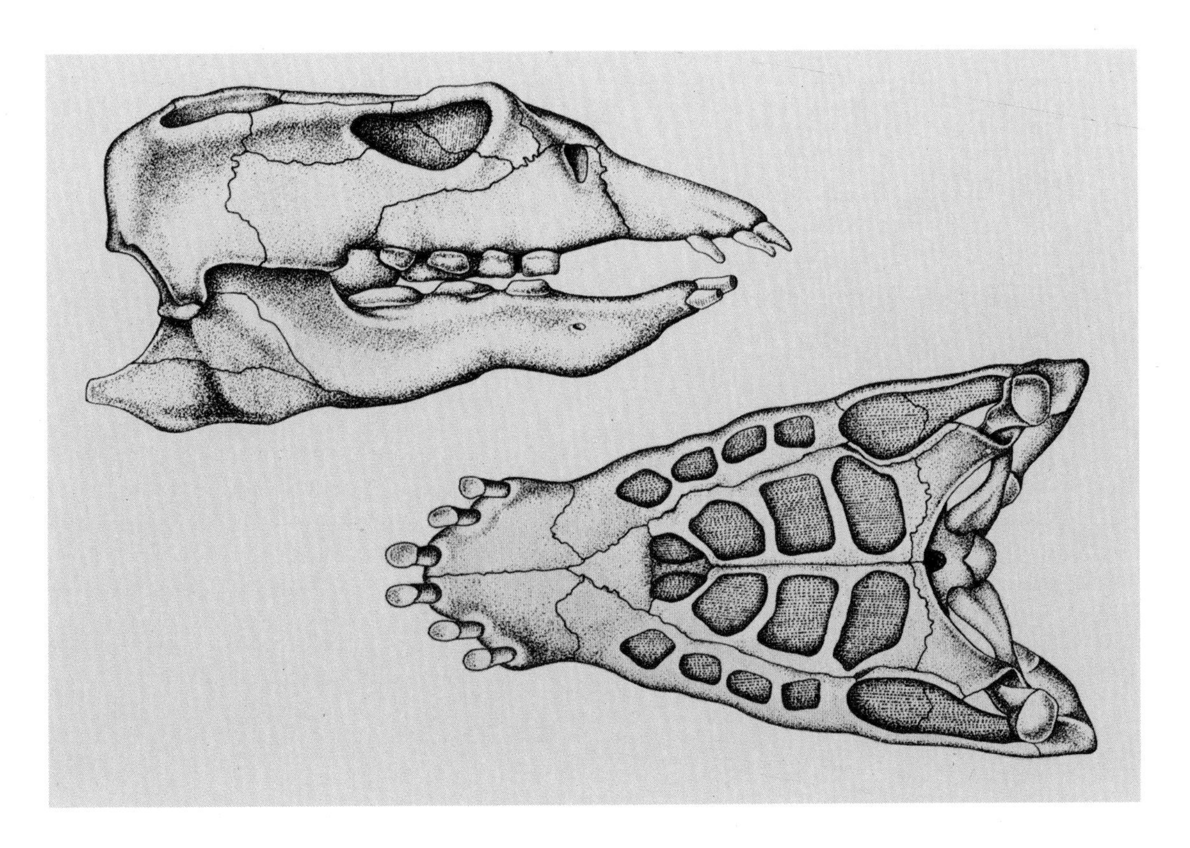

became completely pelagic, living on the open seas. They arose from forms similar to the nothosaurs, and evolved in two very different directions. The first and most surprising of these types has often been described as a cross between a tortoise and a snake: the body was stubby with a relatively short tail, and the very long neck ended in a tiny head. These traits became even more extreme during the Jurassic and Cretaceous Periods, and specialized types such as *Elasmosaurus* of the Upper Cretaceous Period (see below) are quite extraordinary. Specimens nearly 13 metres ($42\frac{1}{2}$ feet) long have been found of this genus, which had no fewer than 76 neck vertebrae! The four flippers allowed the creature to execute quite complex on-the-spot turns and even reversing motions, too. Thanks to their extremely flexible necks these reptiles did not have to pursue their prey but rather darted their heads into the water to grab unsuspecting fishes.

These creatures are among the strangest fossils ever to have been found. In the last century the American palaeontologist E.D. Cope mistakenly reconstructed one by placing its cranium at the end of its tail, a fact that his rival O.C. Marsh did not fail to publicize, and this further increased their intense dislike of each other. The plesiosaurs have always fired the imagination of Man: Jules Verne in his novel *Journey to the Centre of the Earth* wrote about these creatures and from time to time, actual sightings have been claimed, from the chilly waters of Loch Ness in Scotland to the blue deeps of the Pacific Ocean where some years ago a Japanese vessel picked up what was at first thought to be a rotting plesiosaur carcass. Our knowledge of their biology is poor, but it is thought that they had to drag themselves out of the water to lay their eggs: this must have been a spectacular sight. They were sluggish creatures and were very often attacked by a more ferocious type of plesiosaur: the pliosaur. These creatures had a stubby neck and a very long skull, and a huge mouth bristling with dagger-like teeth often as much as 10 centimetres (4 inches) long. The oldest were a modest 3.5 metres ($11\frac{1}{2}$ feet) in size, but in

Elasmosaurus of North America was a giant among plesiosaurs, and by looking at its cervical vertebrae we can see that it could flex its neck very much like the body of a snake.

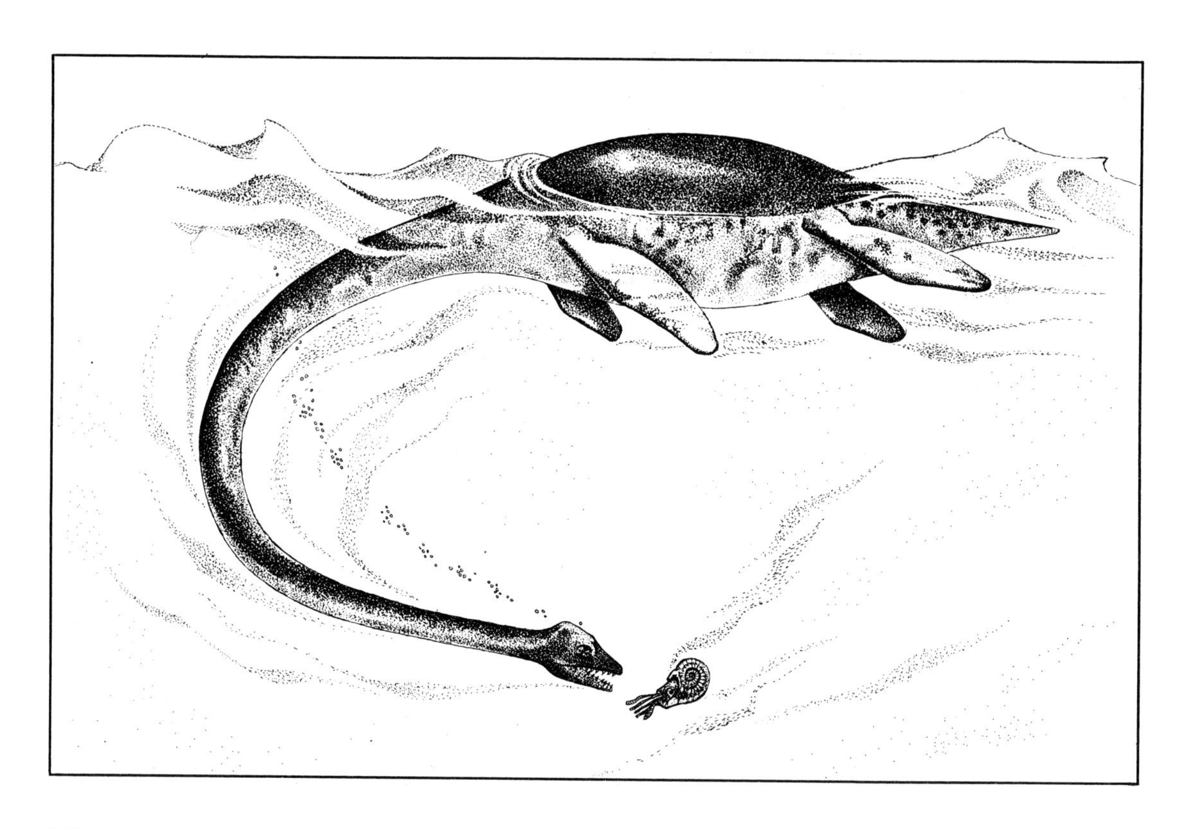

Pliosaurus ferox, a sauropterygian from the Jurassic Period of Britain had a length of 4.8 metres ($15\frac{3}{4}$ feet). With its streamlined shape, this predator was a faster swimmer than the long-necked plesiosaurs.

the Lower Cretaceous Period of Queensland, Australia, a real monster has been found. This was *Kronosaurus*, which grew to a length of 12 metres (39 feet). All the plesiosaurs died out at the end of the Mesozoic Period.

Rhynchocephalians

On land the great reptilian conquest of the Mesozoic Era was mainly due to two groups of diapsids (see page 157) which also colonized the waters. While the mammal-like reptiles, which later gave rise to the mammals, evolved most rapidly during the Permian and the beginning of the Triassic Periods, the remainder of the Mesozoic Era was dominated by the sauropsids. This swapping of dominant roles is often explained by climatic effects. It is claimed that the mammal-like reptiles rapidly perfected a way of regulating their body heat and this adaptation made them better suited to withstand the period of cooling that marked the end of the Palaeozoic and the beginning of the Mesozoic Eras. The sauropsids, more sensitive to temperature changes because they could not maintain their internal temperature at a constant level, had to wait until the climate warmed up in the middle of the Triassic Period before they could evolve into other forms. The archosaurs – the group of diapsid reptiles including the dinosaurs and pterosaurs – dominated the world during the Jurassic and Cretaceous Periods. It seems that some of them also succeeded in making their bodies 'homeothermic' (warm-blooded).

Today the lepidosaurs are only represented by the squamates (snakes and lizards), whose lower temporal opening is not bordered by a bar of bone, and by the Rhynchocephalia, which has only a single species, *Sphenodon*, a real living fossil! *Sphenodon* has kept the primitive arrangement of the two temporal windows characteristic of the diapsids. Both of these groups have clearly descended from the captorhinomorphs, (forms without any 'holes in the skull'). The

The only rhynchocephalian still living is the *Sphenodon* (tuatara) which survives on a few scattered islets of New Zealand, where it has been isolated for millions of years. Free from competition with other reptiles, it had never encountered mammals until the arrival of Man in the region.

first reptiles with temporal openings were small lizards from deposits of the Permian Period in South Africa. The lower opening appeared first, followed by the upper. *Youngina* of the Upper Permian Period from the same region is the first known diapsid, and like all these primitive genera, has been included with the eosuchians, a group that contains the ancestors of both the lepidosaurs and the archosaurs.

The modern *Sphenodon* (see above) is descended from a line that appeared in the Triassic Period and has hardly developed since. Although they were very widespread during the Mesozoic Era, today *Sphenodon* only survives on certain islands off the coasts of New Zealand. It looks like a sort of iguana about 60 centimetres (2 feet) long, living mainly on insects and other small creatures. It is called 'Tuatara' by the Maoris which means 'the spiny one' and although it looks ordinary, its appearance is deceptive, for it has some astonishing characteristics. It has a well-developed pineal eye, located in the centre of the forehead, which is an outgrowth of the brain covered in scales but sensitive to light, and its thoracic cage extends as far as the pelvis. It also has some

Like all rhynchosaurs, *Paradepedon huxleyi* is a distant cousin of *Sphenodon*, although its size was far more impressive since it grew to 2 metres ($6\frac{1}{2}$ feet) or more.

other surprising features – it grows very slowly and may live for as much as a hundred years. There is a gap of two years between egg fertilization and hatching! It also has some strange habits for it hunts by night and often shares its burrow with a shearwater (a type of seabird) whose eggs it does not appear to attack.

Two other rhynchocephalian groups should be mentioned. The rhynchosaurs, (see above) lived during the Triassic Period and spread to almost all parts of the globe. They had a very specialized skull: the front teeth had disappeared and were replaced by a powerful beak. This was probably used to crush roots and rhizomes which they dug out of the ground with their clawed feet. The champsosaurs lived from the Upper Cretaceous to the Lower Eocene Periods and two examples, *Champsosaurus* and *Simoedosaurus*, look very like modern-day crocodiles and gavials. This is another example of a convergence in evolution.

The strange fossils of Monte San Giorgio

Two famous fossil deposits are Monte San Giorgio in Switzerland and Besano near Lake Lugano in Italy, where bituminous schists of the Triassic Period contain reptile skeletons that, while often complete, are flattened on slabs of rock. This often makes them difficult to study and X-ray photography is often used instead of normal methods.

Three genera from these sites, *Askeptosaurus*, *Macrocnemus* and *Tanystrophaeus*, are similar to the lepidosaurs although their exact relationship to them has yet to be fully clarified. *Askeptosaurus* (see page 173) was an animal adapted to life in the sea. It had a highly elongated profile, with a narrow snout which tells us it probably ate fishes. Its legs were very short, and it probably propelled itself in water by whipping its tail from side to side.

Macrocnemus is much smaller and probably lived on land. The hind legs were much more developed than the forelegs, from which we can guess that the creature ran on two legs in the same way as some modern lizards do. It is placed in the same group as the primitive lizards and snakes, as is the third of these genera, *Tanystrophaeus* – a very strange animal indeed. The first remains of *Tanystrophaeus*, discovered around 1830, were neck vertebrae found in the limestone rocks near Bayreuth in Bavaria. They were so long and thin, however, that they were first thought to be limb bones. This mistake was soon corrected, but it was a century later that the first complete specimens were found

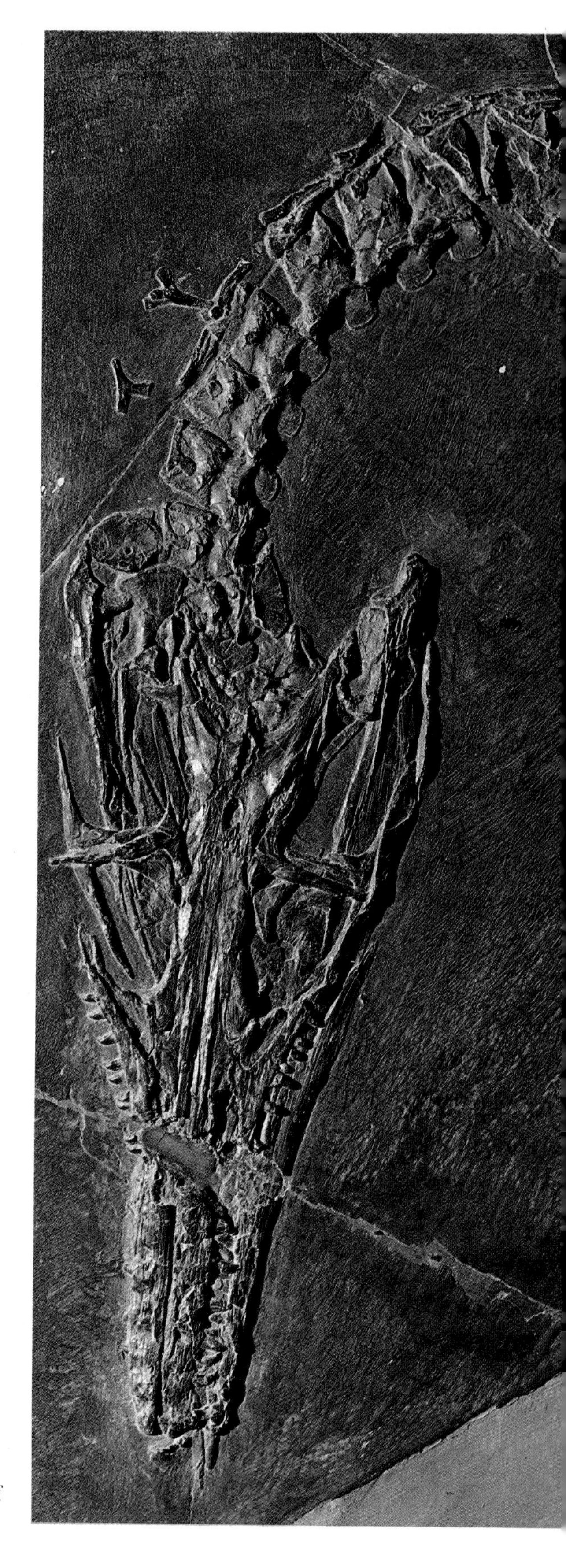

One of the specimens of *Askeptosaurus italicus* from the deposits of Besano on display in the Museum of Natural History, Milan (× ½ approx.).

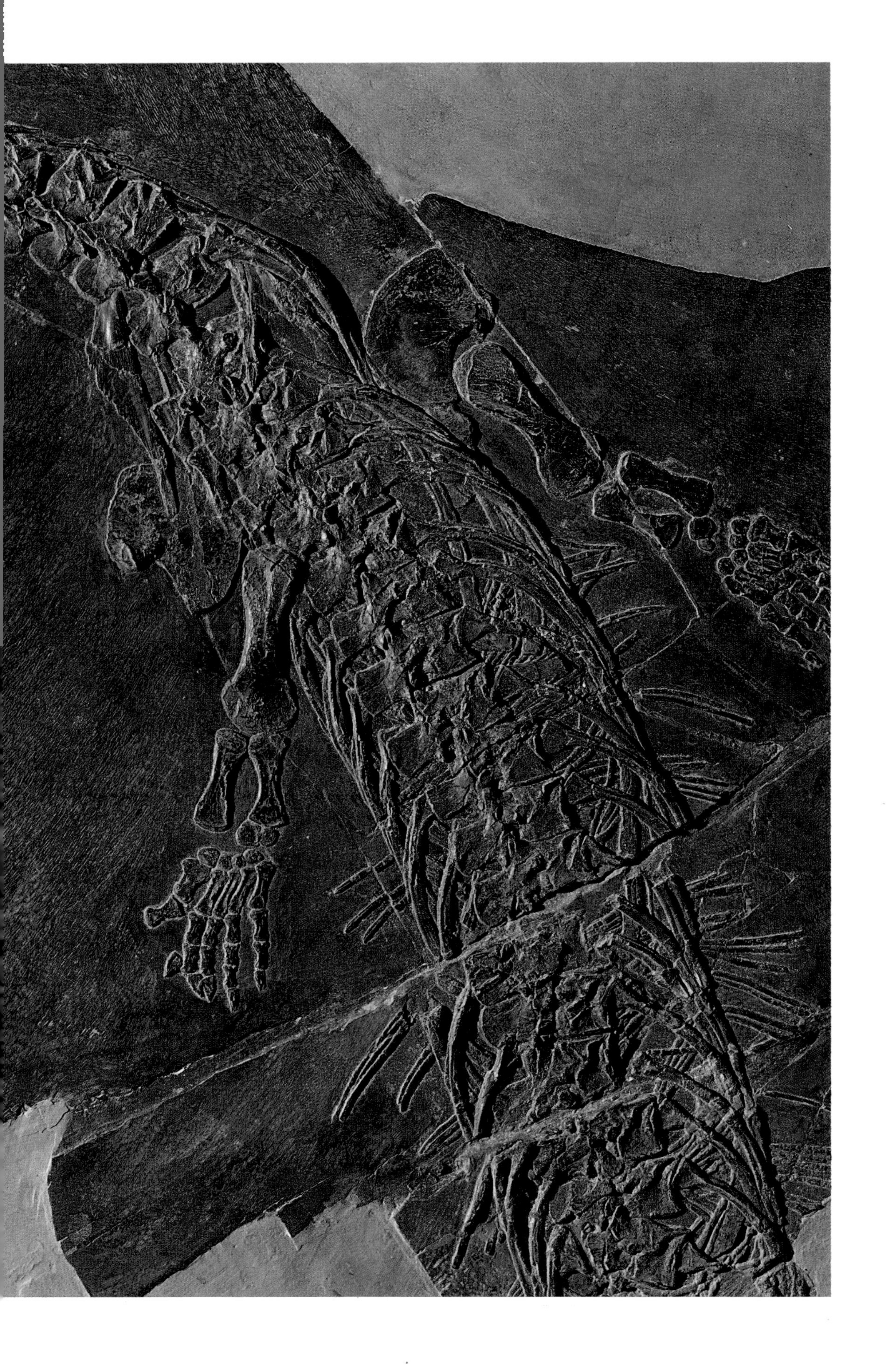

at Monte San Giorgio. The animal grew to some 6 metres (19½ feet) in length with a large but rather rigid neck. Although the neck had only 12 vertebrae, it was the great length of each which made the neck so long. The exact life-style of *Tanystrophaeus* is still not really known, but it is generally thought to have lived on the shore or in shallow waters, keeping under cover and using its long neck like a fishing rod to hunt its prey. The fishes on which it fed must have lived in the surface waters since the zone that contained the bituminous schists of Monte San Giorgio was covered by waters that were calm and contained hardly any oxygen at depth. These conditions produced very well preserved fossils and some reptile skeletons from the genera *Askeptosaurus* and *Macrocnemus* have not been found anywhere else. These creatures show that the lepidosaurs evolved into several different types soon after their first appearance on earth and this makes them a rather difficult group to classify accurately.

Two extreme specializations among the lepidosaurs. Right: *Tanystrophaeus*, an 'angler' reptile of the European Triassic Period. Left: *Tylosaurus*, one of the many mosasaurs from the Upper Cretaceous Period of North America. They are not to scale, however, since the biggest *Tanystrophaeus* was never larger than 6 metres (19½ feet) while *Tylosaurus* could be as long as 9 metres (29½ feet).

Lizards

Although the lepidosaurs did not give rise to as great a variety of types as did the archosaurs, they produced species which colonized every environment. The flying lizards of the Triassic Period such as *Icarosaurus* (see Chapter 7) demonstrate how strange some lepidosaurs could be. Although most were quite small, a few lepidosaurs, such as the mosasaurs (see below and opposite), grew very large. The 'giant lizard of Maastricht' (see page 25) was one such animal. They measured up to 14 metres (46 feet) long and are among the giants of the Mesozoic Era. While many archosaurs died out at the Mesozoic-Cainozoic Era boundary, the lepidosaurs managed to survive. Two main groups had appeared by then: the snakes and the lizards. They have remained up to the present-day, and are now the dominant two groups of modern reptiles. The squamates were characterized by their lack of a bar of bone underneath the lower temporal skull opening. From the Lower Triassic Period onwards, an eosuchian from Karoo in South Africa, called *Prolacerta*, already showed this bone to be reduced. Eventually, the bar completely disappeared and this enabled

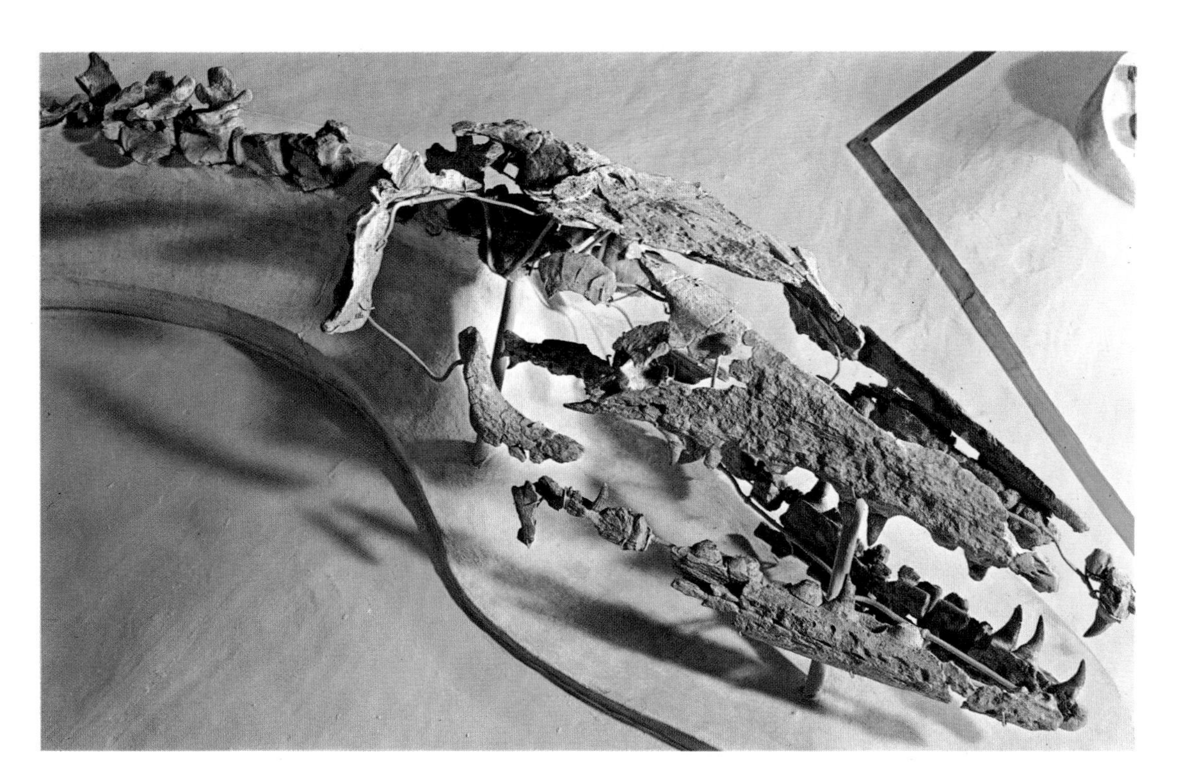

another bone called the quadrate, found at the back lower corner of the skull, to lie next to the mandible. Since the quadrate itself could move against the skull, the whole mouth of the animal could now be opened much wider than before. This is a characteristic of modern lizards and snakes.

Several families of lizards are found in the Jurassic and Cretaceous Periods. The monitors evolved at the beginning of the Cretaceous Period, their most famous living representative being the Komodo dragon of Indonesia. Although land-living monitors up to 7 metres (23 feet) long are known from the Upper Pliocene Period of Australia, the sea-living forms are certainly the more spectacular. The mosasaurs lived only in the Upper Cretaceous Period when they evolved into many species, probably replacing the other marine reptiles such as the ichthyosaurs. The Cretaceous seas contained plenty of food for these great predators. As well as the food already there, such as the cephalopods, many new types of teleost fishes were appearing in the seas at the time. The body structure of the marine monitors was quite similar to those that lived on land, but they had of course modified their legs into flippers for swimming and their necks were shorter. During the Upper Cretaceous Period the number of bones in the hands and feet increased as did the surface area of the forelimbs. The fingers closed up and the hands turned into a paddle (see page 174). They must have been able to swim fast using the paddles and the strong tail. The lower jaw was hung from a moveable quadrate bone and the mandible was hinged along its length. This arrangement is especially well developed in the mosasaurs, which were capable of tracking down and swallowing very bulky quarry. While some hunted on the surface, others dived deep in search of food – we know this because those that dived had a very bony eardrum similar to that of modern whales and dolphins. This protected the ear against the huge pressures created when the mososaurs were diving in deep water.

The teeth of the mosasaurs are those of a ferocious predator, and usually consist of hooked spikes that are sometimes flattened on the side. In the genus *Globidens*, however, the crowns were very rounded and swollen. This is because the animal fed on molluscs, crustaceans and echinoderms in coastal waters, catching them on the shallow sea-bed.

Snakes

The snakes constitute an even more specialized group of squamates. Their lack of paired limbs is their most obvious characteristic, but many burrowing lizards also have either very small legs or none at all. There is another strange, worm-like, group of living squamates called the Amphisbaena. They have the extraordinary ability to move either forwards or backwards in the underground galleries where they hunt for termites and ants.

Because their bones are so fragile, snakes do not fossilize easily and so their origins are poorly known: most snake fossils are vertebrae, of which there can be several hundred along the whole body. The oldest specimens go back to the Lower Cretaceous Period and belong to the genus *Lapparentophis* discovered in southern Algeria. We know, however, that the history of the snakes began much earlier than that. The skull of the snakes has some features in common with the monitors, although this fact is not generally regarded today as proof of close kinship, but has come about because both groups of animals feed in a particular way on similar prey – particularly vertebrates. The snakes have gone much further in perfecting these methods, however. The cranium is very light and its bony components are very moveable. The second temporal arch – that which separates the two skull openings – is breached, and this increases the mobility of the quadrate even more. The two branches of the mandible, already capable of flexing in the centre, are in snakes only connected at the front end by elastic tissue, and are thus able to move interdependently. Finally, the snout is itself hinged on the brain-pan.

Despite the fact that many reptilian species have declined or disappeared altogether through the course of evolution, this is far from being the case with the snakes – or for that matter the lizards. Since their arrival in the Cretaceous Period they have spread rapidly to find a niche in almost all environments.

Thecodonts

The archosaurs are without doubt the group of reptiles that dominated the scene during the Mesozoic Era. The oldest among them were the thecodonts. The name is a reference to the fact that the teeth were fixed in sockets, a general feature among the archosaurs, and one which is even found in other reptiles.

The beginning of the Triassic Period saw the emergence of diapsids whose cranium bore an opening in front of each orbit, and which was to become the trade-mark of all the archosaurs. The exact function of this pre-orbital opening, which subsequently disappeared in several other groups as evolution progressed, remains obscure. It may have been the site of a gland, the passage of a nerve or important vessel, or even the insertion of a muscle, but none of these explanations is entirely satisfactory by itself, probably because it had a number of different functions.

Chasmatosaurus is the most ancient of the thecodonts. It was a curious animal that looked something like a small modern crocodile, and probably lived an amphibious lifestyle. Its sharp, notched teeth were curved backwards and leave us in no doubt that it was a carnivore, but its oddest feature is surely the shape of its muzzle, whose tip was pointed downwards to form a toothed hook that was used to seize its victims.

Later on in the Triassic Period of Karoo in South Africa, we find the remains of a genus of thecondont called *Euparkeria*. This much lighter animal was capable of walking in a bipedal fashion on hind legs which were much more developed than the forelimbs. Its tail was very long, and the eye sockets in its skull contained rings of bony plates. *Saltoposuchus* of the Upper Triassic Period of Germany, shown on page 179, belongs to the same group, but was more advanced in terms of its bipedal ability, as demonstrated by rear limbs far larger than those at the front, coupled with a stronger pelvis now connected to three, not two, vertebrae. This bipedal method of locomotion is a feature towards which many archosaurs subsequently tended.

The aetosaurs and other related forms were decidedly quadrupedal, on the other hand. These thecodonts are primarily characterized by the substantial bony covering which protected their backs. In a genus such as *Desmatosuchus* (see page 178), the neck region was also protected by spikes that reached lengths of up to 45 centimetres (18 inches). In some species the teeth gave way to a kind of beak.

The venomous snakes are the most advanced forms of ophidians. The family Viperidae to which this rattlesnake belongs first appeared in the Lower Miocene Period. The hook-like fangs that inject the venom lie flat when not in use and extend when the mouth opens to strike.

While certain aetosaurs had a somewhat crocodilian way of walking, the convergence is even more striking when we come to consider the phytosaurs. This group, the most characteristic of the thecodonts, reached their peak during the Upper Triassic Period with a form that was very similar to that of modern crocodiles both in appearance and habits (see page 179). They were in no way the ancestors of the latter creatures, however, and belonged to a completely separate line of archosaurs. They frequented fresh waters such as rivers, lakes and swamps, and their remains constitute an important part of Triassic Period fossil vertebrates.

In the various environments they inhabited they developed a number of different specializations very similar to those of the crocodilians. Some had a very elongated snout like the present-day gavials of the River Ganges in India. Also like them, their diet was chiefly fish-based. Other forms, however, had heavy heads with shorter muzzles and stronger teeth, and lived by ambushing land animals along the shores. As frequently occurred among the thecodonts, the phytosaur body was protected by a bony covering which heightened the crocodilian look of the reptiles and covered the entire back as well as parts of the belly in certain species. Nevertheless, there is at least one major feature which enables us to tell a phytosaur from a crocodile without hesitation: in a phytosaur the nostrils were positioned way back along the snout just in front of the eyes, and were often raised in relation to the rest of the skull. This was, of course, a marine adaptation that allowed the animal to submerge almost completely while continuing to breathe normally. Although crocodiles also breathe by means of nostrils placed on the top of their snout, the openings are positioned at the tip.

The phytosaurs were capable of growing to great lengths – sometimes to as much as 5 metres ($16\frac{1}{2}$ feet) – and were quite aggressive, as is witnessed by the frequent wounds found on their skulls which must have been caused by fighting.

Crocodilians

A number of thecodonts, and the phytosaurs in particular, have shown once again how successful the crocodile 'design' was for the reptiles. During the course of evolution several groups of animals evolved this particular body shape and life-style, and the best proof of its success is that among the reptiles the crocodilians are the only order of archosaurs that have survived to this day. We should also note that this type of shape comprises a number of variations, one of the better examples of which is provided by the long-muzzled forms which feed on fishes. Not only has this ecological niche been successively occupied by different types of reptiles, but even within the crocodilians themselves several families flourished at various times during the course of evolution by adopting this shape and life-style.

It was probably the spread of the true crocodiles which led to the decline of the phytosaurs at the end of the Triassic Period, at which time the former animals were still quite small: not more than 1 metre ($3\frac{1}{4}$ feet) in length. Their snout was still short, but their legs on the other hand were longer than those of their modern descendants. Although some had already adapted to an aquatic way of life, the basic body shape was in fact that of a small terrestrial carnivore.

One of the chief characteristics of the crocodiles is the presence of a secondary palate. This is a horizontal bony wall that separates the mouth from the nasal cavities, enabling the animal to open its mouth under water without flooding its respiratory passages. In the protosuchians, the crocodiles of the Upper Triassic Period, this secondary palate was only partly developed, and must still have been partially membranous. More highly evolved families with a much more substantial secondary palate are known from the end of the Lower Jurassic Period, and are grouped in the suborder of mesosuchians. This group is far better

Desmatosuchus is a thecodont in the group of aetosaurs which had heavy skin armour. In this particular genus, the 'horns' grew very large indeed. Its overall length was some 4 metres (13 feet).

Above: Two specimens demonstrating the variety of the thecodonts (the stock group of the archosaurs) – a phytosaur, looking like a crocodile with its long muzzle, and a small bipedal reptile, *Saltoposuchus*. These creatures lived around 200 million years ago during the Upper Triassic Period in southern Germany. Despite its appearance, *Saltoposuchus* was probably a lot closer to the crocodilians than the phytosaur. In fresh waters the phytosaurs were the chief predators of the age, but competition from the true crocodiles probably led to their downfall.

known than the protosuchians. It is highly likely that several lines of protosuchians evolved in the same direction to constitute this second unit, but our lack of knowledge of the crocodiles of the Lower Jurassic Period does not allow any confirmation of this hypothesis.

The mesosuchians were a prolific group, and produced a whole range of species which never diverged far from the basic body shape but which were nevertheless still quite diverse. Thanks to the abundance of fossil material and the skills of the experts, their history has been largely unravelled, even though some points remain unclear. The Jurassic Period essentially saw forms adapted to life in the oceans. The metriorhyncs – a group of mesosuchians that persisted down to the Lower Cretaceous Period – were the group most modified by this very constricting environment (see page 181), undergoing an almost complete 'facelift' to become very hydrodynamic (highly modified for life in water). The armour of bony plates which the mesosuchians had inherited from their predecessors disappeared in these marine animals and their skin became soft and smooth. The limbs were transformed into flippers while the tail, too, developed a kind of fin comparable to that of the ichthyosaurs.

More recent periods saw species of mesosuchians whose ecology was similar to that of modern varieties. In the Lower Cretaceous Period of Niger, Africa, P. Taquet has discovered a giant among these species: *Sarcosuchus imperator* which measured 11

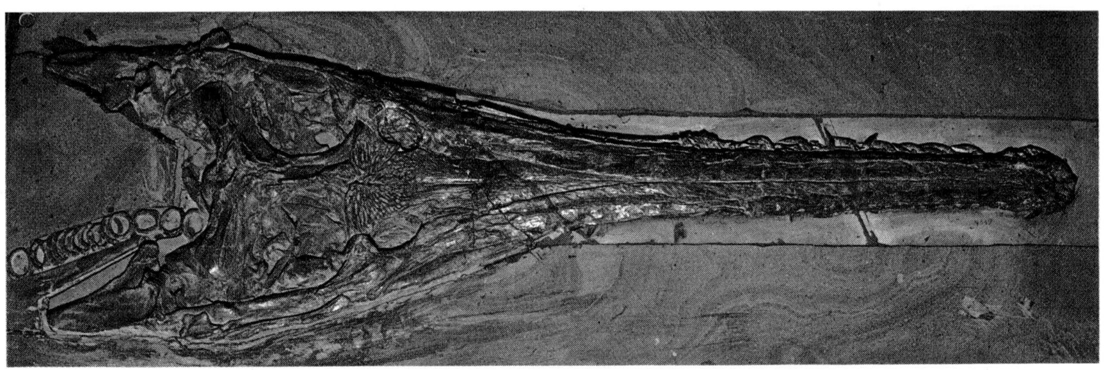

Top to bottom: Three stages in the evolution of the crocodiles.

Protosuchus of the Upper Triassic Period was a small protosuchian about 1 metre ($3\frac{1}{4}$ feet) long. It was a basically terrestrial animal and was found in Arizona, U.S.A.

The long, thin skull of *Steneosaurus bollensis* from the Lower Jurassic Period of Germany was well adapted to catch fishes and is a classic form from the Holzmaden deposits in Wurttemberg, Germany.

Asiatosuchus, from the Eocene Period of Monte Bolca, Italy, belongs to the group of eosuchians which includes the crocodiles of today. It lived in Europe at the beginning of the Tertiary Period and so the climate must have been very different from what it is today.

metres (36 feet) long and looked like a gigantic gavial (a kind of long-nosed crocodile still living today). In the meantime very different crocodiles were developing in decidedly terrestrial environments. These types – referred to as 'runners' – have since disappeared, but at the time were mainly represented by small carnivorous forms whose bodies were carried higher on their legs than the amphibian genera. *Araripesuchus* which we saw earlier was one of these animals. In South America, a continent that became totally isolated during the Tertiary Period (see illustrations on pages 232–233), two families of land-living reptiles enjoyed great prosperity: in the absence of placental

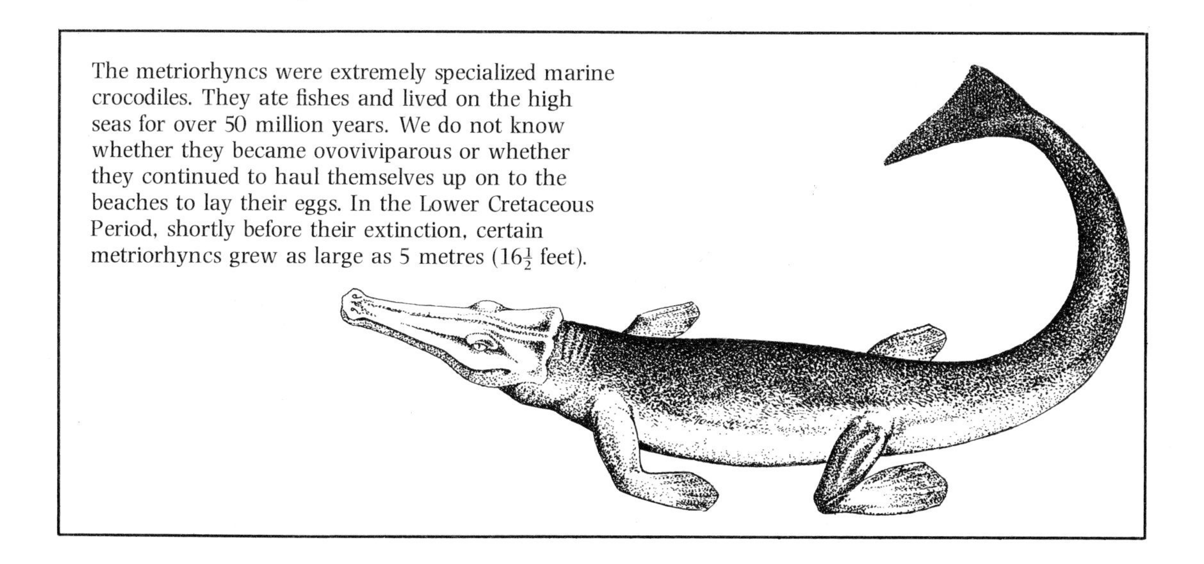

The metriorhyncs were extremely specialized marine crocodiles. They ate fishes and lived on the high seas for over 50 million years. We do not know whether they became ovoviviparous or whether they continued to haul themselves up on to the beaches to lay their eggs. In the Lower Cretaceous Period, shortly before their extinction, certain metriorhyncs grew as large as 5 metres ($16\frac{1}{2}$ feet).

mammals in that area the baurusuchids and then the sebecids shared the role of the great terrestrial predator alongside the marsupial mammals and the running birds. Their flattened, notched teeth are noteworthy, since fossil crocodile teeth hardly ever bear any particular feature that enables us to identify them.

When the evolutionary spread of the mesosuchians was at its height, the branch of modern crocodiles – the eosuchians – split off from the main stem. Characterized by vertebrae that are convex at the rear, they also have a more extensive secondary palate. Two out of the three modern families, the Crocodilidae and the Alligatoridae, were already living at the end of the Cretaceous Period, while another family, the Gavialidae, appeared a little later on at the start of the Tertiary Period. Competition from these new arrivals gradually put an end to the mesosuchians whose very last representatives, the sebecids, nevertheless survived into the Pliocene Period.

Dinosaurs

If any extinct creature is famous in the minds of most people it must surely be a dinosaur, whose image springs to the mind as soon as the word 'prehistoric' is mentioned. This group of creatures, extinct for over 65 million years, is paradoxically more familiar to most of us than are many living animals. The palaeontologist might well rejoice in this popularity were it not for the other side of the coin: there are more false ideas about the dinosaurs than about any other fossil animals. It is unlikely that these ideas will ever be put straight for as long as these creatures continue to evoke our most deep-rooted fears about 'monsters' and remain an important part of the film-maker's equipment. We like to imagine, for example, that in far-off days our own ancestors had to do battle against these frightful reptiles, a scene that has flashed across the cinema screen so many times. However, a glance at the chronological table on pages 302 and 303 will show straight away that several tens of millions of years separated the death of the last dinosaurs at the end of the Mesozoic Era from the development of Mankind during the Pliocene Period and, more particularly, the Quaternary Period.

The term 'dinosaur' was first coined by Richard Owen in 1841, when, on the 2nd of August of that year, the English palaeontologist presented a memorable report at Plymouth dealing with the discoveries of the first *Megalosaurus*, *Iguanodon* and *Hylaeosaurus*. The translation from Greek of the expression dinosaur means 'terrible lizard', and it is indeed true that they sometimes attained monstrous proportions. But – and here is another misconception of the dinosaurs – this was not always the case. As the illustration on pages 182 and 183 show, some species were of a size comparable with that of modern mammals, while the smallest genus, *Compsognathus* from the Upper Jurassic Period, was no bigger than a farmyard chicken.

These animals, practically unrivalled during the Mesozoic Era, in fact reflect an extreme degree of diversity, and we must look for their origins within the stock group of the thecodonts. The boundary between the two groups is not easy to detect, however, and some of the thecodonts, such as *Euparkeria*, are not all that different from certain small bipedal dinosaurs. These origins are therefore complex, and at least two separate lines of 'terrible lizards' evolved from the primitive archosaurs. These two lines, to each of which we attribute the rank of order, have little in common apart from the fact that each had giant representatives. To sum up, therefore, we can state quite firmly that the dinosaurs probably did not constitute a natural group.

Saurischians and ornithischians The fundamental characteristic separating the two orders of dinosaurs is the structure of the pelvis. This part of the skeleton, whose shape is directly linked to the method of walking, consists of three bones on either side. These are the ilium, which is attached to the vertebral column (and more precisely to the sacrum); the pubis, which forms the forward

part of the pelvis; and the ischium, which faces the rear.

In the saurischians, or lizard-hipped dinosaurs, the pelvis was arranged in a fashion similar to that found in the lizards (see page 184). The pubis was simply extended to the front and the ischium to the rear; both bones were elongated, and were attached to very

Although we often think that dinosaurs were enormous creatures, some dinosaurs were quite 'small' compared to the great mammals. *Brachiosaurus* was the heaviest dinosaur at 80 tonnes – the weight of a whole herd of elephants – but the modern blue whale exceeds this. There are signs, however, that the blue whale may soon be beaten by a truly gigantic 'supersaurus'.

powerful muscles.

The ornithischians, or bird-hipped dinosaurs, on the other hand, featured a forked pubis with two expansions, one elongated towards the front – parallel to the ilium which is especially well developed in this direction – and the other shaped like a stick and pointing to the rear. The question was long debated as to which of these two bones represented the true original pubis. Some experts claimed that the posterior 'stick' had been a later addition, while their opponents held that the pubis had turned to the rear in the course of evolution and that a 'prepubis' had developed in front. To make matters worse, the regression of one or the other of these bones in species in certain lines of evolutionary development clouded the issue even further. Nevertheless, the discovery of primitive ornithischians that did not have a prepubis seems to confirm the second theory.

Within the two lines of development there were numerous animals which demonstrated clearly the bipedal characteristics we have already seen in some thecodonts. Still, the quadrupeds or four-legged forms held their own in both orders, and it is within the saurischian order that the most colossal forms appeared some of whose representatives, weighing in the region of 80 tonnes, were the largest creatures that have ever walked the Earth. Their bodies had to undergo far-reaching strengthening modifications before being able to support such huge masses, let alone move them (see page 185).

Coelurosaurs The saurischians are subdivided into two clearly distinct groups. The first includes the basically bipedal, carnivorous forms, while the second, which emerged

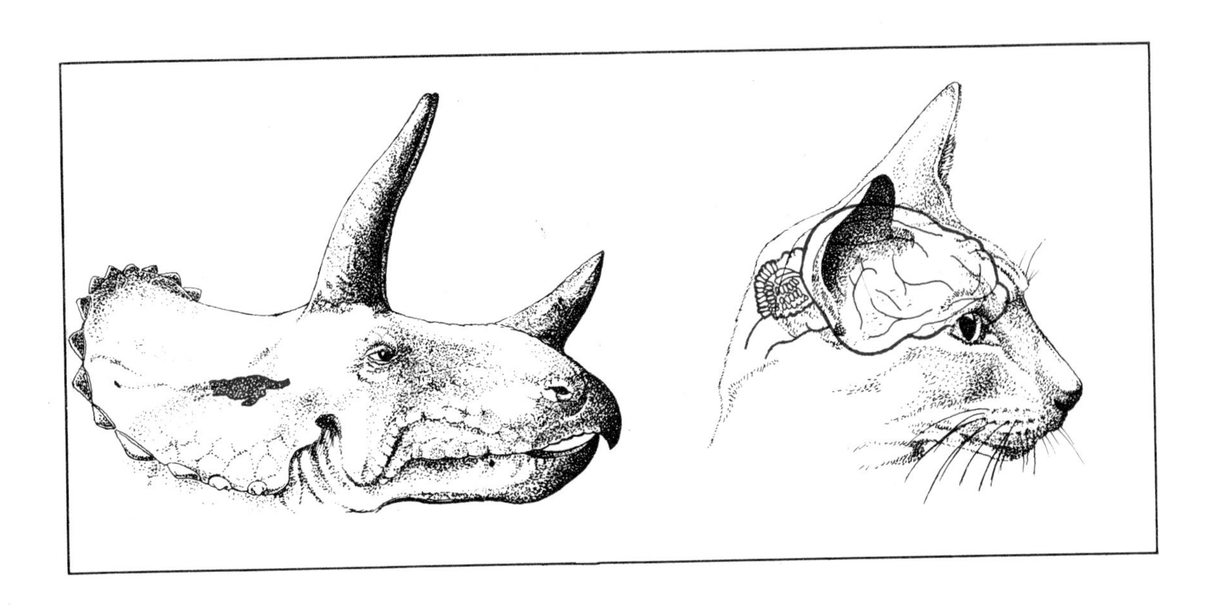

Above: People have often noted that dinosaurs had small brains. The brain of this *Triceratops* (an ornithischian of the Cretaceous Period), compared to a modern cat's is very poorly developed. This was not a feature unique among dinosaurs, however, and was shared by the other reptiles, a characteristic all the more astonishing when we consider the huge bulk of the saurians. Still, the contrast is not always so evident, especially in the carnivorous dinosaurs.

Below: Right-hand lateral views of the three-branched pelvis of a saurischian (left) and the four-branched pelvis of an ornithischian dinosaur. Bones similar to these can be found in modern crocodiles and birds, and this explains why the saurischians are called lizard-hipped and the ornithischians bird-hipped dinosaurs.

a little later on in time, contains the herbivores which began by being bipedal or semi-bipedal, but eventually became firmly quadrupedal. Of the two orders of dinosaurs, the saurischians are those whose thecodont roots are the more evident, especially when we consider the small bipedal carnivores known as coelurosaurs. The very oldest of these, such as the *Saltoposuchus* shown on page 179, are reminiscent of the thecodonts. They are known to us from the Upper Triassic Period onwards and are generally regarded as the most primitive of all carnivorous saurischians. Nevertheless they survived as far as the end of the Mesozoic Era,

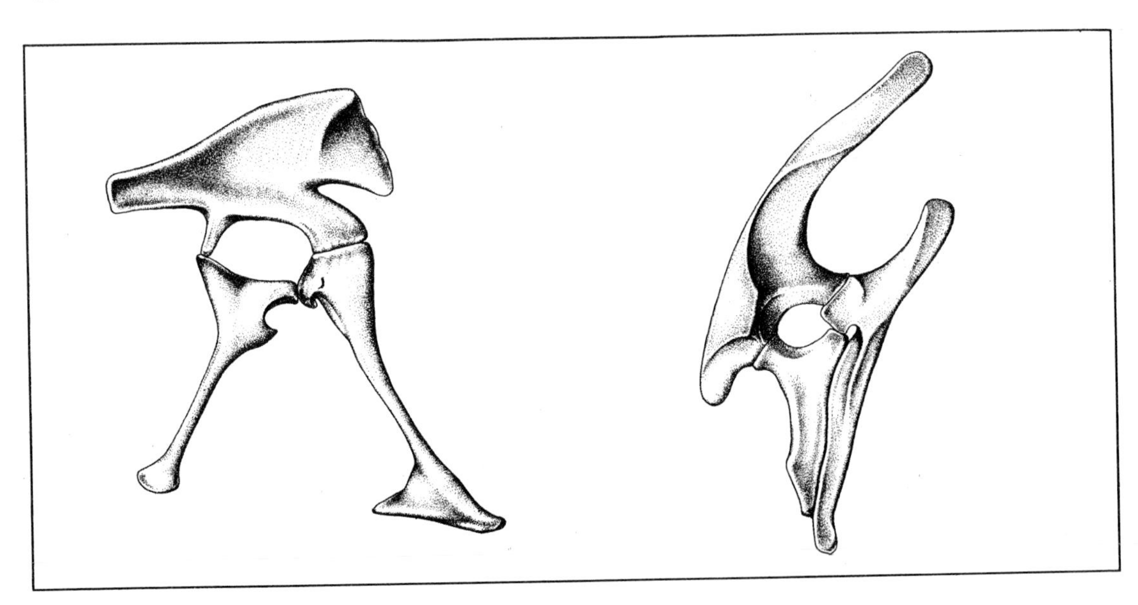

demonstrating their great success at adapting to the environment. As already stated, they were relatively modest in size, with very light bones, and had the slender build of an animal adapted to running and leaping. Their head and neck were both long, and a substantial tail counterbalanced their body weight. The hind limbs were much more developed than the front legs.

The Upper Triassic Period saw the arrival of several genera of coelurosaurs, such as *Coelophysis* for instance (see page 186) which was discovered in New Mexico. Everything about this dinosaur suggests speed and agility. Its rear legs resembled those of some gigantic bird (it measured some 2.5 metres/ 8 feet long) and it is not difficult to imagine it running after its prey. The smaller amphibians and reptiles, not to mention the mammals whose evolution was just beginning, must have had a difficult time trying to elude its clawed hands and long jaws. Its speed and the flexibility of its neck hardly gave their quarry a chance when taken unawares.

Other, much smaller, coelurosaurs lived during the same period but their overall appearance differed little from that of *Coelophysis*. Similar forms are also found in the Jurassic Period, where we find *Compsognathus*, the smallest of all the dinosaurs (see page 182). For a long time this genus was known only from a single but almost complete skeleton from the Upper Jurassic limestone of Bavaria. Strata in the Plain of Canjuers to the south of the gorges of Verdon in France have also yielded up remains, however. There is little doubt that the fragility of the skeletons of this type of reptile makes their fossilization difficult, especially for the smaller among them. In fact only the more exceptional deposits like the lithographic limestones at the end of the Jurassic Period where *Compsognathus* was discovered are capable of preserving them to any satisfactory degree. These are extremely fine rocks laid down in a calm environment, and are best able to fossilize such small skeletons. Surprisingly though, these fossils, while rare, are magnificently preserved when we are fortunate enough to find one.

During the Cretaceous Period the coelurosaurs gave rise to more specialized forms, one of the most astonishing of which was uncovered in the Lower Cretaceous Period of Montana, U.S.A. in 1964, and has been called *Deinonychus*. Reconstructions from its remains have produced a bipedal creature some 2 metres ($6\frac{1}{2}$ feet) in height and 3 metres (10 feet) long. Its rather massive head bore eyes which were much larger than average among dinosaurs. Its forelimbs were thin, but ended in three fingers armed with powerful claws. There were also three claws on the even more powerful hind legs. The three principal toes were equipped with these claws but the first toe, shorter than the other two, had an enormous claw over 12 centimetres ($4\frac{1}{2}$ inches) in length, and it is this monstrous hook which has given the animal its name of *Deinonychus* (meaning 'terrible

This clawed foot belongs to a gigantic sauropod dinosaur. The limbs of this creature, like those of the largest mammals (elephants for instance) are like straight pillars which give the best possible support to the tremendous weight of the body.

claw' in Greek). This dinosaur was perfectly adapted for running, and had a relatively long tail which it carried high and rigid. Once it had caught up with its quarry, this terrifying predator threw it to the ground and jumped on it, disembowelling it with a backwards kick whilst holding it fast with its forelimbs.

A quite different but no less remarkable type of specialization is demonstrated by *Ornithomimus* of the Upper Cretaceous Period, a genus that measured 4 metres (13 feet) in length – a size never exceeded by any other coelurosaurs. As its name indicates, this animal was a kind of 'bird imitator', resembling the ostriches in particular. The discovery of the first complete skull of this reptile was something of a shock, since it showed that this minute and completely toothless head was attached to the body of a bipedal carnivore. The snout in fact was covered by a horny beak similiar to that of an ostrich, and the very long neck made the resemblance even more marked. Its hind legs showed adaptations similar to those of the great wingless birds, with a marked elongation of the terminal bones. These features were undoubtedly those of a high-speed

Coelophysis from the Upper Triassic Period was a hunter of small vertebrates that must have caused havoc among the lepidosaurs. Young skeletons have been found inside adult specimens which seems to indicate that *Coelophysis* was sometimes a cannibal.

runner capable of outpacing any other dinosaur. Its highly developed forelimbs ended in long hands with three fingers.

Many questions have been asked as to the significance of this strange combination of characteristics that made *Ornithomimus* the most original creature it certainly was. Its mode of living and in particular its diet have aroused much speculation, and scientific theories have made this dinosaur, which like the modern ostriches must have inhabited vast open expanses of terrain, assume a number of varied life-styles. Some experts say it fed on molluscs, while others claimed for it a diet of ants; still others state squarely that it was a herbivore. Of all these imaginative theories, however, the most attractive is without doubt that which suggests that this dinosaur was a nest robber. We may assume that dinosaur eggs were large and represented food sources that were nutritious and easy to obtain. So *Ornithomimus* may have used its long fingers to dig out the eggs which it then broke open with a sharp blow of its beak – teeth would have been of little use for such a meal. In the event of the unexpected return of the parents, the speed of *Ornithomimus* would have enabled it to escape without difficulty. This theory has been further underlined by giving a related genus the name of *Oviraptor*, meaning 'egg robber'.

During the course of the Cretaceous Period, coelurosaurs belonging to the same family as *Deinonychus* (separated by some scientists from the coelurosaurs) acquired a body form more or less comparable with that of the dinosaur-ostriches. One such example is *Saurornithoides*, which was distinguished from the latter, however, by the size of its head, which was much larger, and by its jaws, now armed with teeth instead of the beak. Like *Deinonychus*, it had a large brain compared with most dinosaurs, and when we look at a reconstruction of this 'intelligent' reptile, which was some 2 metres ($6\frac{1}{2}$ feet) in length, we cannot fail to be surprised by its unusually large eyes. From this, certain palaeontologists have deduced that *Saurornithoides* was nocturnal – a rarity among reptiles. Perhaps it had adapted to a life in pursuit of mammals, which during the Mesozoic Era were small animals forced to lead a secretive life in the shadow of the giant reptiles.

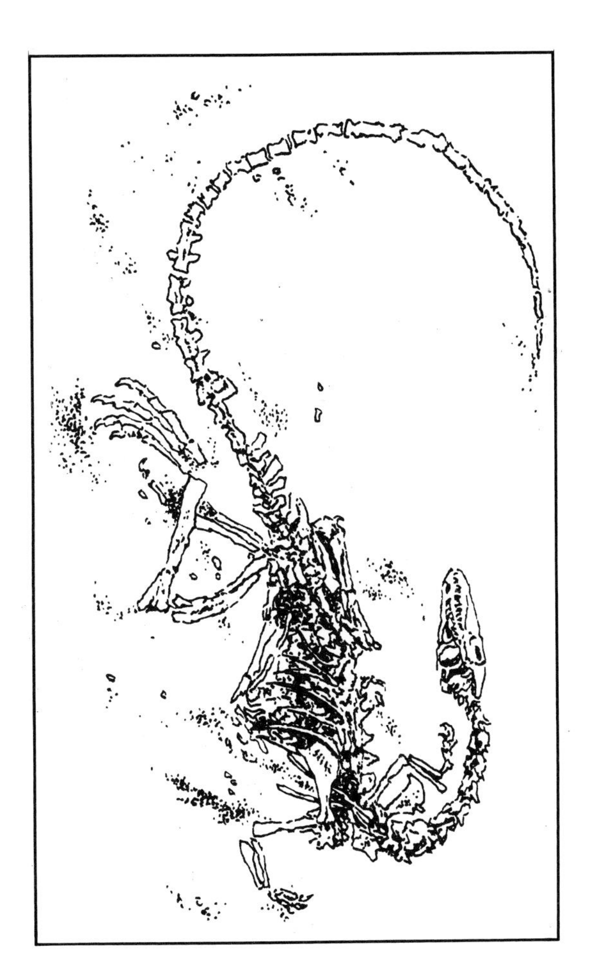

Skeleton of *Coelophysis* from the Upper Triassic Period of Ghost Ranch (New Mexico).

Carnosaurs The carnosaurs comprise the dinosaur genera that truly deserve the name of 'terrible lizards'. Even the smallest among them were terrifying beasts, and it is hard indeed to imagine fully these super-predators of the Mesozoic continents. As their name suggests, they were in fact the main meat-eaters of the dinosaur world.

Among these animals there were none with delicate skeletons, and none with bones lightened, like those of the coelurosaurs, by cavities and thin walls. These creatures weighed several tonnes, and continued to increase in weight until the end of the Mesozoic Era. While the coelurosaurs fed upon lizards and mammals, and were even capable of attacking the first birds from the Upper Jurassic onwards, the carnosaurs, by their sheer weight, could never be satisfied with such small prey. Their structure in-

The enormous skull of *Tyrannosaurus rex* had vast openings, especially the orbits and ante-orbital apertures which were very large. The jaws contained rows of terrible teeth and a single tooth was as much as 20 centimetres (8 inches) long.

dicates that they were adapted for predation on the great vertebrates which alone were capable of supplying them with adequate amounts of meat to satisfy their huge energy requirements. Thus it is that the herbivorous saurischians and ornithischians provided their main source of food.

By the Jurassic Period, the carnosaurs had already attained remarkable sizes. *Megalosaurus* measured as much as 7 metres (23 feet) from nose to tail, rearing its head over 4 metres (13 feet) from the ground (see page 27), while *Allosaurus* from the Morrison Formation (in the Upper Jurassic Period of North America) was 11 metres (36 feet) long and weighed some 5 tonnes. These colossal proportions were surpassed at the end of the Cretaceous Period, however, by representatives of the genus *Tyrannosaurus*, the giant of the group with an overall length of 14 metres (46 feet), a height of 5.5 metres (18 feet) and a weight of up to 8 tonnes.

The carnosaurs were completely bipedal, their hind legs being extremely powerful and able to support the entire body weight. The three- or four-toed feet looked like those of some monstrous bird, and were armed with huge claws, the blows of which must have been fearsome. Despite their size, the powerful muscles of these terrible predators enabled them to move quickly in pursuit of their quarry, the body held almost horizontally and the tail raised above the ground in counterbalance. The forelegs were very short indeed and of little use; as the carnosaurs evolved, these limbs continued to diminish in size so that in late evolved genera like *Tyrannosaurus* they were nothing more than stumps ending in two small clawed fingers. When at rest, therefore, these animals could not have used them as support, and so sat on their tail or even lay down. Unlike the coelurosaurs, their neck was short and massive, since it had to support a huge skull with very powerful jaws (see left): the teeth which lined the jaws looked like flattened, notched daggers. The well-preserved skulls of certain specimens have revealed devices that are comparable with those we have encountered in the monitors and snakes. The quadrate bone on which the mandible is hinged had acquired a certain degree of mobility. The branches of the lower jaw were slightly flexible, and it would seem that the snout was capable of some degree of movement independently of the cranium. All these adaptations were again intended to aid the swallowing of bulky victims. Once the carnosaur had finished its prey off with blows from its feet and mouth, it tore out huge chunks of flesh from the carcass and swallowed them with ease.

Some carnosaurs differ from this 'average type' in certain original ways. *Ceratosaurus* from the Upper Jurassic Period of North America bore a small horn on the end of its snout, a puzzling feature which may have had a secondary sexual nature. The far more surprising genus *Spinosaurus* was discovered in the Upper Cretaceous Period of Egypt. This big meat-eater had forelimbs that were more developed than those of its cousins, and was partly quadrupedal. On its back it had a membrane of skin, held vertical by long vertebral projections, some of which were nearly 2 metres ($6\frac{1}{2}$ feet) long. It is likely that *Spinosaurus*, which inhabited a very warm region, was able partly to regulate its own body temperature with the aid of this 'radiator', which it used to cool or heat its blood according to whether it was directed parallel or perpendicular to the sun's rays. This device – already utilized by members of the therapsid line (as we shall see in Chapter 8) – was also put to use by herbivores living at the same time as *Spinosaurus*.

Sauropods The second branch of the saurischian order is that of the sauropods. Without doubt the most famous of the dinosaurs, they are characterized by bulky bodies, huge tails and long necks ending in minute heads. Their direct kinship with the coelurosaurs and the carnosaurs is far from being proven, however; the structure of the lizard-hipped pelvis is rather primitive, and it is not hard to imagine that it was a feature of many totally independent types of archosaurs. We should also add that the origins of the sauropods are a lot less clear than those of the carnivorous saurischians. The most widely supported belief is that these creatures had their origins in a group known by the name of 'prosauropods', but this theory is by no means generally accepted.

The infra-order of prosauropods includes rather primitive dinosaurs of which the best known must be *Plateosaurus* of the Upper Triassic Period (see pages 190 and 191). In the 1920s a magnificent series of remains belonging to this genus was found in Germany and studied in depth by Professor F. von Huene of the University of Tubingen. His work made it relatively easy to form an idea of this animal's appearance: it was quite large (7 metres /23 feet long) with a compact

Around 80 million years ago predators as monstrous as this tyrannosaur ruled the World. Only very fast creatures like the ostrich-dinosaur *Ornithomimus* were able to escape from these voracious hunters.

body. The rear limbs were more powerful than those at the front, but without the disproportionate size of the bipedal carnivores we have just seen. Its neck was long and carried a small head. *Plateosaurus* was probably partly bipedal, standing up on occasion to graze upon the leaves of trees. The spoon-shaped teeth of the prosauropods give rise to the belief that most of them were herbivores, although some types with sharper teeth must have been flesh-eaters. It was from similar forms that the sauropods may have emerged, the substantial increase in weight meaning that there was no chance of them walking by bipedal motion.

The reign of the sauropods started in the Lower Jurassic Period. They quickly attained colossal proportions and corresponding weights. To estimate the weights, experts proceed as follows: starting from the most precise possible reconstruction of the skeleton, a scale model is made, showing the major muscles. By immersing this model in a calibrated vessel it is possible to measure its volume and to obtain that of the living animal by applying a simple mathematical formula. Since we know that the density of a modern reptile, a young alligator for instance, is 0.9, we are able to calculate the weight of our gigantic saurian with reasonable certainty. We have thus been able to 'weigh' a genus like *Apatosaurus* (more commonly known by its older name of *Brontosaurus*), and have come up with a mass of 30 tonnes maximum for a length of 27 metres ($88\frac{1}{2}$ feet). Compared to this, *Diplodocus*, weighing in at 10 tonnes, was a Jurassic Period featherweight, all the more so when we consider its contemporary *Brachiosaurus* from North America and East Africa (see page 205), which attained a record weight of 80 tonnes. Isolated remains recently unearthed in North America have been quickly attributed by the press to a '*Supersaurus*' that may even have smashed this record. At the other end of this dizzy scale we find creatures such as *Camarasaurus*, a 'small' animal just 6 metres ($19\frac{1}{2}$ feet) in length.

It is generally agreed that the habitat of these monstrous animals was swampy areas or on the shore of lakes, and in regions where the vegetation on which they fed was plentiful and where the water helped in supporting their weight. With the exception of size, all the sauropods appear to have resembled each other, but a closer look reveals notable differences in the proportions of their limbs. To take the examples already quoted, we find in *Diplodocus* or *Apatosaurus* that the hind legs are larger than the forelimbs, while the situation is exactly the reverse in *Camarasaurus* and in *Brachiosaurus*. Additional differences can be seen in the shape of the

Plateosaurus was one of the great land vertebrates of the Upper Triassic Period and sometimes measured as much as 8 metres (26 feet) from nose to tail. It may have stood upright on two legs but the strong forelegs suggest that it may have used all four.

skull and the arrangment of the teeth, which in part at least correspond to a different way of life.

While there can be little doubt as to the more or less amphibian life-style of the sauropods, these aquatic tendencies have been overestimated, especially in the case of *Brachiosaurus*. It is highly unlikely that this animal lived as is shown in most classical reconstructions: totally immersed in water, merely raising its head out of the water to breathe, or even showing only its nostrils above water, curiously situated as they are at the top of the cranium. At a depth of over 10 metres (33 feet) the pressure exerted by the water on the creature's rib cage would have made breathing very difficult, and it is probable that it spent its time only up to its shoulders in water so that its trunk could be partly supported by the water.

At certain times the giant sauropods

would venture on to land, and the wet surfaces of the shorelines have preserved their footprints. A study of these trace fossils seems to point to the beginnings of a social life, since numerous individuals belonging to the same species have left their traces in a relatively brief space of time. From this, it may be assumed that they moved in herds in search of food. The sheer giant size of these animals is not without its own puzzles for the biologist. What for instance was their period of growth? The relative rarity of fossilized juveniles indicates rapid growth during the first years of life. In this context it has been noted that these animals, whose brain was very small, possessed an enormous pituitary gland at the base of the brain, and it is this gland which produces the hormones that are responsible for growth in modern animals.

The lack of development of the nervous system indicates a low rate of activity and it is likely that the sauropods were peaceful beings which economized on movement. The metabolism of these heavyweights, although relatively low, still required a huge supply of energy. This forced the tiny-headed creatures to swallow enormous amounts of vegetation, a rather poor source of food by any standards. It is, however, wrong to draw comparisons by using information gathered on the feeding habits of the great mammals (such as elephants), whose biology is of course very different, and then to try and apply it to animals about which we know much less.

The weight and volume of the sauropods perhaps enabled them to develop a certain degree of body heat control, enabling them to maintain the body temperature at a constant level. This ability, which makes an animal's activities independent of climatic change, has even been attributed to all dinosaurs from time to time. It has also been noted that the running carnivores must have had a high rate of metabolism. Furthermore, the structure of the bones of dinosaurs is nearer to that of the mammals than to modern lizards.

Ornithopods Although there were many different species of herbivorous saurischians, they were for the most part quite similar. Ornithischians, on the other hand, were much more diverse, giving rise to a wide range of very different biological types. Although rather smaller than the sauropods they must certainly count as the strangest representatives of the dinosaur world.

The ornithiscians were represented by both bipedal and quadrupedal herbivores, although their rear limbs were always better developed than the forelimbs. It is generally agreed that all the ornithischians arose from bipedal forms. The most ancient forms belong to the group known as the ornithopods, a name which means 'bird-foot', and refers to the shape of their hind legs. This group, which comes in the centre of the division of ornithischians, is represented in the Upper Triassic Period by genera like *Heterodontosaurus* and *Fabrosaurus* from South Africa, and *Pisanosaurus* which was discovered in northern Argentina and which appears to be the oldest-known ornithischian. *Fabrosaurus* was a small bipedal animal about 1 metre ($3\frac{1}{4}$ feet) long whose snout ended in a horny beak, a structure that was a common feature in many of the ornithopods.

Even later on in the Mesozoic Era we still come across small forms of ornithopods. *Hypsilophodon* (see page 194) from the Lower Cretaceous Period of Europe was still rather primitive, having retained the teeth in the front of its mouth, at least on the upper jaw, for its mandible ended in a 'predentary' bone which formed the framework for the beak in the ornithischians and which is not found anywhere else among the reptiles. This agile-looking creature never exceeded 1.3 metres (4 feet) in length. Its forelegs ended in five fingers although the fifth digit was very small, while the hind legs still had four fingers. Initial studies indicated that these feet and hands were able to seize branches, and the creature was even likened to a reptilian version of the tree-eating kangaroo. More recent work has come up with quite different conclusions, however. *Hypsilo-*

Apatosaurus (commonly known as *Brontosaurus*) spent most of its life in the margins and lakes and swamps, fairly safe from attack by the carnosaurs which, despite being much smaller, could use their terrible jaws to break its long neck or crush its tiny skull.

Hypsilophodon (so-called because of the enamel folds in its teeth), was a small ornithopod whose back was covered in hard bony outgrowths. Originally it was thought to have lived in trees but now we think it was more 'terrestrial'. The ossified tendons which ran along its spine enabled it to run with its tail held up almost horizontal, acting as a counterweight to the body.

phodon was without doubt a fast runner which found refuge among thick branches only very occasionally.

The best-known family among the ornithopods must surely be the iguanodontids. Discoveries made by people such as Bernissart did much to boost the popularity of this group, which occupied a very important place among the herbivorous fauna of the Lower Cretaceous Period. *Camptosaurus* of the Upper Jurassic Period can be thought of as the average type of ornithopod; this was an inoffensive herbivore which, although bipedal, still possessed forelegs that were more developed than those of the carnosaurs, and which were probably used as support when the animal moved slowly or searched for food on the ground. The genus *Iguanodon* of the Lower Cretaceous Period was about twice the size of *Camptosaurus*. *Iguanodon bernissartensis* measured 8 metres (26 feet) in length and could stand to a height of 5 metres ($16\frac{1}{2}$ feet).

One of the most curious specializations found in this animal was the conical, spiked shape of its thumb (see page 22). As a result of a study of the discoveries of Bernissart by the Belgian palaeontologist L. Dollo, this spiked bone, positioned on the end of its nose as a horn by early scientists and immortalized in this incorrect way by the statues of Waterhouse Hawkins, was repositioned correctly to form these strange spurs. Their precise significance is uncertain: the theory that they were a defensive weapon seems the least extravagant, although the usefulness of such a weapon against an attacking carnosaur seems doubtful. The fierce predators were evidently better adapted for running, but escape on foot may still have been the best form of defence for these peaceful herbivores.

The deposits of Gadoufaoua in the Republic of Niger in Africa have revealed a very special iguanodontid. This is *Ouranosaurus nigeriensis* which, like the carnosaur *Spinosaurus*, had very long vertebral outgrowths supporting a vertical membrane. This again was probably a device for regulating body temperature in a very warm climate.

In the Upper Cretaceous Period the ornithopods known as the hadrosaurs were the most remarkable group, and are usually referred to as the duck-billed dinosaurs owing to the flattened shape of their muzzle. The great majority of these reptiles have come from North America: not only have complete skeletons been discovered, but 'mummified' specimens have also been dug up. The first of these extremely rare finds was made by the Sternberg family who, in 1908, uncovered two complete specimens of *Anatosaurus* in southern Wyoming. The 'mummies' are in actual fact natural moulds of individual creatures that became dried up following death. Similar imprints have enabled us to form a very accurate idea of the shape of the body and the appearance of the skin, which was covered in tiny knobbly outgrowths. An analysis of the gut contents has even revealed the ingredients of the animal's last meal: twigs, seeds, pine cones and pine needles. The post-cranial skeleton of the hadrosaurs was little different from that of an *Iguanodon*, yet the head itself was highly unusual, and above all very variable from one species to the next (see below and page 196). The feet had three toes and ended in short claws, as did the upper limbs where only the three central fingers were developed. The 'mummified' specimens show that the fingers were interlinked by a skin membrane and these paddle-like hands and feet seem to point to a life-style that was at least partly aquatic. In addition, the laterally flattened tail could have been an excellent means of propulsion in water. It is likely that the duck-billed dinosaurs lived on riverbanks where they grazed on foliage, taking to the

The hadrosaurs like this *Anatosaurus* grew up to 10 metres (33 feet) long. It is called a duck-billed dinosaur because of the shape of its snout. Although the front part of the mouth was only armed with a horned structure, inside the mouth there were literally hundreds of foliaceous (leaf-like) teeth arranged in several rows on each half-jaw. These tightly fitting and constantly renewed teeth were very efficient at crushing and pulverizing vegetation.

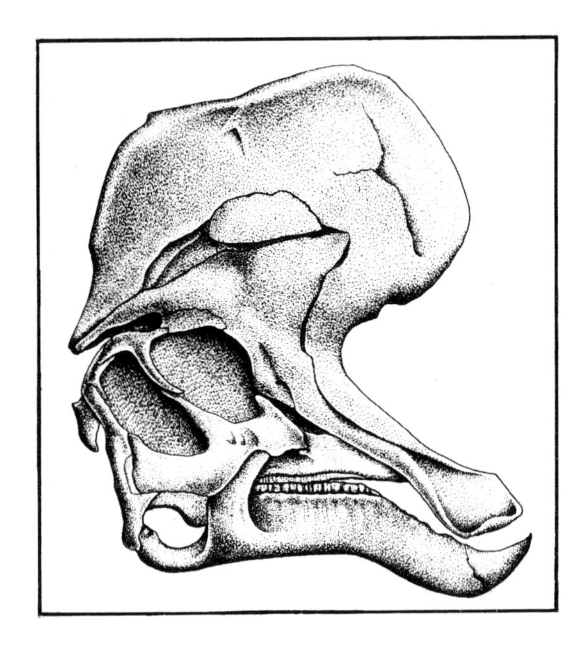

The duck-billed dinosaurs showed an odd tendency to develop lumps of bone on the skull that formed a kind of crest or even a long horn that pointed backwards. Two examples are shown here: *Lambeosaurus* (left) and *Corythosaurus* (below). These ornamentations, which varied in size with age and sex, were perforated by nasal ducts. The reasons for these are unexplained but there have been several theories. It has been suggested that they were a 'snorkel' permitting breathing during a dive, or a reserve of air. This is obviously untrue as their anatomy is not consistent with such ideas. Other theories claim that these devices helped to improve the creature's sense of smell or were a way of amplifying their cries. This last idea is more likely, but difficult to test!

water when danger threatened.

Even more surprises were in store for the scientists who first investigated the pachycephalosaurs, which also lived during the Upper Cretaceous Period. Here again we find the same bipedal outlines which varied little from one ornithopod to the next; but it is their skull which is of particular interest. The illustrations on page 197 show two specimens which are astonishing because of the 'intelligent' look of their domed foreheads. The appearance is deceptive, however, for beneath this dome there was only a small brain covered by skull bones which grew to thicknesses of up to 22 centimetres (8½ inches). This enormous growth of the cranial bones is hard to explain, especially as there was precious little worth protecting inside the head itself! Of the various theories put forward, one of the more interesting suggests that these heavy boned skulls were related to a form of combat carried out between members of the same species. If we try and liken these reptiles to mammalian equivalents, therefore, the pachycephalosaurs become a kind of bad-tempered ruminant. If we continue the comparison we might imagine these creatures living like wild sheep, in flocks led by a dominant male, in which case violent battles (see page 198) would take place from time to time between rivals bent on attaining flock supremacy. The club-like head might also have been used as a defence against predators, much in the same way as

modern dolphins deal terrible blows to the bellies of sharks with their beaks in order to ward them off. These theories presuppose a relatively complex social pattern, however, and are hardly compatible with the smallness of the pachycephalosaurian brain.

Although most pachycephalosaurs have been discovered in North America and Asia, they seem to have been present in Europe and have even been found on Madagascar. *Yaverlandia*, discovered in the Lower Cretaceous Period of the Isle of Wight in Britain, possessed a cranial vault that was rather less domed, and the characteristics of this fossil demonstrate clear similarities between the 'bone-headed dinosaurs' and *Hypsilophodon*. We should also note that – as in the latter – the pachycephalosaurs, whose denture was otherwise weak, had retained the teeth towards the front of the mouth.

Stegosaurs With the stegosaurs we have arrived at a group of ornithischians totally different from the ornithopods. Although large, the ornithopods still give the impression of being relatively agile animals, whereas the stegosaurs can only be described as a kind of living fortress. As powerful creatures too slow and bulky to escape the unwelcome attentions of the carnosaurs and thus forced to meet them face to face, the stegosaurs, ankylosaurs and ceratopians perfected strategies which, though rather different, nevertheless derived from the same basic requirement, that of staying alive.

Stegosaurus of the Upper Jurassic Period is well known thanks to the complete skeletons extracted from the Morrison Formation in the U.S.A. This was a massive quadruped some 6 metres (20 feet) or more in length and weighing in at 1.8 tonnes. Its familiar appearance (see page 199 top) featured large dorsal plates which were biggest around the pelvis, plus two pairs of spikes on the tail. There is little doubt that the latter had a defensive role, and a powerful blow from a *Stegosaurus* tail must have often been lethal, even for a big carnosaur. The function of the dorsal plates is less obvious, however. They may have protected the spinal column from being bitten, but the creature's flanks were still open to attack. In some reconstructions they are flattened on the back of the animal to form a more efficient type of armour, but there is no evidence to prove that this was how they really were arranged. It has also been noted that the plates exaggerated the height of the animal when seen in profile, making it all the more impressive – and fearsome. Yet another potential function was that of temperature regulator: the large masses present at the base of the plates and inside them as well might have constituted sinuses in which the blood was able to warm up or cool down according to the outside temperature.

The three-toed hind legs were much longer than the five-fingered forelegs, a fact which heightened the very arched look of the *Stegosaurus* back. It is also one reason for searching among bipedal animals for the ancestor of the stegosaurs, and *Scelidosaurus* from the Lower Jurassic Period of England has often been thought of as the first representative of the group; still, its affinities with the ankylosaurs are more marked. The head of *Stegosaurus* was very small, ending in a beak, while the mouth had just a single row of teeth in each half of the jaw. It is thought to have fed mainly upon soft vegetation, and it is likely that this dinosaur inhabited lowland regions of a more or less swampy nature where it would find an abundance of such food. A slightly different form of stegosaur has been found in the Tendaguru deposits of East Africa, again in the Upper Jurassic Period. *Kentrosaurus*, from East Africa (see

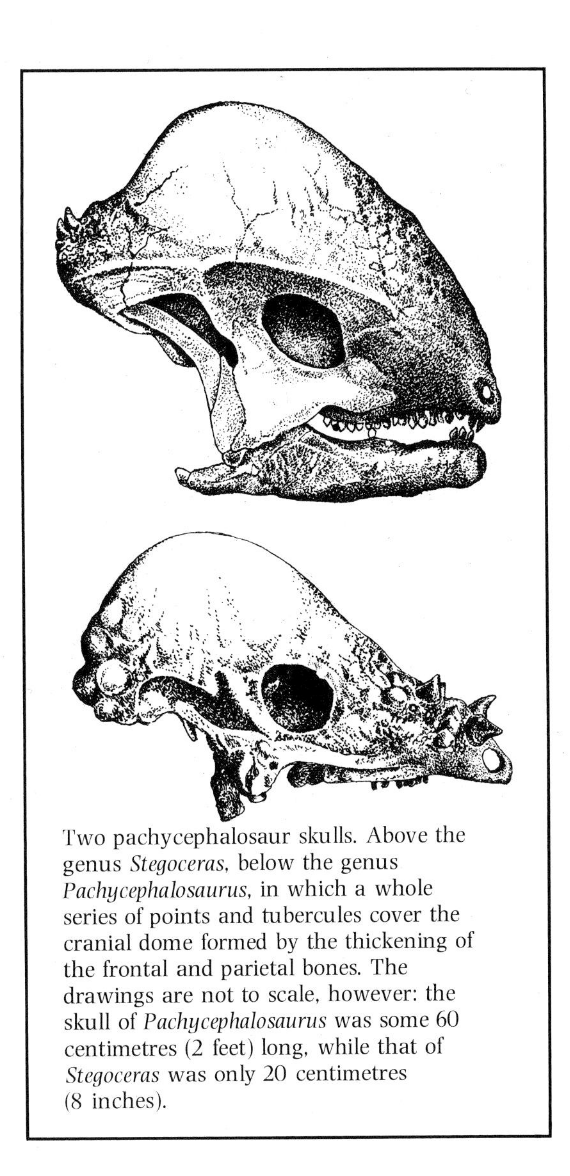

Two pachycephalosaur skulls. Above the genus *Stegoceras*, below the genus *Pachycephalosaurus*, in which a whole series of points and tubercules cover the cranial dome formed by the thickening of the frontal and parietal bones. The drawings are not to scale, however: the skull of *Pachycephalosaurus* was some 60 centimetres (2 feet) long, while that of *Stegoceras* was only 20 centimetres (8 inches).

Stegoceras in a battle for supremacy – assuming that the dinosaurs of the Cretaceous Period behaved like modern rams!

page 199, bottom), was somewhat smaller, with a back lined with spikes rather than plates. It must also have occupied a similar ecological niche to *Stegosaurus* in North America. Yet another genus, *Omosaurus*, has been found in Europe.

Ankylosaurs During the Cretaceous Period the stegosaurs were gradually deposed by another group of armour-plated dinosaurs, the ankylosaurs, to which they are related. There is an animal which scientists believe is ancestral to both the stegosaurs and the ankylosaurs, and this creature is called *Scelidosaurus*. This animal was the first ornithischian to be described in 1861 by R. Owen on the basis of remains found among the marine deposits of the Lower Jurassic Period in Dorset, England. These remains had obviously been washed up at some time in prehistory, and *Scelidosaurus* was no doubt a land animal. Its exact appearance remains obscure, but it is best described as a heavy, low-slung reptile about 4 metres (13 feet) in length. Its limbs were rather short (especially at the front) and massive, while its back was covered in a series of bony plates that constituted its sole means of protection. In the course of their evolution the ankylosaurs perfected the technique of passive defence, an art at which they were far more expert than the stegosaurs.

Oddly enough, there is a long gap in the fossil record that separates *Scelidosaurus* from the ankylosaurs proper, whose remains have only been found in Cretaceous strata. It may be that the ecology of the intermediate forms was unfavourable to fossilization; the fact remains that the first fossils attributable to this group are from the Lower Cretaceous Period, and it did not assume any major significance until the Upper Cretaceous Period, in other words towards the end of the reign of the dinosaurs.

Polacanthus, discovered in the Lower Cretaceous Period of the Isle of Wight, somewhat resembles the stegosaurs with the rows of huge spikes on its back and tail. Nevertheless, the region around the lumbar vertebrae and sacrum was protected by a large shield of rigid bone.

The Upper Cretaceous Period saw the first appearance of much heavier and much more powerfully armoured types, and a great variety of these forms inhabited North America in particular. The back of *Scolosaurus* was entirely covered with hinged bony plates, a mode of defence from which its head and tail also benefited. To this were added longitudinal rows of tough spines, making an almost impregnable 'fortress'. The most exposed portion of the forelegs was also provided with similar armour. When attacked by a carnosaur it is unlikely that

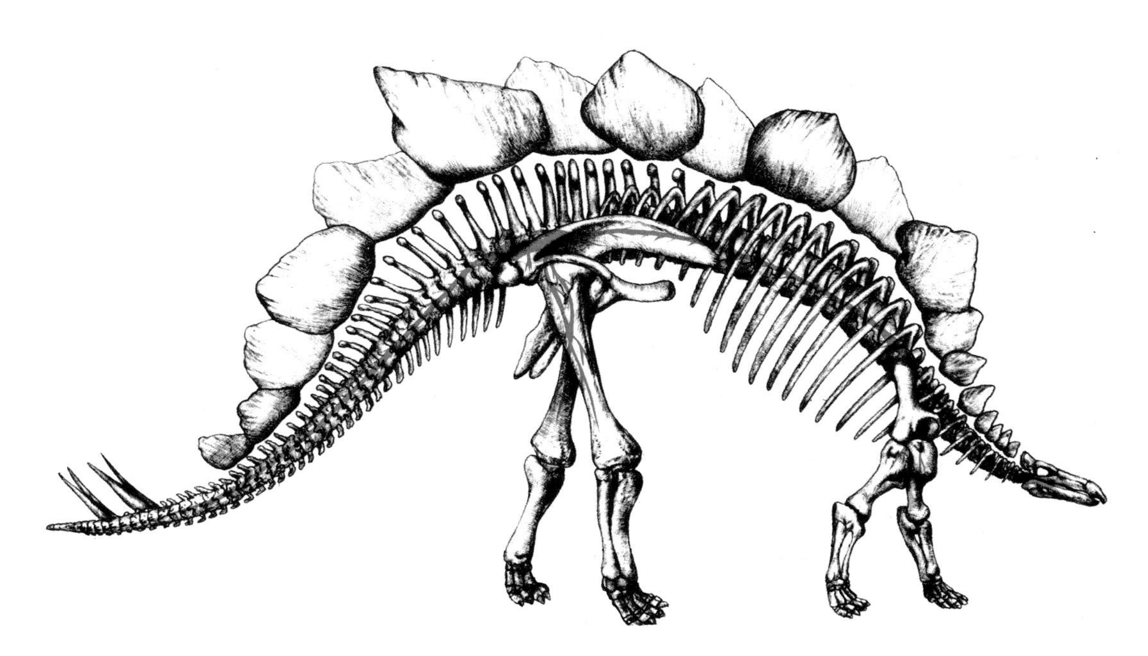

Scolosaurus would have been content merely to crouch on the ground. With a body weight of 3.5 tonnes and a tail which ended in a pair of strong spikes it is likely that this creature used to strike out at its enemies. In some ankylosaurs the tail vertebrae were even welded together and reinforced by bony tendons, with the tip weighted like a heavy rounded club. Alongside such beasts – often compared to living tanks – a genus like *Acanthopholis* (see page 200) looked almost elegant, with its much smaller plates and spikes.

In many ways the ankylosaurs were very similar to the stegosaurs, having weak teeth, or in some species even no teeth at all, forming a complete contrast to the impressive array of teeth which we saw in the

Above: *Stegosaurus* has a reputation as a stupid plant-eater with a very small brain in strong contrast to its massive body. Nevertheless, in its lumbar region the animal had a swelling of the spinal marrow (a sort of second brain), that was much larger and of a size more in keeping with its muscular masses. Today, the ostrich has a very similar structure, and its head is relatively small, too.

Below: *Kentrosaurus* is the East African cousin of *Stegosaurus*. Its head was more flattened and it had more spines along the back and tail. It was also smaller; 5 metres ($16\frac{1}{2}$ feet) and weighed about a tonne. Because *Stegosaurus* was so heavy it had to feed on the ground, but the lighter *Kentrosaurus* could have got up on two legs and grazed on tree foliage from time to time like the ornithopods. It was a particularly well-protected creature and its powerful tail bristled with long spikes, forming a fearful weapon against an aggressor.

ornithopods. Because they ate soft vegetation, the ankylosaurs competed with the stegosaurs for very similar habitats and sources of food. It is unlikely, however, that the stegosaurs were able to resist this competition, possibly because of the much greater efficiency of the protective devices which the ankylosaurs had adopted. There is no doubt that in order to withstand the attacks of the ferocious carnosaurs, such forms of protection had to be really effective, and in this respect only the turtles and certain giant armadillos, which we shall look at in chapter 8 (Mammals), can compare with the ankylosaurs.

Ceratopians This last group of ornithischians appeared, flourished and died out during the Upper Cretaceous Period. They were a new type of animal totally different from any that had gone before, and the group evolved rapidly. We have been able to reconstruct the various stages of their development fairly accurately. The ceratopians were a group of horned dinosaurs – a sort of rhinoceros of the reptile world. The more primitive forms were, however, totally devoid of horns.

The ceratopians evolved from the ornithopods at a fairly late stage in evolution. Their origins are to be found close to the arrival of the genus *Psittacosaurus*, a parrot-beaked dinosaur. This essentially bipedal animal has been discovered in Asia in layers dating from the Lower Cretaceous Period; about 2 metres ($6\frac{1}{2}$ feet) long, it was heavily built and especially remarkable by the shape of its skull, which started in a pointed beak that curved downwards like that of a parrot's. The head then broadened out to form a kind of bony collar protecting the front section of the neck. This is basically the form of head found among the very first ceratopians.

Thanks to the abundance of remains found in the Gobi desert of Mongolia, the genus *Protoceratops*, the first in this line of curious dinosaurs, is especially well known. Not only have a number of adult skeletons been found, we have also discovered eggs and young specimens at all stages of development, and we are thus able to study the animal's progress from birth onwards in the greatest detail. Some embryonic remains have even been found in certain shells. From all this evidence we can say that *Protoceratops* was a rather small creature, not more than 2.4 metres (8 feet) long, but unlike *Psittacosaurus* it was already perfectly quadrupedal. Its neck collar or 'frill' was made lighter by large openings and was much more extensive. Although practically absent at the moment of hatching, this frill pro-

Acanthopholis from the Upper Cretaceous Period of Britain is one of several European ankylosaurs. We do not know very much about it though, because only incomplete skeletons have been unearthed.

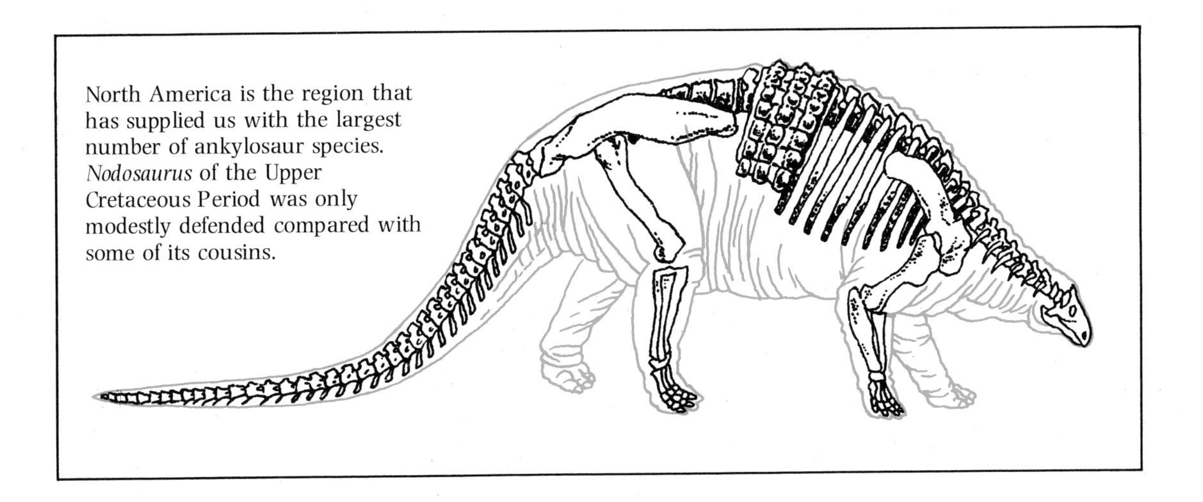

North America is the region that has supplied us with the largest number of ankylosaur species. *Nodosaurus* of the Upper Cretaceous Period was only modestly defended compared with some of its cousins.

gressively grew in size towards the adult stage. Later on it helped to protect the neck region, but its most important function was to serve as an attachment for the powerful chewing muscles. The whole head assembly is very massive by contrast with the same small cranium we noted in the stegosaurs and ankylosaurs, and this was in fact a constant feature of all ceratopians. Their bodies were thick-set and low-slung, and their general appearance varied little in the course of the group's evolution, apart from a considerable increase in size.

The fabulous discoveries of Mongolia have brought to light various aspects of the life of *Protoceraptops*. It was a small dinosaur which apparently led a gregarious life. When laying, the female would deposit twenty or so eggs in a depression in the ground which she then covered over with sand. Like modern crocodiles, they inspected their nests from time to time, and defended them against possible predators. This much at least has been deduced from the presence of the crushed skull of an *Oviraptor* above one such nest. But not all such confrontations turned out in favour of the herbivores: in 1971 a Polish-Mongolian palaeontological expedition to the Gobi brought back two intermingled skeletons, one belonging to a *Protoceratops* and the other to *Velociraptor*, a small carnivorous saurischian. Both creatures had obviously died locked together in battle and were found as they had fallen.

The horn of *Protoceratops* was not supported on a bony shaft, as it was in later ceratopians, although certain specimens do have an irregular area on their snout that might well have formed the base of a small horned outgrowth. In the later periods of the Upper Cretaceous the ceratopians evolved into a variety of forms which were chiefly reflected in all sorts of arrangements of bony points and spikes on their skulls. These heavy animals, with their necks covered by a bony 'ruff', were thus able to defend themselves actively against the great carnivores by striking out with their frontal and nasal horns. The frequency of wounds observed on their heads and vertebrae demonstrates that these herbivores must have led quite an eventful existence. It has even been suggested that there was frequent combat between members of their own species. Whether that is true or not, there is no doubt that, when attacked by one of the great carnosaurs, the more powerful of the ceratopians made the aggressor pay dearly with its life.

Several groups of ceratopians arose from the *Protoceratops* stock, two of which were particularly distinctive, one having a short collar and the other a long collar. The long-collared forms, which included genera such as *Pentaceratops* and *Torosaurus*, probably split off from the short-collared forms. Right at the beginning of the short-collared line, *Monoclonius* (see page 203) represents a fairly simple stage. The edge of the collar was scalloped, and two small outgrowths rose above the eye orbits, while the main feature of the head was a single nasal horn. In *Triceratops* this horn was smaller in size, while the two horns above the eye orbits had

Thanks to their powerful armour and a tail that frequently ended in a 'club', the ankylosaurs could hold off attacks from the great carnosaurs.

increased in proportion making the animal look like a kind of prehistoric cow (see page 36). Indeed, the first fragment of this fossil genus was believed to be that of some kind of ancient cattle, for one of the horns discovered in 1877 was said to belong to an extinct bison! *Triceratops*, which we have been able to reconstruct in perfect detail, was a colossal reptile of up to 8 metres (26 feet) in length, the head alone measuring 2 metres ($6\frac{1}{2}$ feet), while its weight of as much as 8 tonnes was a record among all the ornithischians. Its powerful horned beak was reinforced by rows of numerous teeth which interlocked vertically when closed; and when one crown was worn and fell out, it was replaced immediately. This great battery of teeth on either side of the mouth enabled the animal to deal with even the toughest plant material.

Styracosaurus, illustrated on page 203, is an example of the great complexity of the bony formations found on the skulls of certain genera. Nevertheless, there were some ceratopians, such as *Pachyrhinosaurus*, that were completely without horns. The upper portion of this animal's snout was thickened to form a flat bony hump. Mention should also be made of *Leptoceratops*, again one of the members of a separate line of ceratopians. This genus had retained a cranial structure close to that of *Protoceratops*, but with a less pronounced beak. This more lightly built creature was just 2 metres ($6\frac{1}{2}$ feet) long, and is the only ceratopian that might possibly have walked on two legs.

The death of the dinosaurs

The boundaries between geological eras are usually marked by major events that frequently had a great effect upon the face of the Earth and the living world. One of the most spectaclar of such upheavals occurred during the transition from the Mesozoic Era to the Cainozoic Era. At this time there was a major 're-modelling' of terrestrial and marine life, and many animal groups never in fact survived. Of the species that disappeared dinosaurs are all the more noteworthy since it was they who so dominated the Earth before their sudden extinction. Other groups died out completely at the same time, and vast numbers of invertebrates which had flourished during the Mesozoic Era disappeared almost instantaneously. These included the ammonites, the belemnites and the hippurites (see Chapter 3). The largest of the marine vertebrates – the mosasaurs, plesiosaurs and ichthyosaurs – suffered the same fate, leaving only the crocodiles. In the air, the pterosaurs (see

Chapter 7) were added to the long list of creatures that became extinct, while in the rest of the reptile world – so dominant during the Mesozoic Era – only the squamates, turtles, crocodiles and a few rhynchocephalians were spared. The mammals, which up to now had been represented only by small, shrew-like creatures living in the shadows of the mighty reptiles, were suddenly able to become the dominant group of land-living animals.

This abrupt break in the course of biological evolution has for a long time been the cause of much speculation, and many efforts have been made to try and fathom its reasons. It is one of the greatest mysteries of nature, and one about which many theories have been put forward, often without any connection between them. Of these theories, very few are acceptable in the light of our present knowledge of events.

If we discount the more unlikely causes of the mass extinctions at the end of the Mesozoic Era, such as extra-terrestrial invasion, there is little choice but to look to natural events to provide the answer. The most bizarre situations have been imagined: having become so gigantic in size, for example, the dinosaurs died out for sheer lack of room, or rendered our planet completely sterile by their devastation of it. One widespread theory attributes their extinction to a sort of 'senility' of their lineage. Against this theory one can argue that the Upper Cretaceous was a period of great vitality for the giant saurians - nearly all the divisions of dinosaurs were present at that time, while

Monoclonius was a fairly undeveloped North American ceratopian and less well armed than its descendants. Its remains are common in the Upper Cretaceous Period deposits of western Canada.

Styracosaurus must surely have been one of the oddest of all the ceratopians. Unlike *Monoclonius* it had many long bony spikes on its 'ruff' and these protected its neck region from the terrible teeth of the carnivorous dinosaurs. Both these genera were excavated from the same deposits.

new groups had just made their appearance and were in full expansion, especially among the ornithischians.

Some authorities have attempted to detect some variation in the environment that might have brought about the elimination of the species, and the emergence of the flowering plants has been suggested as the cause. Indeed some of these plants did produce toxic alkaloid substances which would have been fatal for the dinosaurs. The arrival of the Lepidoptera (butterflies and moths) which followed that of the flowers might also have constituted a threat, since butterflies produce caterpillars, and these larvae have been accused of devastating the food sources of the great herbivores. Other accusations have been levelled at the mammals whose egg-stealing habits may have put an end to the dinosaurs – yet the mammals had been present since the Triassic Period! All of these causes, to which we may add various assorted epidemics, come in for heavy criticism, not least because they claim to explain the extinction of a specific group without being capable of having the slightest effect on marine invertebrates such as the ammonites – yet they too disappeared. Even if we accept some large-scale catastrophe, it is still hard to conceive how creatures inhabiting very different environments were struck down at the same time, especially since other major groups from the self-same environments crossed the Cretaceous-Tertiary boundary without harm.

Of the various causes that might have affected the entire planet, we might mention the explosion of a supernova or the impact of some gigantic meteorite. Proof of such astronomical disasters has sometimes been affirmed by the sudden richness in marine deposits of rare metals such as iridium, but this does not explain the effects of their action upon the World's fauna. Perhaps the most likely theory of all is that which links the spectacular disappearance of species with the exceptional shrinkage of the seas which marked the end of the Cretaceous

Period. This theory has been strongly argued by L. Ginsburg. The underwater plateaux which bounded the continents underwent an extensive reduction in surface area, and the frantic competition which resulted from the shrinkage of this densely populated habitat brought about the destruction of whole groups, whilst changes in a number of food chains led to the extinction of one species after another. On the continents only certain reptiles of moderate proportions and the warm-blooded animals (the mammals and birds) were able to withstand a climate that had become more variable and rather cooler.

We must of course remember that the end of the dinosaurs – so often presented as the inevitable result of creatures which attain such huge sizes – came more than 150 million years after their first appearance, and the mammals which took over from them are still a long way from attaining such longevity. Still, as the next chapter will show, we might well ask the question: have the dinosaurs really disappeared?

The reasons why the dinosaurs died out are uncertain and several original theories as to why the dinosaurs disappeared at the end of the Cretaceous Period have been suggested. One rather simple theory says that they died out because the predators became too good at catching their prey; when the prey became extinct they starved to death. Shown here: *Allosaurus* in a clinch with *Brachiosaurus* from the Upper Jurassic Period.

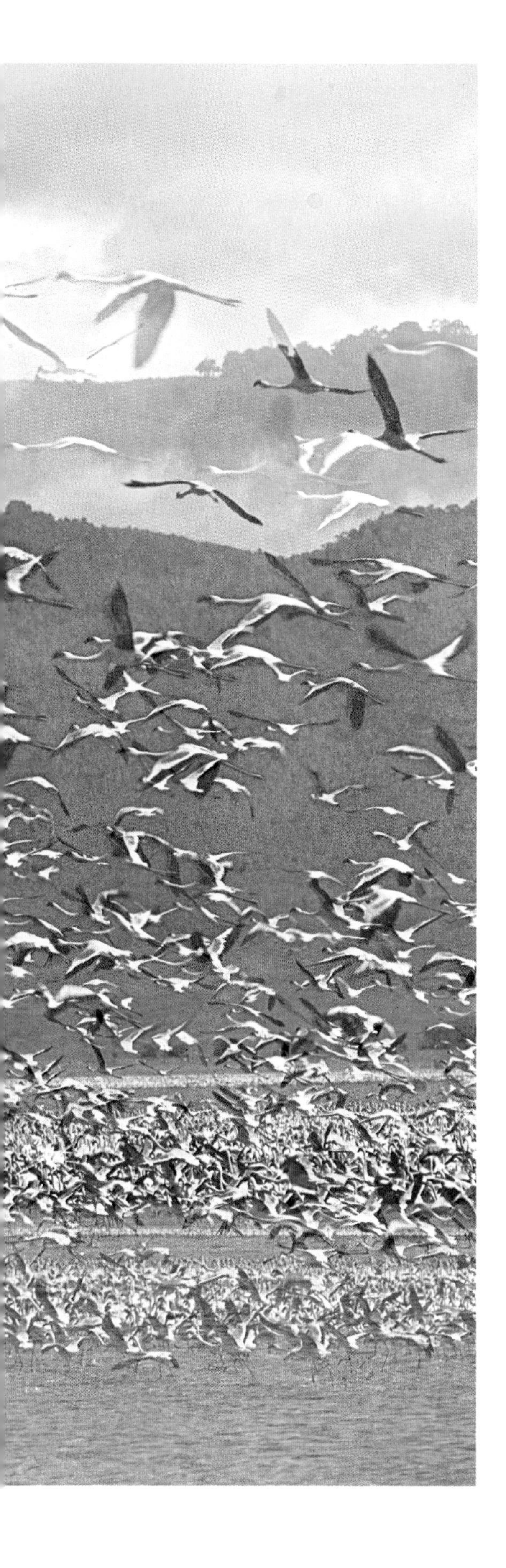

Chapter Seven

Adaptation to Flight

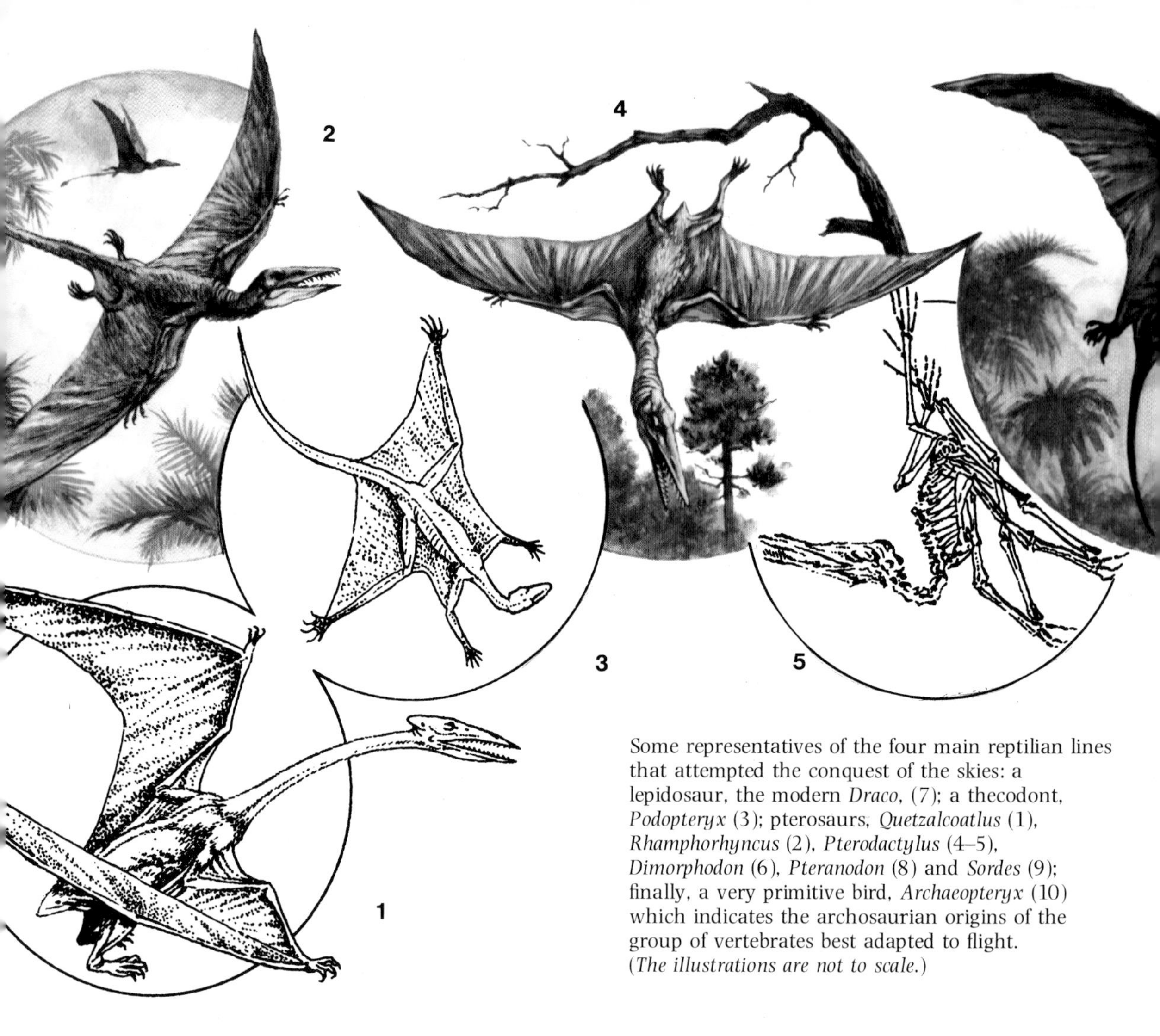

Some representatives of the four main reptilian lines that attempted the conquest of the skies: a lepidosaur, the modern *Draco*, (7); a thecodont, *Podopteryx* (3); pterosaurs, *Quetzalcoatlus* (1), *Rhamphorhyncus* (2), *Pterodactylus* (4–5), *Dimorphodon* (6), *Pteranodon* (8) and *Sordes* (9); finally, a very primitive bird, *Archaeopteryx* (10) which indicates the archosaurian origins of the group of vertebrates best adapted to flight. (*The illustrations are not to scale.*)

Adaptation to flight

The conquest of the skies

The birds have been the most successful group of animals to take to the air. Other animals have the power of flight – the bats for instance – but the birds are the most varied and widespread of all flying animals. It is fascinating to study the solutions which different groups found to the problem of flight.

It is important for us to remember that the first flying creatures which lived on the Earth were the insects, evolving in vast numbers among the forests and swamps of the Carboniferous Period, while the vertebrates were only just beginning to move from the water to dry land. So the arthropods (jointed animals) could fly long before any vertebrate had wings. The flying insects were probably a new and attractive food supply for the land animals, but in order to capture them they had to become skilful fliers.

Not all the flying vertebrates ate insects, however; in fact some of them used flight to escape from their enemies, rather than catch their prey. These small animals were rather poor fliers but avoided being killed by other terrestrial animals by gliding for short distances in the same way that the first fishes capable of leaving water were able to flee from the fearful jaws of their natural enemies. Flight can be seen as the best adaptation to life among the trees: an environment that was gradually conquered by the vertebrates.

Several types of egg-laying reptiles started to develop the power of flight at about the

same time, including the lepidosaurs and more especially the archosaurs, the group from which the birds evolved. Flying mammals appeared much later on and even some amphibians could glide. In the forests of Borneo today lives a tree frog, which cannot fly properly but can at least glide using the webbing between its toes as a sort of parachute.

Flying lizards

Limestone from the Carboniferous Period area around Bristol in the west of England has gradually been dissolved by running water, leaving holes and cracks in the ground. The remains of animals that lived on the surface, and sometimes even the animals themselves, are often found in these holes as fossils. This is a common type of rock formation which usually contains very promising fossils. A number of different vertebrates from the Triassic Period have been found, including a strange animal that is very similar to the modern 'flying dragon' that lives in South-east Asia (see page 210). It was called *Kuehneosaurus* in honour of the German palaeontologist W.G. Kuehne.

Kuehneosaurus is a squamate with a characteristic free-moving quadrate bone (see page 175). Despite being very old this was already a highly specialized animal. It was about the same shape as a lizard with a very long tail and measured some 60 centimetres (2 feet). The hind legs were larger and stronger than the forelegs, and this suggests that the animal ran fast on two legs. *Kuehneosaurus* had very strange ribs, quite unlike those of a normal lizard. The vertebrae in the middle of the trunk stuck out sideways with long bony outgrowths attached to them: these were erect, elongated ribs and were built very lightly with a large central cavity.

This tubular framework was used to carry two triangular 'wings' on each side of the body which the creature was able to fold away backwards when not in use. It is easy to imagine that these wings could be used for flight. Launching itself at full speed, the animal would glide off into the air, its 'wings' held away from the body. *Kuehneosaurus* was probably a more efficient flier than the modern flying lizard, *Draco*. Once in the air it may have caught a few insects using its many sharply pointed teeth. However, *Kuehneosaurus* probably mainly used flight to escape from its enemies. Another rather similar creature called *Icarosaurus* has been found in the Upper Triassic rocks of New Jersey, U.S.A. (see page 211). This reptile is very similar to some British fossils, which is not surprising because these two deposits, separated today by the North Atlantic Ocean, are located in regions which were very close to each other in the Triassic Period (see pages 232 and 233).

The aerial glides of *Kuehneosaurus* or *Icarosaurus* cannot be considered as true active flight, and it is most unlikely that they were ever able to twist and turn in the air like modern-day birds. The lepidosaurs such as the Malay or Indonesian *Draco* are just as primitive as *Kuehneosaurus* and *Icarosaurus* and therefore could not be descended from them. They probably evolved the same gliding ability on their own. In fact the resemblance between the two types is only superficial as their skeletons are very different. There are fewer modified ribs in *Draco* and the surface area of the membrane which they support is much smaller.

The flying geckos are not particularly good fliers either. The folds of skin attached to their limbs, trunk and tail form a natural parachute along with the webbing in between their fingers. Using these as an airbrake, the geckos can make steep glides but are incapable of long leaps. Even one tree-living type of snake, the snake of paradise, can glide. This animal lives in southern Asia and can flatten out its body to give it lift in the air. This probably only prevents it from crashing to the ground should the animal lose its grip on the branches, and cannot be described as true flight.

There are several species of 'flying dragons' (genus *Draco*). This animal lives in the trees of the luxuriant forests of tropical Asia. Its dullish colouring makes it practically invisible when resting, hooked to the trunk of a tree. When it launches itself into the air, however, its 'parachute' shows a brilliant orange and black pattern. *Draco* is capable of soaring over 20 metres (65 feet).

Flying thecodonts

The Soviet palaeontologist A.G. Sharov has found two of the oldest flying vertebrates. Both are from the Lower Triassic Period of the Republic of Kirghiz, right in the heart of the Asian continent, and are thought to be thecodonts.

Icarosaurus of the Upper Triassic Period of North America was slightly smaller than its European contemporary, *Kuehneosaurus*, and probably cast itself from high branches to soar in the air. Its long tail acted as a stabilizer. Like *Kuehneosaurus*, *Icarosaurus* possessed huge orbits denoting the importance of sight to these animals, a rule that applies to all diurnal flying vertebrates.
In 1978 a similar flying reptile was found in Madagascar, *Daedalosaurus* from the Upper Permian. Its ribs extended from the body like a pair of wings. If this age is confirmed, *Daedalosaurus* can claim to be the oldest known flying vertebrate.

Longisquama (see below) was a small tree-living saurian with some odd features. Its eyes were protected by a ring of bony plates and its shoulder-blades were joined together to form a fork, as in a modern bird. The strangest feature of all, however, was the series of appendages which stood out from its back. These structures, which looked like primitive feathers, were arranged in a single line down the back, although some palaeontologists think that there were probably two rows. They seemed to have acted like a type of parachute on the animal's back, and were formed by very long raised scales. The reptile scale is thought to have given rise to the bird feather but maybe Sharov was not quite correct to call these structures 'proto-feathers'. *Longisquama* was probably not an ancestor of the birds but it does show some features which may have helped the animal to glide. A complete skeleton of *Longisquama* has not been found so we do not have a complete idea of what it looked like.

Podopteryx (see page 208) looked like a kind of small lizard about 20 cm (8 inches) long, with a very long tail. The fossil is very detailed, and Sharov has been able to see the outlines of a skin membrane which helped the creature to glide like a modern flying squirrel (see pages 228 and 229). The skin membrane or *patagium* was connected to the long hind legs, the root of the tail and the rear of the forelimbs.

Both these species, although difficult to relate to more recent forms, show some of the ways in which the archosaurs adapted to life in the trees. They show the beginnings of the conquest of the air by the thecodonts, which eventually gave rise to the birds and the pterosaurs.

Pterosaurs

The pterosaurs were the first real archosaur success in the air. These creatures ruled the skies during most of the Mesozoic Era and were not replaced until the true birds evolved. 'Pterodactyls' are popular today in books and films for they were spectacular creatures. Some palaeontologists think that they evolved from a thecodont similar to *Podopteryx* but the layout of their wings, which were modified front legs, is very different from the *patagium* of *Podopteryx* which was arranged around its hind legs.

It is amusing to note that the first pterosaurs discovered were thought to be sea-

Longisquama was an archosaur adapted to flight, but there are still many questions left to answer about its anatomy and way of life.

This small pterosaur belongs to the genus *Pterodactylus*. When it was first identified in the 18th century by Collini it was thought to be an amphibian but in 1801 Cuvier recognized it as a flying reptile. The fossil below comes from Eichstätt in Bavaria, about 20 kilometres ($12\frac{1}{2}$ miles) from the famous deposits of lithographic limestones at Solnhofen.

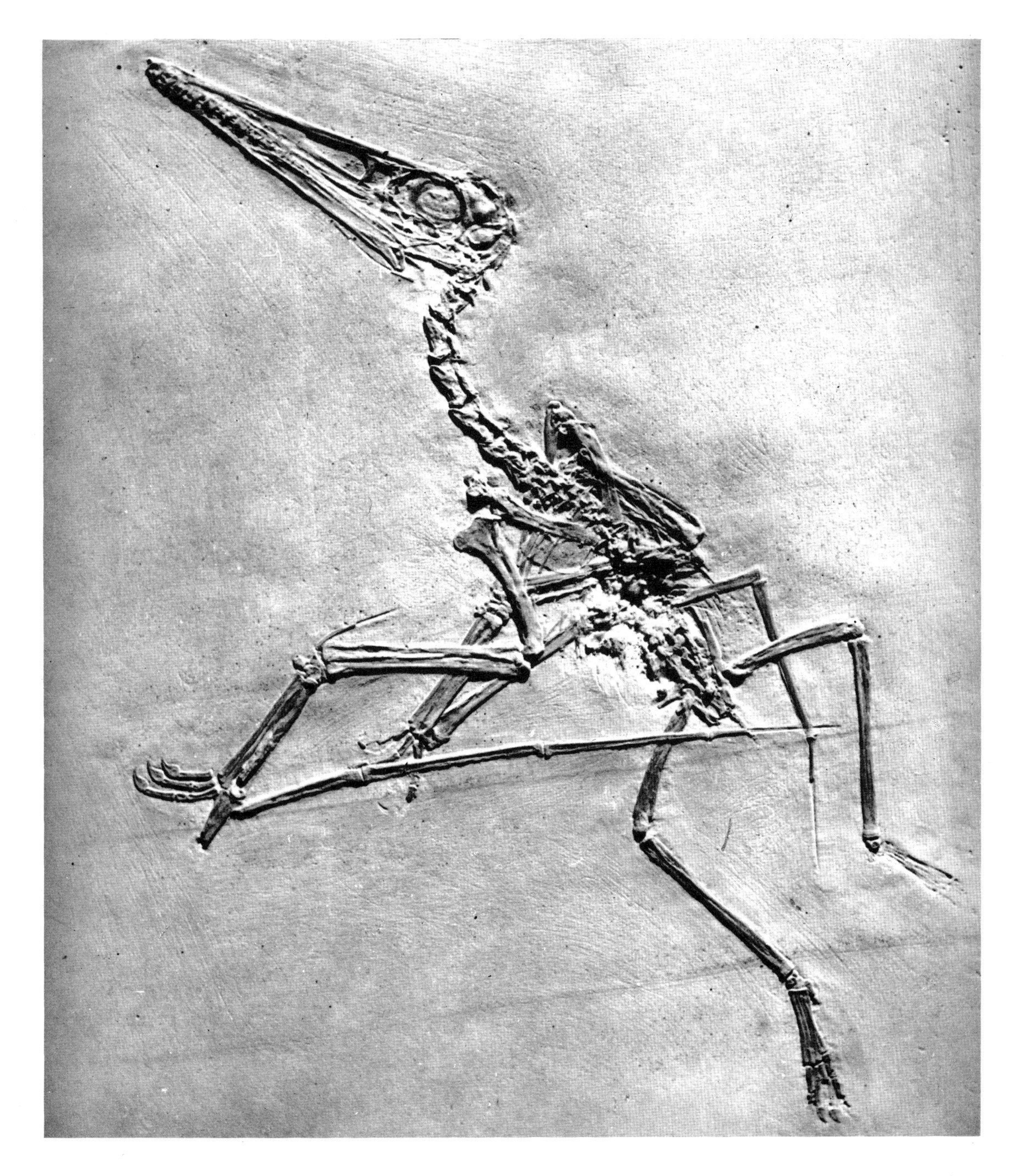

creatures – probably because the remains of these animals were most commonly found in marine sediments. They were present from the Upper Triassic Period in northern Italy until the end of the Cretaceous Period, when the largest forms appeared. Although some people in the last century thought that they were closely related to the birds or the Chiroptera (bats), it was soon realized that they were very different and could not be ancestors of either of the other groups. The wing of the flying reptiles (see pages 208 and 209) had no feathers, and was constructed like that of a bat by means of a skin membrane. However, it showed several differences from the bat wing. The wing of the bat is stretched over four fingers of the hand while that of the pterosaur was held on one only, the fourth, whose finger bones had become extremely long and thin. The *patagium* also ran from the base of the neck to the wrist (perhaps even to the wingtip) and was spread out to the rear on either side of the tail and hind legs. There is still some debate about this arrangement, however, for it may have created problems for the animal when it moved over the ground.

The exact way in which the pterosaurs flew is still not known. Palaeontologists and aircraft designers have joined forces on many occasions to try to understand how

Dimorphodon from the base of the Jurassic Period (above left), and *Rhamphorhynchus* from the Upper Jurassic Period (below), were two pterosaurs of similar size, about the size of an eagle. *Pteranodon* (above right) from the Upper Cretaceous Period was a giant with a wingspan of 7 metres (23 feet) but only weighed 15 kilograms (33 pounds).

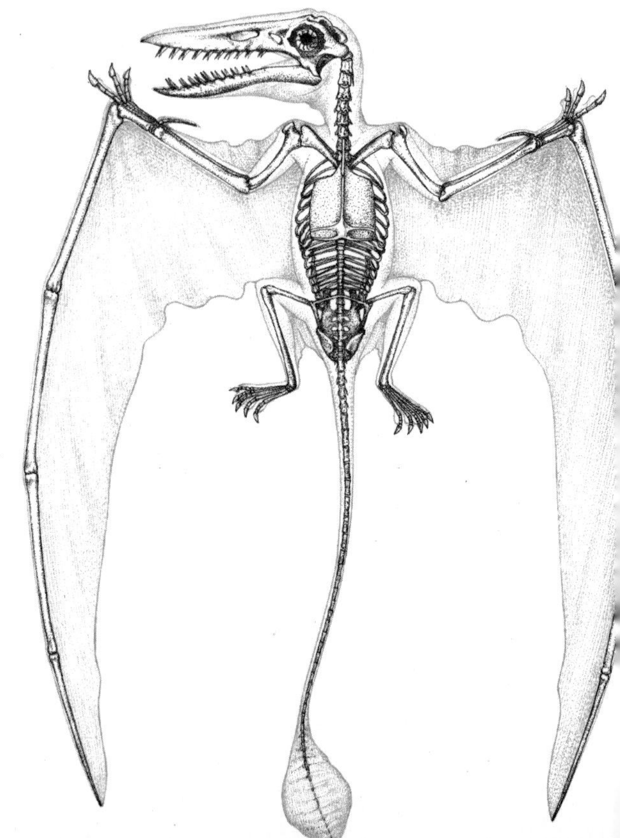

these incredible flying machines worked, but there were so many different types that it is impossible to find any general rules. It is very unlikely that a *Pteranodon*, with a wingspan of 7 metres (23 feet), flew in a similar way to little *Pterodactylus*, the size of a sparrow. However, all the pterosaurs had a certain number of common features, especially those which made the animals' skeletons as light and strong as possible. Their bones were hollow like those of birds, with very thin walls. The head looked far too big for the trunk and the bones of the skull were often fused together. The snout had a large mouth often armed with long, sharp teeth, but it was sometimes completely toothless, as in *Pteranodon*. The brain had certain features also found in birds – another example of convergent evolution. The enormous importance of sight and the sense of balance for these animals explains why the optic lobes and cerebellum were so very large, while the trunk carried a huge sternum (breast-bone, see page 214) in which the wing muscles were inserted. These were rather weaker than those found in birds whose overall skeleton is a lot more robust.

Therefore the winged reptiles were rather frail and weak, but this was made up for by their low weight and large wing area. While the smaller species like *Pterodactylus* flapped along, the larger ones were excellent gliders and hardly moved their wings at all, using the ascending air currents for lift. These giants had a *notarium* welded to several of the back vertebrae. This was a bony plate which absorbed a large part of the enormous stresses transmitted by the wings to the shoulder girdle. The tail was quite long in the more primitive forms but gradually became smaller and eventually almost completely disappeared. This, together with the elongation of the metacarpals (the bones between the wrist and fingers), is one of the main differences between the suborders of pterosaurs known as the rhamphorhynchoids (from the Upper Triassic–Upper Jurassic Periods) and the pterodactyloids (from the Upper Jurassic–Upper Cretaceous Periods).

The tail of *Rhamphorhynchus* (see page 214) ended in a flap of skin which was used as a rudder in flight. The bony 'horn' on the back of the head of *Pteranodon*, which sometimes doubled the length of its skull, was once thought to be used for steering the animal in flight, but modern tests in a wind-tunnel have shown this to be untrue. It was in fact used to balance the weight of the enormous beak and so relieve the stress on the neck muscles. This strange helmet also helped to streamline the air currents flowing over the head of the animal, helping it to fly faster through the skies. Some people think that it may have acted as an air-brake which slowed the creature down when the head was turned to the side, or perhaps it was a sexual feature, as it is not found in all members of the species.

As we have already seen, it is quite possible that some dinosaurs could regulate their body temperature, but we are absolutely certain that some pterosaurs were warm-blooded. Even in the last century, it was realized that a flying vertebrate could only be a warm-blooded animal, as is the case with the birds and mammals, for in order to fly a very high rate of metabolism is needed. In 1927 a fossil of *Rhamphorhynchus* was found which included traces of hair. Then another fossil was discovered by A.G. Sharov in 1971 which he called *Sordes pilosus* (see page 209). This magnificent fossil from the Upper Jurassic Period of Khazakhstan has an imprint of dense fur covering the whole body except for the tail. It is likely that many pterosaurs had some kind of fur covering, even though we do not often find traces of it in their fossils.

Exactly how the pterosaurs walked on land is not known. Did they walk on the hind legs or scramble about on all fours? Certainly the heavier species with their weak hind legs could not have walked upright. Maybe they were forced to crawl on their bellies. There is some evidence for this in a trace fossil found in the Jurassic of Arizona, U.S.A. which is thought to belong to a pterosaur: both hands and feet have left their mark in the sediment as the animal dragged itself along.

It is hard to imagine how a huge *Pteranodon* could move over the ground if it was not able to fold away its wings completely like a bird. It is even more difficult to understand how these creatures launched themselves into the air. Despite its impressive size, *Pteranodon* was not the giant of the group, *Titanopteryx* discovered in Jordan, and *Quetzalcoatlus northropi* of the Upper Cretaceous Period of Texas, U.S.A. (see page 208) with a wingspan of some 12 metres (39 feet) were much larger. The largest known bird (a fossil vulture from Argentina) only had a 7 metre (23 feet) wingspan and so *Quetzalcoatlus* is the largest creature ever to have flown on Earth. In order to take off the big pterosaurs would have hauled themselves to the top of a tree or cliff before casting themselves off into the air. They were certainly not adapted to climb well, and any damage to their wings must have meant their certain death. It seems almost impossible that these creatures were ever capable of taking off from the ground under their own power, and this is why the windy conditions along coastal cliff-faces must have been particularly favourable for them. Remains of *Pteranodon* have been found in marine sediments laid down 100 to 200 kilometres (62 to 124 miles) off the coast and the contents of their stomachs tell us that they lived mainly upon fish. Flying close to the waves, they must have used their long, toothless beaks to seize any fish that came close to the surface, and – like the pelican – were able to keep their catch in reserve in a kind of pouch located beneath the throat. *Quetzalcoatlus* on the other hand has been found in completely inland environments and is often thought to have been some sort of reptilian vulture living on carrion. The teeth of the smaller species suggest that they ate insects although some, like *Pterodaustro*, were highly specialized animals whose lower jaw bore a fine comb of hundreds of needle-like teeth, which it may have used to feed upon plankton.

Like so many other species, these extraordinary vertebrates became extinct towards the end of the Mesozoic Era after a history of over 140 million years, and in comparing them with the birds it would be wrong to think of them as a failure. They were no doubt adapted to a very different flying technique and through this lived successfully in many habitats.

The small pterosaurs could only flap their wings and glide. The complex aerial manoeuvres which the birds can execute due to the great mobility of their wings, the strength of their muscles and the rigidity, suppleness and lightness of their feathers, were not possible for these reptiles.

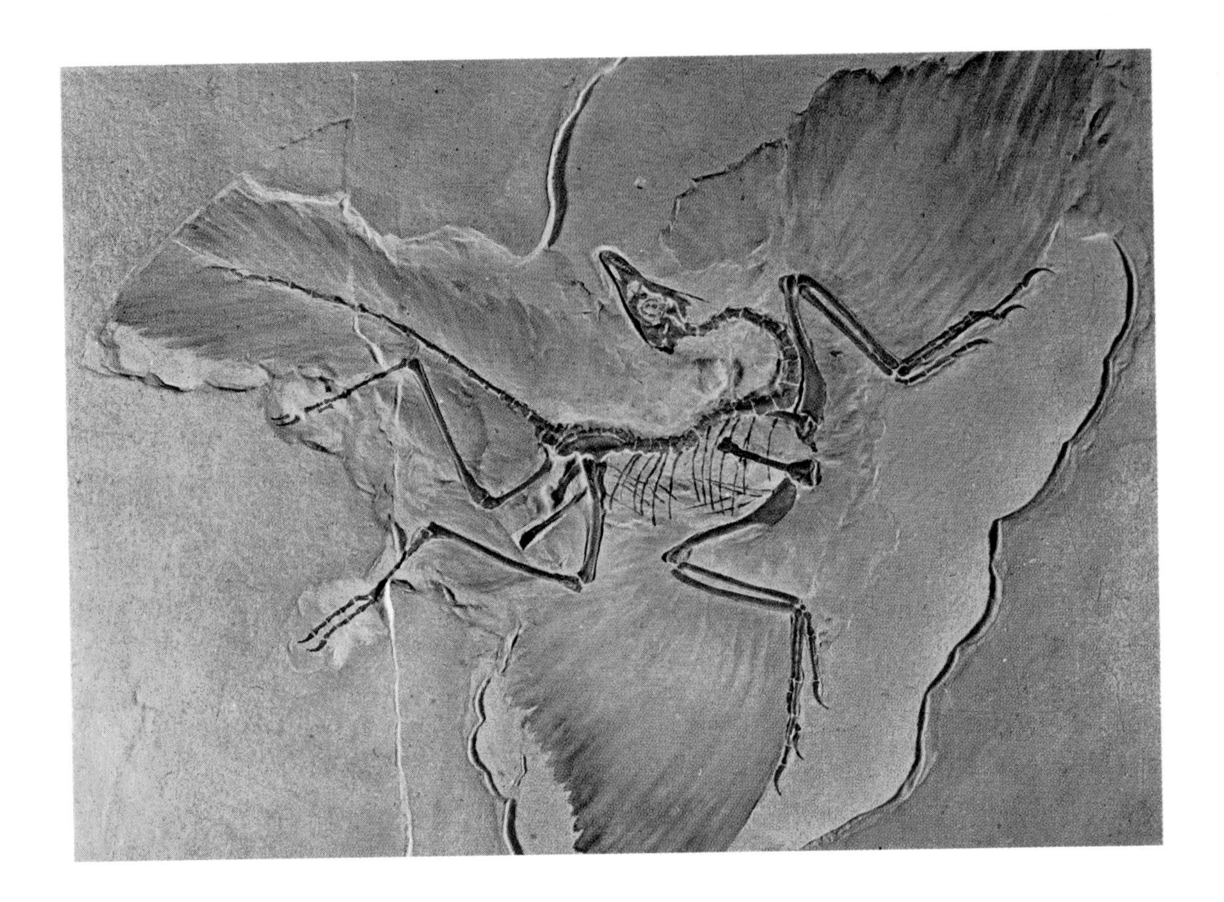

Only the lithographic limestones of Bavaria (Upper Jurassic Period) have so far revealed skeletons of *Archaeopteryx*, a primitive bird the size of a crow. Only five examples are known (the specimen above is owned by the Berlin Museum). A few feathers have been found in the Barcelona province of Spain, but we do not know what kind of bird they belonged to.

1861: A sensational discovery in Bavaria

As we have already mentioned concerning the origin of different divisions in the phylum of the vertebrates, the first representatives of an important group nearly always appear quite suddenly and already equipped with quite a few of that group's characteristics. Even when the group's ancestry is questionable, it is rare that we can find fossil forms which are intermediate between groups. This situation is understandable when we consider that evolution is not a regular phenomenon, for a species that has remained unchanged for thousands of years may undergo a sudden crisis which leads to its modification or disappearance. Very stable and very numerous species will, of course, form the bulk of fossil collections. Still, at the base of an important phylum we nearly always find unstable species represented by populations that are small in number and evolving fast. The likelihood of such species ever being discovered is therefore very remote, and we should remember that fossils can only give us a very incomplete idea of the World's fauna.

There is, however, at least one splendid exception to what is practically a rule: *Archaeopteryx*, a fossil which links the birds to their archosaurian ancestors. This is all the more remarkable since birds usually make poor fossils. Our knowledge of this animal is due to the excellent fossils found in Solnhofen in Bavaria. In the 19th century the extremely fine limestone quarried in this area was cut into slabs intended for use in lithographic printing. These rocks were deposited during the Upper Jurassic Period in a very calm shallow sea or lagoon, and the workmen examining each slab for possible defects occasionally discovered magnificent

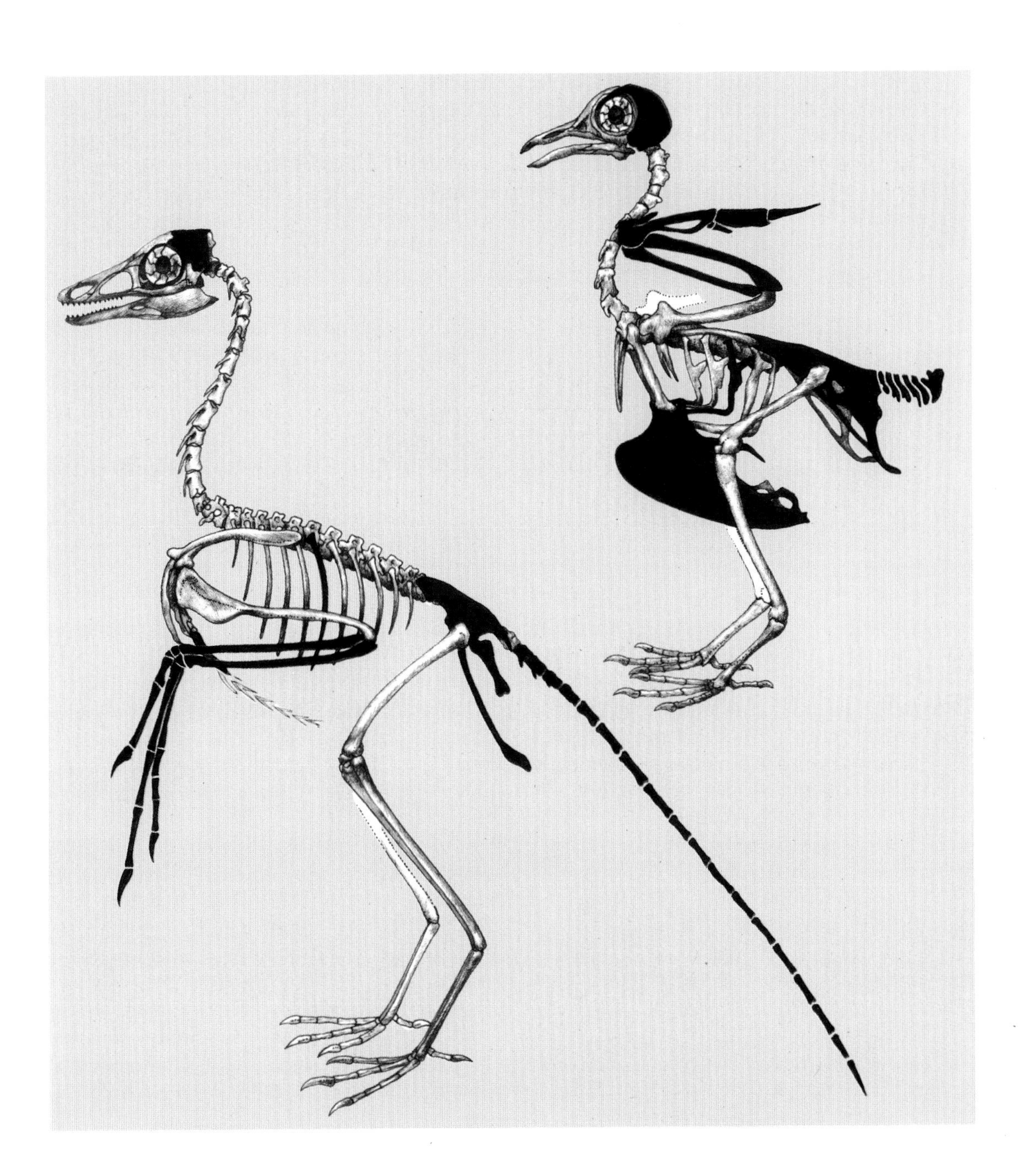

The skeleton of a pigeon (top) compared with the reconstruction of an *Archaeopteryx* (left) shows the extent to which the anatomy of what may well be a primitive bird differs from that of its distant modern cousins. The main points of divergence are marked in red. According to Ostrom, the *Archaeopteryx* should be classed as a small specialized coelurosaurian dinosaur.

fossil plants, invertebrates, fishes and reptiles often showing the imprint of the creature's soft parts (see page 213). Doctor F.K. Häberlein was fascinated by these finds, and frequently traded his medical services for samples of these fossils. He built up a considerable collection of them in the course of many years. In 1861 he was given a fossil which could only have been a bird's feather, and this was followed shortly afterwards by the first *Archaeopteryx* skeleton from the Ottmann quarry near Pappenheim. A less complete specimen had in fact been found several years before but its true identity was not recognized until 1970 after a century of being labelled as a pterosaur fossil! In the year after this amazing discovery, Häberlein's specimen was purchased by the British Museum in London, along with the rest of his collection. Fifteen years later a second fossil bird from Blumenberg was obtained by Häberlein's son, and this fossil, even better preserved than the first (see page 218), was also soon sold, this time to the collection of the Natural History Museum, Berlin. For a very long time *Archaeopteryx* was only known from these two specimens, and it was not until 1951 that a further example was brought to light, although this discovery was not revealed until 1973. It was followed in 1956 by a fresh find and in 1970 by the identification of the very first specimen.

The feathered reptile It has often been claimed that, if *Archaeopteryx* had not been preserved along with its feathers in the excellent deposits of Solnhofen, it would certainly never have been thought of as a primitive bird. This is because apart from its feathers, it has practically no avian characteristics at all (see page 219). The skull had many windows, and showed the opening in front of the eye typical of an archosaur. The jaws were armed with small conical teeth set in sockets and the eye was surrounded by a bony ring like the one found in the birds as well as in many other groups. The size and shape of the brain cavity suggested that the brain was primitive and rather reptile-like. The spine, with its many vertebrae, was not rigid as in a modern bird, and ended in a long tail. The dorsal ribs did not possess the supporting struts which reinforce the thoracic cage of the birds (see pages 219 and 223), and there was still a ventral rib framework. The pelvis was rather primitive and the bones that made it up were not interconnected. The foot did, however, possess the avian feature of a thumb turned to the rear which could be brought up next to the other three digits (the fifth digit had disappeared). At the pectoral girdle the shoulder bones were joined together to form the *furcula* (or 'fork') – the wishbone so prized by those hoping to have a wish come true – but the powerful breast-bone of the birds was completely lacking. Although the upper limbs had started to take on the layout of a wing – the plumage is there to prove it – they still had three separate, clawed digits, and finally, the bones were not hollow.

So *Archaeopteryx* had few skeletal adaptations for flight and it must have flown poorly. Its brain is by no means that of an animal capable of aerial acrobatics, while the absence of large areas of bone for the attachment of the wing muscles suggest that it could not manage prolonged flapping flight. The feathered reptile could not have been a good glider either because of its very short wings. Therefore *Archaeopteryx* probably had poor flight powers and the rather flexible construction of its spinal column and pelvis must have made landing a hazardous business.

There is little doubt therefore that *Archaeopteryx* was unable to compete with the pterosaurs which by this time had become reasonably agile fliers. The success of the first birds was really due to their adaptation to life in a forest environment where competition from the 'winged reptiles' was most unlikely.

Archaeopteryx is traditionally depicted as a tree-dwelling animal capable of progressing along a trunk with its clawed hands and holding on to branches by its prehensile feet. It was able to soar heavily from branch to branch and execute an as yet inefficient flapping flight.

Above are the outlines of *Archaeopteryx* (in blue) and *Compsognathus* (black), a coelurosaurian dinosaur. Ostrom thought that the three clawed digits of *Archaeopteryx* were used to capture prey rather than to climb trees and were similar to the hands of small carnivorous dinosaurs. Perhaps *Archaeopteryx* used its 'wings' to engulf its victims, rather like modern birds of prey.

The origin of the birds

Four groups have been suggested as close relatives of the birds, and the least expected of these must surely be the crocodiles! Other suggestions have included the thecodonts and the ornithischian and saurischian dinosaurs. This last theory, put forward by J.H. Ostrom, suggests that the birds derived from small bipedal coelurosaurs. This is quite an old idea but Ostrom presented some new evidence to support it. Certain skeletal arrangements such as that of the shoulder girdle and the forelimbs are very similar in *Archaeopteryx* and the coelurosaurs and, according to Ostrom, this could only be explained by a close kinship, between the two.

The appearance of feathers has been regarded for more than a century as an adaptation to gliding by a tree-living creature or even as a means of increasing the speed of a bipedal running form. To Ostrom, however, these organs would not have played any part in locomotion to start with: plumage would have been an insulating covering enabling a small animal to regulate its temperature, and later on would have played a part in the development of the forelimbs (see above). The increasingly powerful shoulder muscles might even have allowed very short flapping flights consisting of vertical leaps for capturing the creature's prey.

These theories have been criticized, however, and some people do not think that *Archaeopteryx* ever lived in the trees. Other experts prefer to stick to the more classic theory of the origin of the birds in a species of thecodont similar to *Euparkeria*. Even so, making all the lines of archosaurs derive from a single stock group hardly solves the problem of their relationship to each other.

Birds of the Cretaceous Period

During their subsequent history, the birds acquired the features which characterize them today; as we have seen, the skeleton of *Archaeopteryx* possessed practically none at all. Nearly all these later modifications were due to the two requirements of flapping flight: lightness and strength. The bones were lightened by large air-filled cavities that very much reduced the skeleton's weight. A number of bones were fused together. The joints in the skull disappeared rather quickly, while the brain itself increased in volume and the teeth vanished. The wing became essentially organized around two digits and some of the wing bones were joined together. The strong pelvis was formed by a bony mass while in the lower limbs some of the ankle bones were added onto the base of the tibia while others were included with the metatarsals now fused into a single bone. The ribcage became very rigid, and while the neck vertebrae became quite moveable, those of the back were jointed together and even fused; the tail became much smaller. The ribs were interconnected by means of supporting struts, and at the front of the breast, beneath the *furcula*, the sternum developed an enormous crest where the wing muscles were inserted. All these arrangements were accompanied by physiological improvements which although they left no fossil trace, helped the bird to fly strongly and make safe and easy landings.

It is hard to retrace the steps in the evolution of birds, for so few fossils have been left behind, partly owing to the avian way of life and partly because of their fragile and thin-walled bones. While the discovery of *Archaeopteryx* was very fortunate, it is depressing to note that very few bird fossils have been found since and those that have are much more recent. Lakeland deposits of the Lower Cretaceous Period in the state of Victoria, Australia, have provided us with a few feathers, but more complete remains are not found until the Upper Cretaceous Period. The most famous of these fossils are without any doubt the genera *Ichthyornis* and *Hesperornis* which are found in marine deposits in Kansas, U.S.A.

Ichthyornis (see below) has often been compared to a gull – it was about the same size and probably had similar habits. The strength of its breast-bone indicates that it must have been a good flier, while its overall skeleton greatly resembles that of a modern bird. It still had a few primitive characteristics, however, such as the presence of

After *Archaeopteryx*, the oldest remains of fossil birds do not appear until the Upper Cretaceous Period, and these are only small fragments. So we can only guess at what *Ichthyornis* (below) looked like, and this is one of the better-known creatures!

teeth on the jaws, pointed out by O.C. Marsh at the end of the last century. This observation was based upon isolated fragments and jawbones whose connection with the skull was very deformed. For a time it was doubted if these remains actually belonged to the genus *Ichthyornis*, and they were taken to belong to small marine reptiles. The mosasaurs are in fact commonly found in these layers and the bones of a young specimen could well have been mixed with those of the Cretaceous 'gull'. However, the discovery of an extremely well-preserved example has recently removed the doubts levelled at O.C. Marsh's identification of this creature.

Hesperornis is better known as we have all the parts of its skeleton. It was much larger than *Ichthyornis*, measuring over a metre ($3\frac{1}{4}$ feet) in height. It was a very specialized form and demonstrates a most advanced reduction of the forelimbs which are now limited to a very small humerus. This makes it easier to understand why the sternum of this flightless bird had no trace of a keel. However, the hind limbs, used for swimming, were very powerful and the bird was able to use its toothed jaws to seize the fishes on which, like *Ichthyornis*, it fed. The teeth of *Hesperornis* are also the subject of debate. Some people have doubted whether the upper jaw – toothless at the front – can be included with the rest of the skeleton, and the branches of the mandible attributed to the genus by Marsh have a central hinging mechanism identical to that found in the mosasaurs (see page 175). This may be interpreted, however, as the result of a convergence by which two totally different animals have both adapted to the ingestion of whole fishes.

Other fossil birds have been discovered from the Upper Cretaceous Period and nearly all of these are incomplete skeletons or parts of skeletons belonging to aquatic genera. Of these, some are quite closely related to *Ichthyornis* and *Hesperornis*, while others apparently represent flamingoes or even primitive cormorants. There is no doubt that the birds must have become widely distributed on land as well, but they were seldom fossilized. One mandible from the Upper Cretaceous Period was thought to belong to a large running bird, but is probably only a fossil of a 'dinosaur-ostrich', and other finds are even more doubtful.

Some large running birds

The most important phases in the diversification of the birds must have taken place during the Cretaceous Period, and by the beginning of the Cainozoic Era, the chief divisions in this class had evolved. The birds of these times have some remarkable characteristics such as the considerable importance of running forms which had lost all means of flight. The dinosaurs had suddenly died out and the great carnivorous mammals had not as yet appeared, and so gigantic birds for a time occupied the ecological niches of terrestrial predators. *Diatryma* (see page 226) whose bones have been excavated in Europe and North America, was one of the more spectacular examples. Another such bird was *Gastornis*, a European species which had a longer neck and smaller head, being part wader and growing to the size of a modern ostrich. In Europe, the appearance and expansion of the placental mammals soon caused these curious creatures to die out, but in South America they survived as meat eaters alongside the marsupials and terrestrial crocodiles. During the Miocene Period lived the most advanced forms of these birds including the genus *Phororhacos*, discovered in Patagonia. The skull of this animal, similar in outline to that of *Diatryma*, measured 60 centimetres (2 feet) and, although its wings were small, its hind limbs were long and very powerful. *Phororhacos* must have chased its quarry – medium-sized herbivores – before overcoming them and

Like *Ichthyornis*, *Hesperornis* is known mainly from remains excavated in the chalk of Niobrara from the Upper Cretaceous Period of Kansas U.S.A. It was an aquatic bird and must have been rather clumsy on land; its webbed feet were used for swimming, and its tail helped it to keep its balance on land.

The largest *Diatryma* exceeded 2 metres ($6\frac{1}{2}$ feet) in height, and their huge, parrot-like beaks made them look very fierce. Its body and small wings were covered in very fine feathers similar to the present-day cassowary.

tearing them apart with its enormous beak. The huge proportions of this flightless bird were nevertheless exceeded by some more recent South American genera, such as *Brontornis*. The ratites (see page 227) were another remarkable group of terrestrial birds whose ancestry remains unclear and whose numbers were drastically reduced by the expansion of mankind.

Gliding mammals

During the Cainozoic Era, a long time after the pterosaurs and the first birds, a third group of vertebrates also gained the power of flight. The class of mammals, whose main evolutionary episodes we shall look at in the next chapter, included many groups which became airborne.

Of the tree-dwelling marsupials and placentals that leapt and soared in the air, the phalangers were the most important, for they were ancestors of the many modern flying forms which now live in Australia, Tasmania and New Guinea. There are several degrees of adaptation to flight, from tiny species provided simply with a bushy tail right up to the great soarer which is over 1 metre ($3\frac{1}{4}$ feet) long and whose *patagium* between the fore and hind limbs enables it to soar up to a hundred metres (328 feet) from the topmost branches of a eucalyptus tree. Between these two extremes are the flying squirrels and the flying phalangers. Feeding upon insects, leaves, buds, flowers and even the gum they find on bark, the flying phalangers are nocturnal animals which sleep by day in hollow tree-trunks.

The placental mammals have also given rise to species of 'soarers' which by convergence have acquired a *patagium* very similar to that of the phalangers. They belong to two orders, the rodents and the Dermoptera.

Of the rodents, the flying squirrels are currently found in the forest regions of North America, Europe and southern Asia, and apart from the large Indian species, they are all nocturnal and live on insects and dry fruit. On leaping from one tree to the next

On certain islands safe from enemies, enormous running birds called ratites survived until a few centuries ago. Their most famous living representative is the ostrich. *Aepyornis*, whose eggs had a volume of some 10 litres (above) lived in Madagascar. New Zealand was the home of *Dinornis* (below), which grew up to 3.5 metres ($11\frac{1}{2}$ feet).

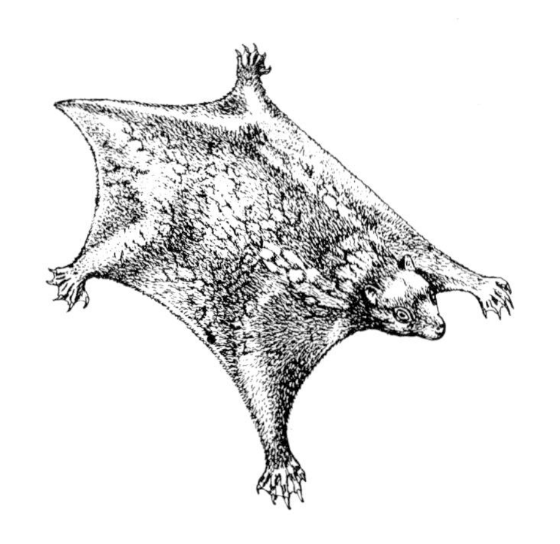

The dermopterans (also called the galeopithecans), are very curious mammals that were once thought to be related to the lemurs or even to bats, but they actually form a totally separate order. They have a fold of skin between their front and hind legs called a patagium which they use to soar from tree to tree (left). Their diet is chiefly vegetable and they rarely reach the ground. They are most active at night and are excellent climbers; by day they remain suspended from a branch, rolled up into a ball (right). The young attach themselves to their mother's breast and stay there even when she glides from tree to tree.

they stretch out the flaps of skin that run between their limbs and form a natural parachute. The long tufted tail is used as a rudder and helps them to steer to some extent. A very similar feature has been evolved by the Anomaluridae, a rodent family that lives in Africa south of the Sahara Desert. The 'scaly-tailed flying squirrels' also have a large *patagium* which is remarkable in being supported by a strong rod of cartilage attached to the forearm. They are so-called because of the curious scales which cover the underside of the tail and may perhaps help these tree-living animals to keep their balance when negotiating tree trunks. It is also found in the flying mice which belong to the same family which unlike the previous herbivorous and fruit eating flying squirrels, are insect eaters. Flying squirrels have been found as fossils in Europe, and the Anomaluridae in Africa, both from the Miocene Period onwards.

The strangest of all the soaring mammals are without doubt the Dermoptera, an order that today is made up of just one genus called *Cynocephalus* living in South-east Asia, Indonesia and the Philippines. They are tree-living creatures which sleep by day and hunt at night (see page 229). The Dermoptera have some very interesting features, the first of which is that their *patagium* is far more developed than in any other gliding mammal, extending from the base of the head – forward of the arms – to the tip of the tail, and reaching as far as the ends of the digits on each limb. The skeleton is rather specialized and they have strange lower incisor teeth which are enlarged in the shape of a comb. The Dermoptera are not closely related to any living group of placental mammals and they have been traced back to the Upper Palaeocene Period as fossils. The genus *Planetetherium* has been excavated in North America, as has *Plagiomene* from the lower Eocene Period. Although these remains show that the group appeared at the beginning of the Cainozoic Era they do not tell exactly when the Dermoptera achieved the ability to soar from tree to tree.

Bats

The Chiroptera, or bats, have the best developed powers of flight among mammals and can flap their wings, which makes the most sophisticated manoeuvres possible. While the soaring mammals are only a little further developed than their leaping ancestors, the bats have evolved true wings and some have perfected an extraordinary sonar system by which they can find their way in total darkness (see page 230).

The bats are usually rather small although certain species can have a wingspan of over 1 metre ($3\frac{1}{4}$ feet). Their *patagium* is stretched between the very modified hands and the rear legs, and is often developed as far as the tip of the tail. Unlike the wings of the pterosaurs and the birds which are supported by a single digit, the bats use all four digits of the hand almost equally and the

This bat is able to steer its course by night, avoiding obstacles and locating its prey by means of its 'radar' system – ultrasonic waves which it sends out in flight and uses its huge ears to pick up the echoes.

metacarpals and phalanges are very much reduced in size. The thumb, separate from the *patagium* and ending in a strong claw, can move in all directions. It is interesting to note that once again the trunk is made rigid by the fusing together of the vertebrae and the breast bone bears a central crest which reminds us of the keel of the birds, though rather weaker. The foot is an astonishing device in which the toe claws are able to hook on to a support while the animal rests upside down, without muscular contraction.

As in all flying animals, the skeleton is as light as possible and easily crushed, therefore conditions in the rocks must be exceptionally favourable for the delicate bones of a bat to be fossilized. Nevertheless, very ancient genera have been discovered, the oldest of which is *Icaronycteris* (see page 231) which goes back some 50 million years yet gives the impression of a perfectly developed bat so similar to modern species that it tells us practically nothing about the group's origins. It has a few primitive features such as the proportions of its limb bones, but this bat must have lived a life very similar to that of many of its modern relatives which hunt flying insects at dusk.

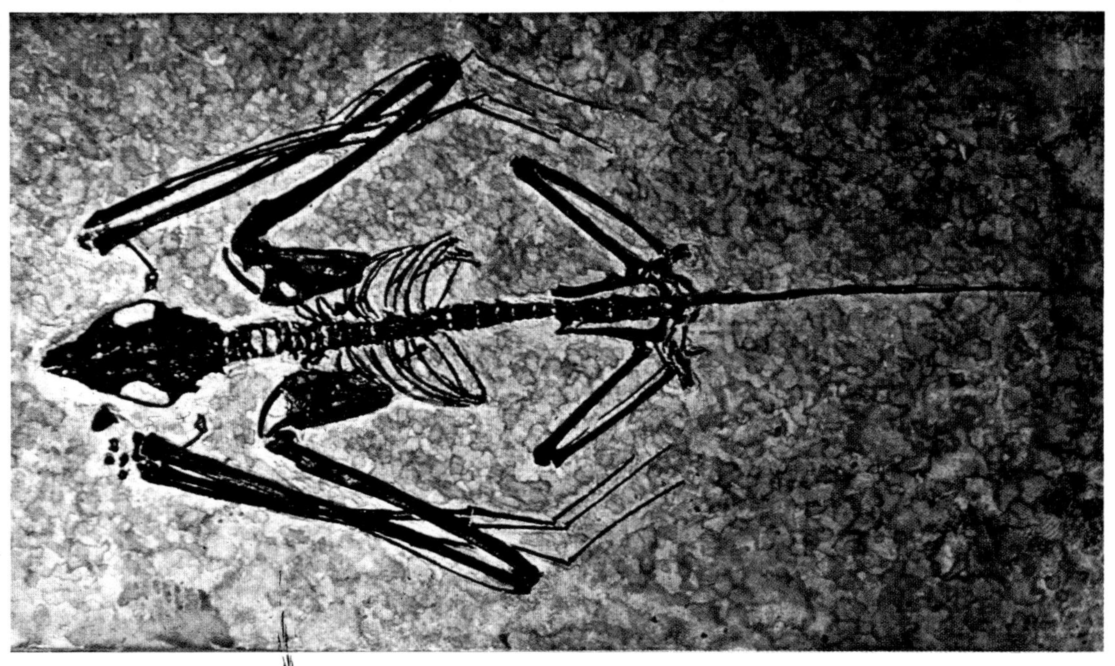

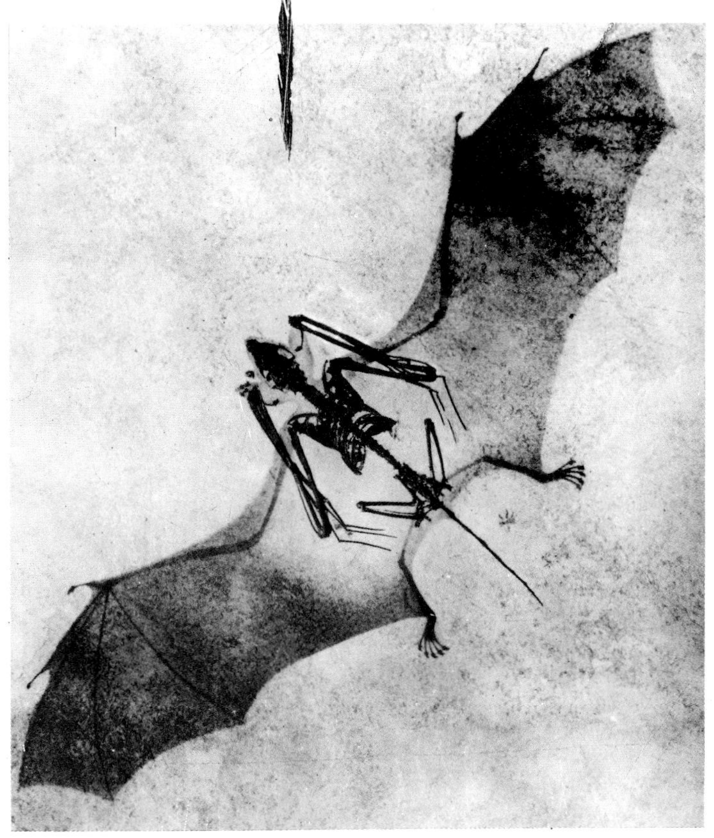

Icaronycteris, the oldest of all known chiropterans, lived in North America about 50 million years ago. It was quite large with a wingspan of 30 centimetres (1 foot) and more. The magnificent fossil above was discovered in lakeland sediments, where it had possibly drowned. The animal had then been buried rapidly, its wings folded, in very calm conditions. Apart from outstanding specimens such as this, most species of fossil chiropterans come from cave deposits where their remains often abound. On the left the skeleton of *Icaronycteris* is shown superimposed over the silhouette of an outstretched pair of bat wings.

Chapter Eight
The Mammals

1 | 2
3 | 4

The Earth 220 million years ago (1), 100 million years ago (2), 40 million years ago (3) and today (4).

The mammals

Reptilian origins

The term 'reptile' covers a number of groups which, although they had a common ancestor, are very different today. Some of the reptiles were the direct ancestors of the mammals and this is why they are included in palaeontological classification along with the mammals under the name of 'theropsid'. They had a particular arrangement of windows in the skull (see page 157): the synapsid type, with a single lower temporal opening. This arrangement is found even among the very first theropsids – the pelycosaurs – which have very few mammal-like features. These forms are very ancient, as they first appeared in the Upper Carboniferous Period and lived throughout most of the Permian Period. It is interesting to see that the theropsids were only really successful when the other reptile groups were uncommon. They were dominant at the end of the Palaeozoic Era but were unimportant during the Mesozoic Era while dinosaurs ruled the World. However, the pendulum swung back, when at the end of the Mesozoic Era many reptiles died out and the theropsids, as mammals, resumed their position of supremacy which of course they still enjoy.

The most primitive of all the pelycosaurs were the ophiacodonts. Resembling large modern-day iguanas, they lived at the water's edge and ate mainly fishes. Their jaws were armed with a large number of sharp, identical teeth. They also possessed a couple of larger and stronger teeth, and these were probably the forerunners of the canine teeth of mammals.

Alongside the ophiacodonts lived the sphenacodonts and the edaphosaurs. These creatures were more massive and had even more powerful jaws. *Dimetrodon* was a carnivorous sphenacodont (see above right), with a high and narrow skull. Its vertebrae had extremely long vertical struts which supported a membrane with a good blood supply, a device probably used to regulate the body's temperature. Most reptiles, unlike birds and mammals, have an internal temperature that varies according to their

surroundings, and this causes a drop in their biological activity when the climate cools. However, in the early morning, by holding its 'sail' sideways on to the sun's rays, *Dimetrodon* was able to heat up its 250 kilogram (550 pound) body weight quickly and attack its still-drowsy prey more easily.

The edaphosaurs show similar characteristics to the sphenacodonts, but their teeth

indicate that they had a very different diet. Their palate was covered by a large number of tiny teeth which formed a sort of 'grater'. They were probably herbivores, although it has also been suggested that they had a diet of molluscs and crustaceans.

Mammal-like reptiles

The therapsids or 'mammal-like reptiles'

Dimetrodon is one of the most common fossil vertebrates in the Lower Permian Period of North America. It weighed up to 250 kilograms (550 pounds) and was 3.5 metres ($11\frac{1}{2}$ feet) long, but was not the largest of the pelycosaurs, since *Cotylorhyncus*, an edaphosaur, exceeded 300 kilograms (660 pounds). The edaphosaurs and the sphenacodonts did not all have this dorsal 'sail' which constituted an initial attempt at body temperature regulation by the theropsids.

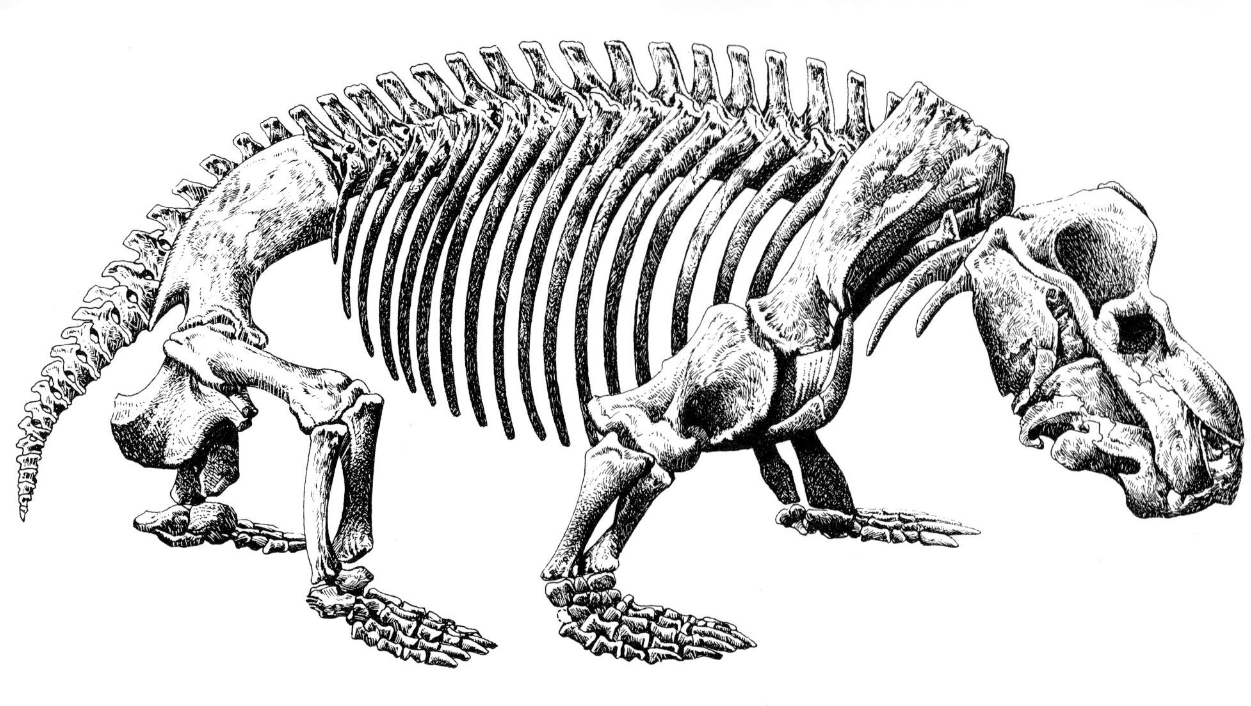

took over where the pelycosaurs left off. Their history is well known, thanks to the many Permian-Triassic Period formations found in South Africa, Russia, China, Laos, the West Indies and Antarctica. By an extraordinary phenomenon of evolutionary convergence, the various families that make up this vast unit all tended more or less to acquire mammal-like structures, independently of each other and each at its own pace. It is possible that some of them even crossed the line that is so difficult to draw between reptilian and mammalian stages of organization.

The change from reptile to mammal was characterized by a series of changes in bone structure, and one of the most significant was the reduction in the number of bones that made up the mandible until there was only one bone: the dentary (see page 237). The main bones which accompany it in the complex reptile mandible did not disappear altogether, however. Instead they were reduced to structures called ossicles, and were incorporated in the middle ear where they formed an acoustic link between the ear-drum and the inner ear. This astonishing transformation can be observed during the development of a mammal embryo. The teeth, which are replaced throughout the life

Above: *Kannemeyeria* was about 2 metres ($6\frac{1}{2}$ feet) long and had a bulky body. It dug up roots and rhyzomes with its powerful claws and cut them up using its horned beak; its mandible was able to slide backwards and forwards to crush and grind its vegetable diet. This was a relatively advanced genus of the dicynodont therapsids, since it possessed the outlines of a secondary palate.

Below: *Cynognathus* is found in the same South African deposits as *Kannemeyeria*. It was an active meat eater with teeth very similar to the first mammals.

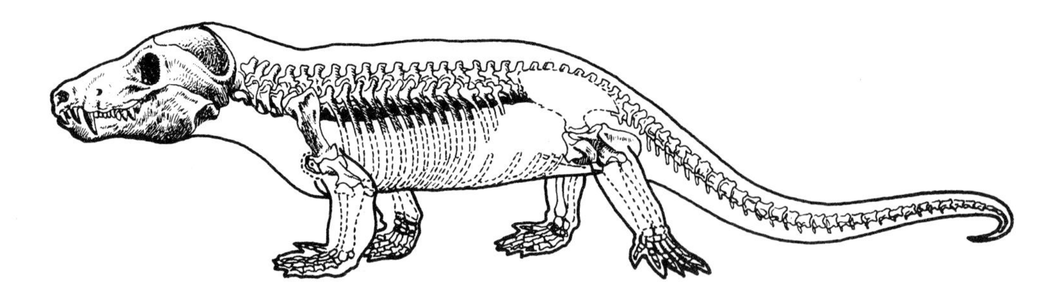

of a reptile, were replaced by fewer and fewer growths until eventually only one replacement took place, and that did not affect the molars. The molars became increasingly differentiated, as did the incisors, canines and premolars, assuming particular shapes and functions.

The skull underwent changes, and was hinged by two condyles instead of just one on the vertebral column. A secondary palate, membranous at first, then bony, gradually separated the respiratory passages from the mouth cavity. The food that fills the mouth of a mammal does not stop it breathing, and so it is able to chew its food without depriving its body of oxygen. The limbs continued to 'straighten out' and the body was no longer dragged along the ground. Growth was more specifically programmed; in mammals growth stops in the adult stage, while that of the reptiles slowly continues. As with the birds and the dinosaurs, an increasingly efficient thermal control maintained the internal temperature of the animals which were able partly to free themselves from outside variations in the climate by a layer of fur. The young were suckled by their mother, and were protected from enemies by their parents, although in the most primitive mammals they continued their embryonic growth within an egg, much like the modern monotremes (platypuses and spiny anteaters or echidnas) which are not viviparous.

The therapsids underwent numerous adaptations. Some were relatively unremarkable vegetarians whose remains abound in certain Triassic Period deposits. They were frequently large beasts like *Kannemeyeria*, shown on page 236, whose skeleton was very heavy. *Lystrosaurus* of the Lower Triassic Period led a life comparable with that of modern hippopotamuses; its teeth had disappeared and had been replaced with a horned beak, except for very powerful canines which probably functioned as defensive weapons. These herbivores were indeed menaced by ferocious predators: lighter therapsids with more mammalian features. This herbivorous way of life was particularly demonstrated by the therocephalans and the cynodonts (see page 236), and it is likely that the cynodonts are at the base of the origin of the class of mammals. The cynodonts in turn diversified, giving rise to new branches of herbivores, in particular the tritylodonts, the only group of mammal-

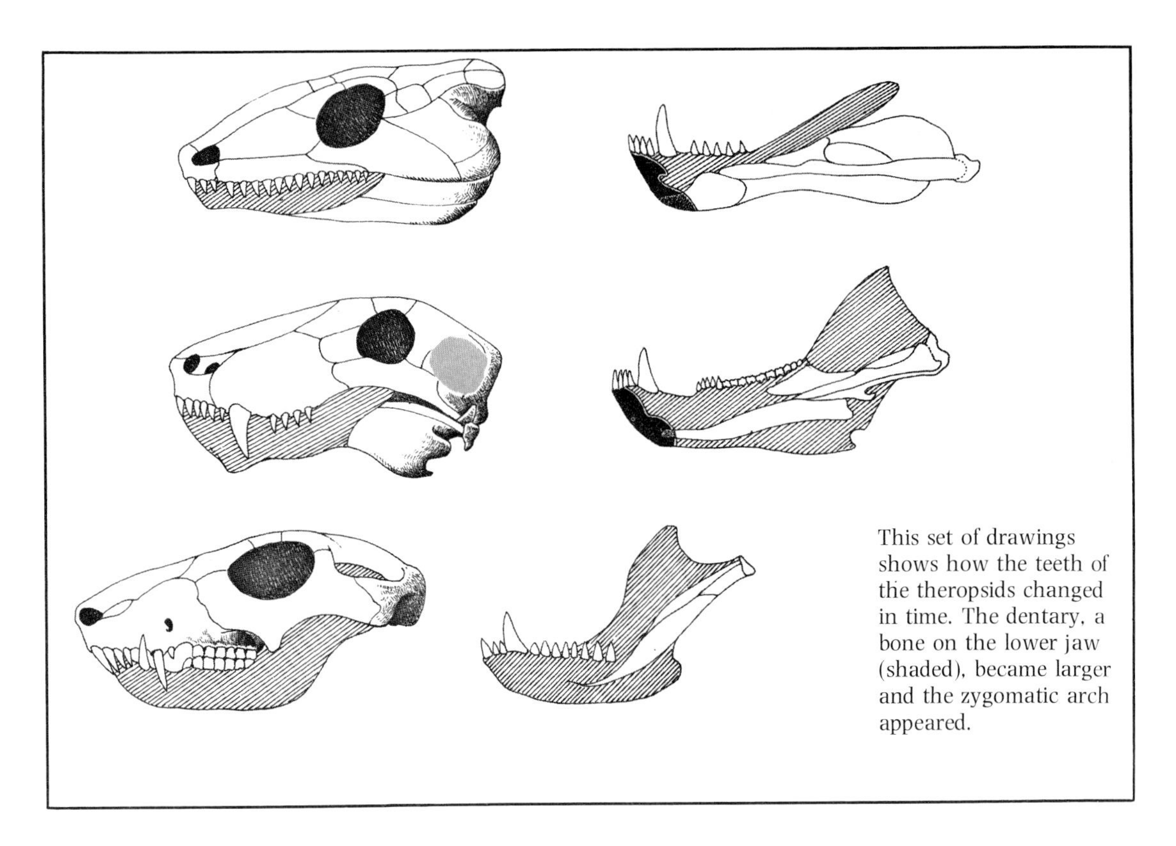

This set of drawings shows how the teeth of the theropsids changed in time. The dentary, a bone on the lower jaw (shaded), became larger and the zygomatic arch appeared.

like reptiles to make it successfully across the Triassic–Jurassic Period boundary.

Nevertheless, the carnivores once again proved to be the more 'progressive' creatures. *Thrinaxodon* and *Cynognathus* already possessed most of the mammalian features despite the fact that their mandibles still consisted of several bones and their teeth were frequently renewed. The presence of small openings in the bones of the muzzle of some cynodonts implies in all likelihood the existence of whiskers, like those sensitive hairs around the mouth of a cat. Still, they were probably covered in hairs, while a secondary palate allowed the warm-blooded meat-eaters to breathe without interruption, an essential requirement of their high metabolic rate.

Below: *Bienotherium* was a tritylodont from the Triassic Period of Yunnan in China. The gap separating the canines from the premolars and the general look of the skull is reminiscent of a mammal, but it is not considered as such, as its mandible is still reptilian.

The order of insectivores is an insignificant group including the most primitive of the placentals and some of their descendants which adapted to this type of diet. *Zalambdalestes* (left) and *Deltatheridium* (above) of the Upper Cretaceous Period of Mongolia are shown here (× 1 approx). The latter was probably not even a placental, but rather a marsupial or the representative of another group of equivalent rank.

The first mammals

During the Triassic Period, the carnivorous mammal-like reptiles that were the forerunners of the mammals were a lot smaller than the sphenacodonts of the end of the Palaeozoic Era. While the theropsids receded to make way for other reptilian branches then flourishing, their representatives diminished in size, and the first mammals turned out to be minute creatures that lived in the shadow of the giant reptiles which we considered in Chapter 6. Their remains are also extremely rare, and it may be necessary for palaeontologists to sift and wash tonnes of sediment just for a few teeth. Much has been written with regard to this rarity of the Mesozoic Era mammals. Does it reflect an equal degree of rarity among all the animals living at the time? This is no doubt part of the story, but it is also not difficult to imagine that the mammals of those ancient times inhabited regions, or at least environments, that were different from those where the dinosaurs roamed. Deposits where dinosaur fossils have been found, and which are the ones most frequently investigated, are not necessarily favourable to the preservation of fossil mammals. It may be necessary to continue our research elsewhere – this at least is suggested by the work of a team of Berlin palaeontologists who recently resumed the working of a Jurassic Period lignite mine at Guimarota, Portugal, and who have uncovered a splendid series of mammals, including some complete individuals.

The very first true mammals appeared on Earth some 190 million years ago. Those first forms, however, gave little hint of the huge diversity of animals which was to be a feature of this class, for they were kinds of shrews or small mice whose teeth indicate an insect diet. As we said, these distant mammalian ancestors were the contemporaries of the great saurians but also lived at the same time as small reptiles whose feeding habits were identical. In the competition which ensued, the mammals had the distinct advantage of being able to regulate their body heat. When night fell, therefore, they were able to continue hunting prey which their cold-blooded rivals were unable to capture, forced to seek shelter by the cold.

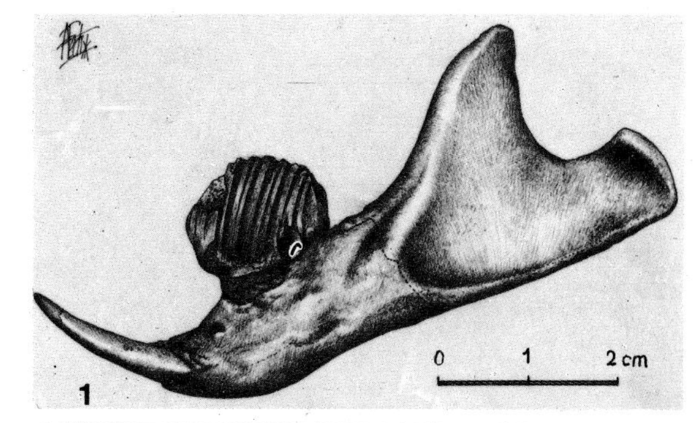

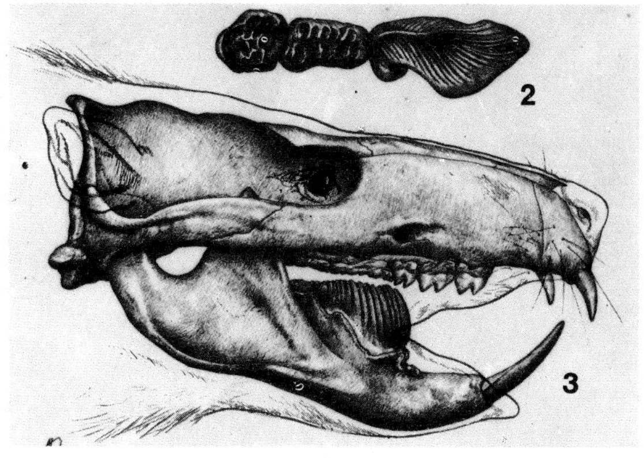

1 Jawbone of *Neoplagiaulax* ($\times$ 3) a multituberculate from the European Palaeocene Period.
2 and 3 Lower jugal dentition and reconstruction of the cranium of *Ptilodus* ($\times \frac{4}{3}$) from the North American Palaeocene Period, another multituberculate.

Mesozoic expansion

Two distinct mammalian groups were present from the Triassic–Jurassic Period boundary onwards. Is their common origin to be found among the reptiles, of which at least two species independently attained the mammal stage, or did they descend from an unknown Triassic Period mammal? The question is still without an answer. The first of these groups so far only contains a single genus, *Kuehneotherium*, whose molars have three points or cusps arranged in a triangle, and it was from this group that the marsupials and placentals (in other words practically all modern mammals) arose much later on. The second group is rich in a number of species, and covers all the animals whose molars bear cusps in rows. This group formed the basis for the prototherians, which are only represented today by the monotremes but which were very common in the Mesozoic Era. The Multituberculata in particular, which appeared at the end of the Jurassic Period and adopted great significance during the Cretaceous Period and the beginning of the Cainozoic Era, were important prototherians (see above) and owe their name to their molars which contained numerous outgrowths. Their powerful lower premolar nearly always grew as a sort of notched blade, and their incisors were highly developed. These vegetarians abounded in a world where rodents were non-existent, and it is probable that the appearance of the rodents and of other herbivorous placentals put an end to the success of the Multituberculata.

Ectoconus majusculus is one of the better-known placentals of the American Palaeocene Period. We have been able to reconstruct its skeleton completely, and its physical proportions are quite representative of the other condylarths, with a length of 1 metre ($3\frac{1}{4}$ feet), excluding the tail.

The prototherians also included the triconodonts, to which *Amphilestes* from the Jurassic Period of Britain belonged. This was the very first Mesozoic Era mammal to be discovered – in 1764 – although its true nature was not realized at the time. The largest triconodonts attained the size of a cat, and were all carnivorous, attacking small reptiles and mammals, and perhaps even dinosaur eggs.

Modern prototherians are relics which are impossible to link directly with any of these Mesozoic Era groups. The spiny anteater, which has completely lost its teeth, and the platypus, which has much reduced teeth, are animals that have retained their primitive way of reproducing. This helps us to understand the ways of life of the very early mammals. Female prototherians lay eggs which are of the consistency of leather. But the eggs are not left to hatch in the ground or in a nest – the female incubates them for some two weeks at the bottom of a closed burrow which she has dug for herself. In the spiny anteater, the eggs are attached to the hairs of a pouch developed by the mother specially for this purpose. After hatching, the young – still barely formed – suck on the milk exuded through their mother's fur. Finally, among all these oddities, we must mention the spur which the male platypuses carry on their anklebone, making them the only venomous mammals.

In the group containing only *Kuehneotherium*, as with the triconodonts, the teeth were very numerous, but diminished in significance in the course of evolutionary time. In the sands of Trinity, Texas, we find a placental (*Pappotherium*) and a marsupial (*Holoclemensia*) from the Middle Cretaceous Period, while another placental (*Endotherium*) is known in Manchuria, more or less from the same time. Contrary to one widely held belief, placentals did not descend from the marsupials – which some regard as more primitive – rather these are two 'sister groups', in other words two groups born simultaneously out of a common ancestor. We shall see that the shift in the continental land masses (see pages 232–233) enabled each of them to develop and flourish independently in their own domains. Today the marsupials are chiefly represented in Australia and New Guinea, lending such a typical stamp to the fauna of these regions inhabited by kangaroos, koalas, flying phalangers and Tasmanian devils. But, a long time ago, the Americas were also the home of numerous marsupial species, of which the opossum is one of the sole survivors.

Marsupials and placentals both possess a particular type of molar which appeared at the beginning of the Cretaceous Period in their common ancestors, the pantotheres. These were distinguished by the number of teeth as well as by several other features, especially by the structure of their internal anatomy and manner of reproduction. Contrary to what is suggested by a rather ill-chosen term, both groups produce a placenta which feeds the foetus (the developing young). In the marsupials gestation is very short and the development of the young embryo is continued 'outdoors' in its mother's ventral pouch. In the placentals gestation is longer and takes place inside the mother's womb, resulting finally in the birth of well-developed offspring.

Alongside the condylarths of the Palaeocene Period lived a group of plant eaters, some of which grew very large indeed. They were called the pantodonts and were commonly found in North America from the Palaeocene Period to the Oligocene Period, when the last of them disappeared from Asia. *Coryphodon* (below) was one of the biggest, about 2.5 metres (8 feet) long. The canines of this bulky animal with powerful limbs were enlarged into defensive weapons, and many pantodonts had teeth of this type.

Condylarths

All the main groups of placental mammals had evolved by the end of the Cretaceous Period. The insectivores in particular are known by several genera from Mongolia (see page 238) and the U.S.A. as are a primate, *Purgatorius* and a primitive ungulate, *Protungulatum*, both of which were discovered in Montana. From the Palaeocene Period, now

that the dinosaurs had finally disappeared, there was a sudden rapid increase in the number of genera. Apart from the Multituberculata which are now at their peak, fairly large herbivores and carnivores suddenly began to occupy the ecological niches vacated by the reptiles. Two placental groups formed the bulk of this new increase in mammal stocks: the 'condylarths' and the 'creodonts'. The condylarths were essentially herbivorous or omnivorous, and the creodonts were almost totally carnivorous.

The condylarths, with their teeth made for crushing rather than biting, in fact covered all sorts of primitive forms that are at the origin of the modern ungulates, and are what we call a 'stock group'. They certainly derive from carnivorous or insectivorous species, and as our illustration on page 240 shows, their general appearance was reminiscent of a cat. They were usually low-slung with a longish body and long tail, and their name reflects a particular feature of their limbs. In a cow or sheep, for example, the astragal (the bone which joins the foot to the tibia) ends at the bottom in a sort of bony pulley. (The old-fashioned game of 'knucklebones' used the astragals of a sheep.) Because of this pulley the foot can only move vertically in relation to the leg. In the condylarths, on the other hand, the joint consisted of a spherical section that moved in a socket and which gave it the flexibility found in a modern cat.

Most condylarths were not much bigger than a dog, although larger types often appeared in the Eocene Period. Two families, the arctocyonids and the mesonychids, moved towards a more carnivorous diet. The mesonychids in particular were thought for a long time to be creodonts – a fact that clearly demonstrates the extent to which the precise evolution of these primitive groups is still unknown.

Andrewsarchus, a mesonychid of the Upper Eocene Period of Mongolia, was gigantic: its skull alone measured almost 1 metre ($3\frac{1}{4}$ feet) in length, but its exact life-style remains a mystery. It is unlikely that its pursued its prey; equipped with rounded teeth, it may have been a carrion-eater or possibly lived on shellfish, or may simply have lived much in the style of a huge omnivorous bear.

More animals that lived alongside the condylarths: the dinoceratans lived in North America and Asia. They looked rather like the pantodonts and also had very elongated upper canine teeth which, extending downwards past the lower jaw, were protected by a bony outgrowth on which they rested. Some unrelated animals shared a similar feature (this is known as convergence). For instance an identical device is found in one pantodont, some felines and a marsupial. The huge *Uintatherium* shown here measured some 2 metres ($6\frac{1}{2}$ feet) at the shoulder. Like other North American genera its skull was covered in strange bony lumps. The dinoceratans became extinct at the end of the Eocene Period.

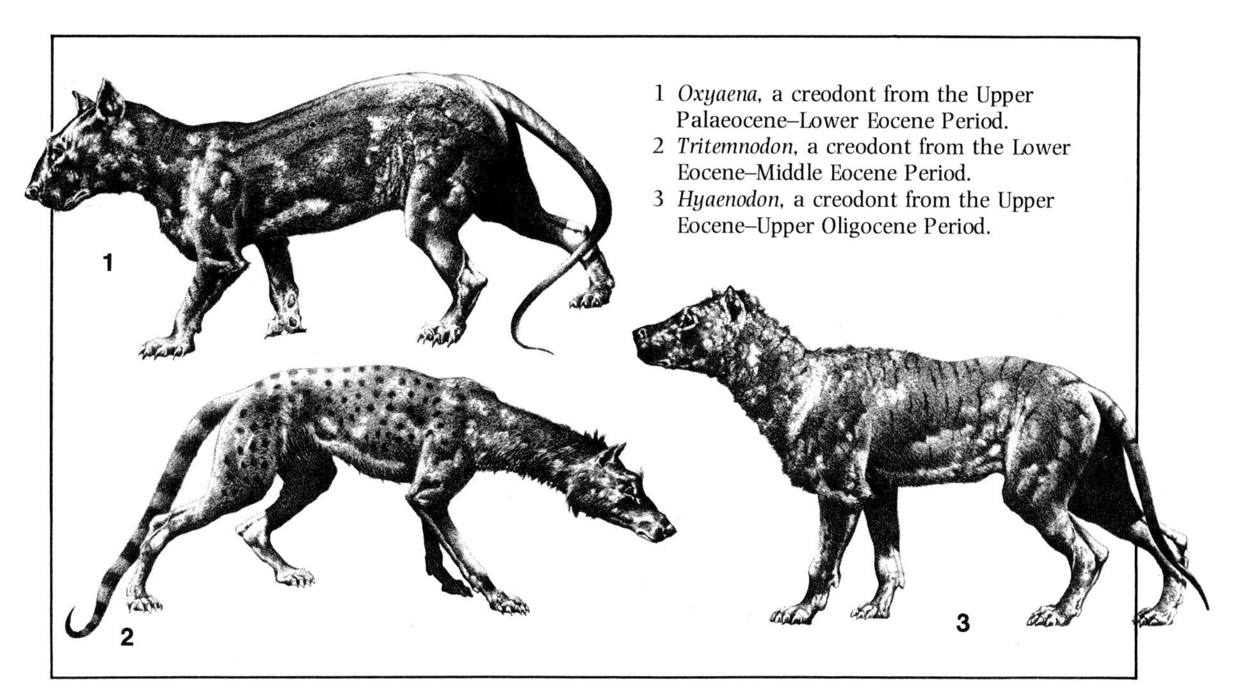

1 *Oxyaena*, a creodont from the Upper Palaeocene–Lower Eocene Period.
2 *Tritemnodon*, a creodont from the Lower Eocene–Middle Eocene Period.
3 *Hyaenodon*, a creodont from the Upper Eocene–Upper Oligocene Period.

Creodonts

The creodonts are not very different from the condylarths, with the same low profile, very long tail, and five digits on each foot. Their skull, although very large in relation to the size of the body, contained only a primitive and rather small brain.

These animals are not the ancestors of modern carnivores, which belong to quite another order, that of the fissipeds. Their common origin must be sought in the Cretaceous Period. Apart from *Protungulatum*, the Upper Cretaceous Period of the U.S.A. has revealed two fossils, *Cimolestes* and *Procerberus*, which show how carnivorous forms split off from the insectivores. But creodonts and fissipeds were already clearly separate during and after the Palaeocene Period, and it is easy to tell them apart by their teeth (see page 244).

Certain creodonts with sharp teeth are a little like small members of the cat family or foxes, while others grew to considerable sizes and fulfilled the roles handed down in modern nature to the wolves, lions or hyaenas. Their primary expansion came during the Palaeocene and Eocene Periods. In North America, one group developed teeth and jaws similar to those of the sabretooth felines (see page 246); the creodonts went into decline during the Oligocene Period when the families of modern carnivores were rapidly expanding, but they were still represented by genera like *Hyaenodon*, which was very widespread. The giant of the group was discovered in the Miocene of Libya: this was *Megistotherium*, with a weight estimated at 1 tonne! This period also saw the last of the condylarths in South America, while the creodonts survived to the early Pliocene Period in the Siwaliks region south of the Himalyas in Tibet.

Fissipeds

Modern carnivores are divided into two main groups. One group contains the Canidae (wolf, dog, fox, etc.), the Ursidae (bears), the Mustelidae (weasel, martens, etc.) and the Procyonidae (raccoons, coatis, etc.). The other group contains the Viverridae (mongoose, civet), the Felidae (lion, panther, cat, etc.) and the Hyaenidae (hyaenas). All are distinguished from the creodonts by a number of features, not the least of which is a larger, more sophisticated brain.

We have said that the first fissipeds appeared in the Palaeocene Period. They belong to the group of miacids which died out before the Oligocene Period; these were small tree-dwelling mammals, and their forest environment was little suited to fossil preservation, so they remain largely unknown. What we do know is that they possessed a long tail and short, supple legs. At the

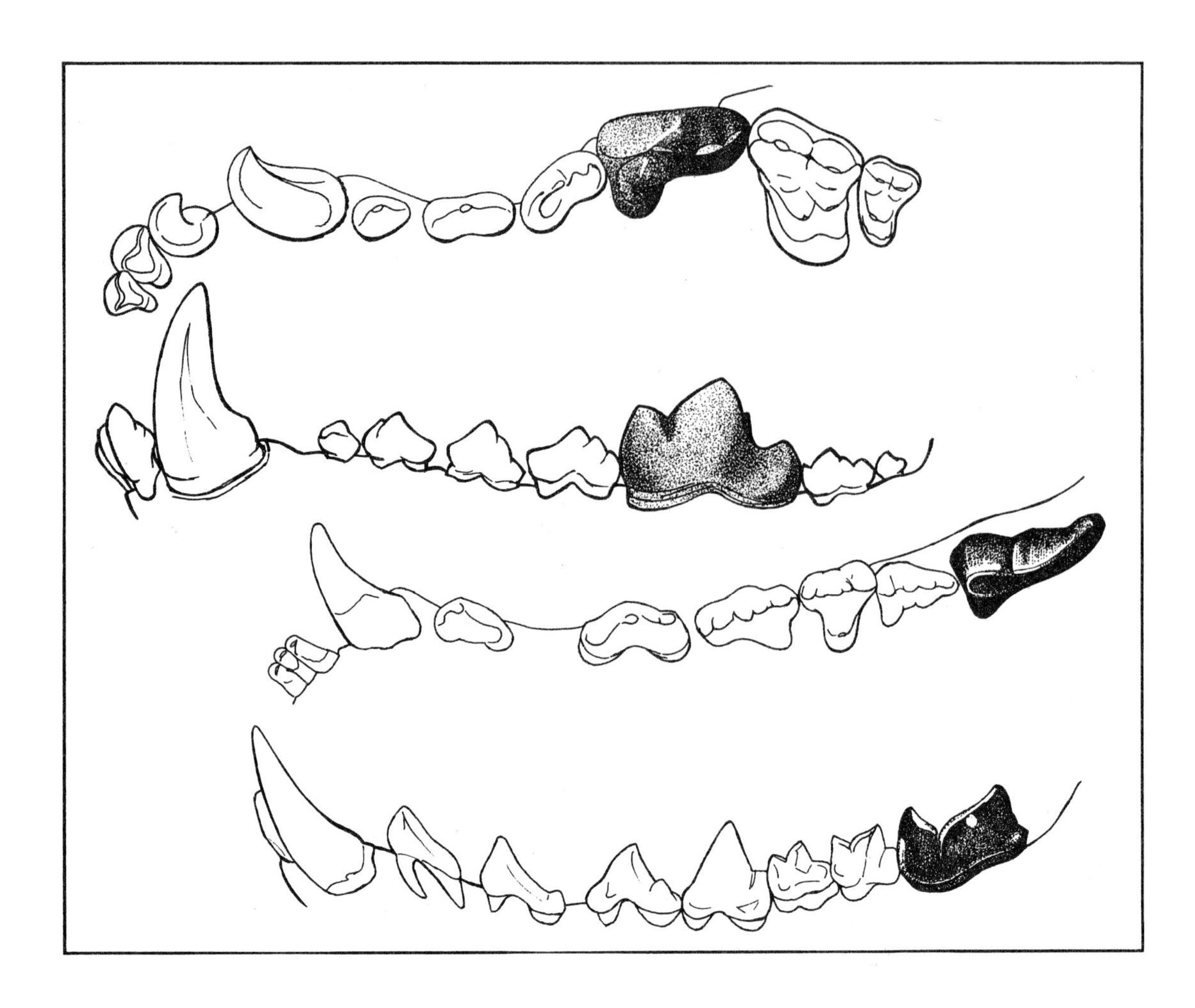

Top: the teeth from the upper and lower jaws of a fissiped, the wolf.
Above: The same teeth of a creodont, *Hyaenodon*. The adaptation to a carnivorous diet is shown here by the sharp cheek teeth. The fissipeds are characterized by the pair of carnassial teeth: the fourth upper premolar and first lower molar which were used for slicing meat. In the creodonts these carnassials were in a different part of the jaw and sometimes several teeth performed the same function at once.

beginning of the Cainozoic Era they were overshadowed by the creodonts which were then at the height of their development, and by the great carnivorous birds which maintained their status as super-predators after the extinction of the dinosaurs. It is not until the end of the Eocene Period, some 40 million years ago, that it becomes possible to distinguish the two groups of fissipeds which were gradually to force out their competitors.

In the Oligocene Period the Mustelidae were represented by species that resembled modern weasels and ferrets. The lake deposits of the end of this period have revealed a kind of 'otter-seal' called *Potamotherium* that was already quite adapted to aquatic life. Wild dogs hunted the large herbivores in packs; bears had not as yet appeared, although *Amphicyon* (see page 245), was physically similar to the bears. The hyaenas, too, were still absent, but their ecological niche was occupied at this time by other carnivores, in particular the creodonts which were thus able to maintain their existence amongst the newly evolving fissipeds. The picture is rounded off by the Viverridae and the Felidae. In the latter group, the big cats which killed their victims in the manner of modern felines were accompanied by a series of creatures which have no equivalent in our own times. Their upper canines were greatly enlarged and, because of other changes in their anatomy they were used to stab through the thick skins of their prey rather than bite them. This odd specialization was to reappear several times in the course of the history of the Felidae (see page 246).

In the Miocene Period the procyonids made their appearance in North America. This represented an attempt by the fissipeds to feed in an omnivorous or even an herbivorous way (in the case of the pandas) but was never very extensive. The Mustelidae continued to thrive: *Megalictis* of the U.S.A. grew to the size of a bear, a record unequalled in this family. The various species of the Canidae flourished, and while some moved towards modern genera such as *Canis*, known from the end of the Miocene Period, others 'imitated' certain groups of fissipeds. Thus it is that we see curious strains of 'feline dogs' and 'hyaena-dogs'. The first bear, *Ursavus*, was the size of a fox terrier but the group increased individually in size during the Miocene Period. *Hemicyon*, whose descendants survived up to the Pleistocene Period, was also a member of the Ursidae. All retained their sharp teeth and fed less omnivorously than the true bears.

Among the 'cats', the Oligocene Felidae were gradually replaced by other forms. *Felis* in particular was present by the end of the Miocene Period, while its contemporary, *Ictitherium*, definitely belongs to the Hyaenidae that had by now separated off from the Viverridae.

The Pliocene Period therefore was the time during which the modern carnivores became established. Among the Ursidae, the genus *Ursus* emerged. Its molars became covered in tubercules and were used to grind its food. This animal showed clearly the omnivorous tendencies of this family. In addition to the fissipeds that have survived up to the present day, there were also some varieties which are now extinct. These included the hemicyonins, 'hyaena-dogs' and the last of the amphicyonids. The hyaenas did not align themselves with other eaters of carrion but, like the Canidae, diversified and

Jaw of *Amphicyon* (Oligocene Period, from the phosphorite deposits of Quercy, France, approx. × 1). These often very large animals probably occupied an ecological niche half-way between that of the modern bears and great cats.

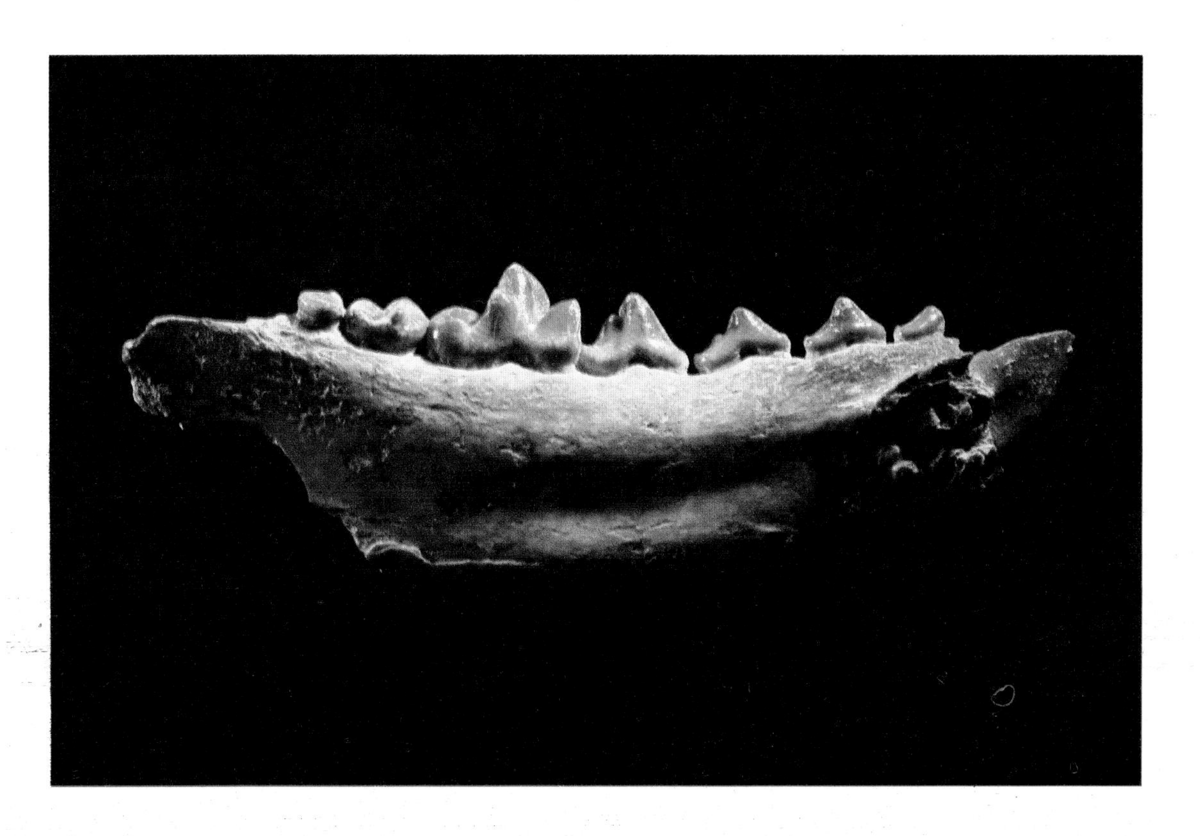

Machairodus from the Upper Miocene–Lower Pliocene Period, had very powerful forelegs and a jaw that could gape open far wider than in a modern cat. This fact, coupled with the extreme development of certain neck muscles, indicates that the animal had quite an original technique of putting an end to its prey: *Machairodus* held down its victims with its forelimbs while dealing them fearful blows with its long teeth.

also produced types which were physically like those of other families. Thus *Euryboas* deserved the nickname of hyaena-leopard because of its short snout and sharp cutting teeth. This category also includes the sole genus of hyaena to have penetrated North America: *Chasmaporthetes.*

The Felidae were still producing sabre-tooth forms: *Machairodus* (see above) of the end of the Miocene Period was succeeded by other genera such as *Megantbereon* and *Homotherium* in Eurasia. *Smilodon* of the American Quaternary Period was the last to die out. It is very well known, for complete specimens have been extracted from the outstanding deposits of Rancho la Brea in California where they were preserved in asphalt. It was about 11,000 years ago in a swamp near to the present site of Los Angeles, where these carnivores attacked bison, elephant and other herbivores that had become bogged down in the mud, and had themselves been sucked under. Over 2000 specimens of *Smilodon* have been found. It would seem that other Felidae which inhabited the same region, such as lions and pumas, were wiser, for their remains are much rarer in this swamp.

This procession of carnivorous animals applies chiefly to Eurasia and North America; South America remained isolated for the major part of the Tertiary Period (see pages 232–233). As we shall see, the marsupials living there had partially filled the niches occupied in our own geographic area by the creodonts and the fissipeds, and the carnivorous flightless birds enjoyed prolonged success. In the Upper Miocene Period only a few procyonids arrived from North America, clinging on to natural rafts of vegetation, and it was not until the end of the Pliocene Period and especially the Pleis-

tocene Period that we find a true invasion of the Canidae, Mustelidae, Ursidae, Procyonidae and Felidae. In Africa, the persistence of the creodonts is similarly explained by the continent's isolation during a large part of the Tertiary Period, and by the late arrival of fissipeds. In the Lower Miocene Period, the collision between the African and European continents allowed first the Canidae, Felidae and Viverridae to go south, followed by the Hyaenidae and the Mustelidae, while the Ursidae were the last to arrive during the Pliocene–Quaternary Periods. As for Australia, its destiny was for a long time linked to that of South America, and the sole fissiped to live there before the arrival of the Europeans was the dog, which had been introduced by the aborigines, and which had returned to the wild state.

Return to the oceans

Like many reptiles before them, certain mammals have returned to a life in water, and this is a development pursued by several families with varying degrees of success.

The otter, for instance, is an animal which although adapted to an amphibious way of life has altered little from the basic anatomy of the mustelids. In the fissipeds, however, a life in water was pursued by animals whose whole body structure was completely remodelled. The seals, sealions and walruses were once united under the name of pinnipeds, but advances in palaeontology have shown that these forms, which seem closely related, actually originated from two totally distinct families of fissiped. While the seals descended from the Mustelidae, the sealions and walruses are related to the Ursidae. These marine mammals are known from the Miocene Period; they were very specialized and greatly resembled their modern descendants. As with the marine reptiles (see Chapter 6) the bones of the limbs became short and very strong, while the hands and feet were transformed into swimming flippers. Their teeth were also greatly modified: the carnassials (the meat-shearing cheek teeth) can no longer be distinguished, since all the post-canine teeth have become simplified and are all identical.

The cetaceans, or whales, became totally marine and so must even give birth to their young at sea. We do not have the benefit of a series of fossils showing the changes from land to marine life, and therefore, the palaeontologist must carefully analyze what he does have in order to discern their origins. The cetaceans' ancestry originated among the condylarths, and more precisely near to the mesonychids which as we have seen had adopted a carnivorous diet. The deposits of Fayyum in Egypt which form the basis of our knowledge about the animal life of the early African Tertiary Period have provided us with the oldest cetacean: *Protocetus*. It still possessed a well-developed pelvis, but this subsequently became smaller. Its tooth structure was little modified. The backward-pointing nostrils mounted on the top of a

Basilosaurus of the Upper Eocene Period measured up to 20 metres ($65\frac{1}{2}$ feet) in length. In the middle of the last century an unscrupulous German collector managed to create a 34 metre (111 feet) long skeleton from the remains of several different specimens. He exhibited this 'great sea-serpent' all over Europe before it was exposed as a fraud.

skull with a very long, narrow muzzle were an adaptation to marine life.

From the end of the Eocene Period onwards the earliest cetaceans became very varied and widely distributed. Some grew to gigantic size. Such a creature was *Basilosaurus* of North America (see page 247), whose hind limbs and pelvis became reduced to little more than bony 'relics'. Its forwardmost conical teeth were used to seize its prey which was then shredded by the molars.

Modern cetaceans are split into two groups. The Odontoceti are first known from the Upper Eocene Period and some were thus contemporaries of *Basilosaurus*. Their teeth were arranged in long straight rows. Today they include the sperm-whales and the dolphins. The Mysticeti arose a little later on. The oldest have been excavated from the Oligocene Period of New Zealand, and their teeth have been replaced by a curtain of baleen, the famous 'whalebone' once used in ladies' corsets. With this baleen curtain they filter the sea water and strain off the tiny organisms on which these huge beasts feed.

Two orders directly related to the proboscidians (see pages 261–263) have also gone back to life in the waters. These are the sirenians and the desmostylians. The sirenians are herbivorous mammals with a method of tooth replacement rather similar to that of the elephants. The first of them, *Prorastomas*, is known from the Eocene Period of Jamaica, and the group is well represented in more recent terrain as well. The modern dugong and manatee (sea-cow) were joined only two centuries ago by the North Pacific Steller's sea-cow. However, this creature was exterminated a mere 20 years after its discovery.

The manatees are able to hold themselves erect half out of the water and females have been seen to suckle their young in this position, clasping them in their pectoral flippers. They may be at the root of many a sailor's 'mermaid' legend; but while the sirenians emit loud and plaintive noises, there is little seduction suggested by their fat bodies and large, moustached heads.

The desmostylians never needed to meet Man in order to become extinct. This was a very strange order restricted to the Upper Oligocene and the Miocene Periods of the Pacific coasts in the northern hemisphere.

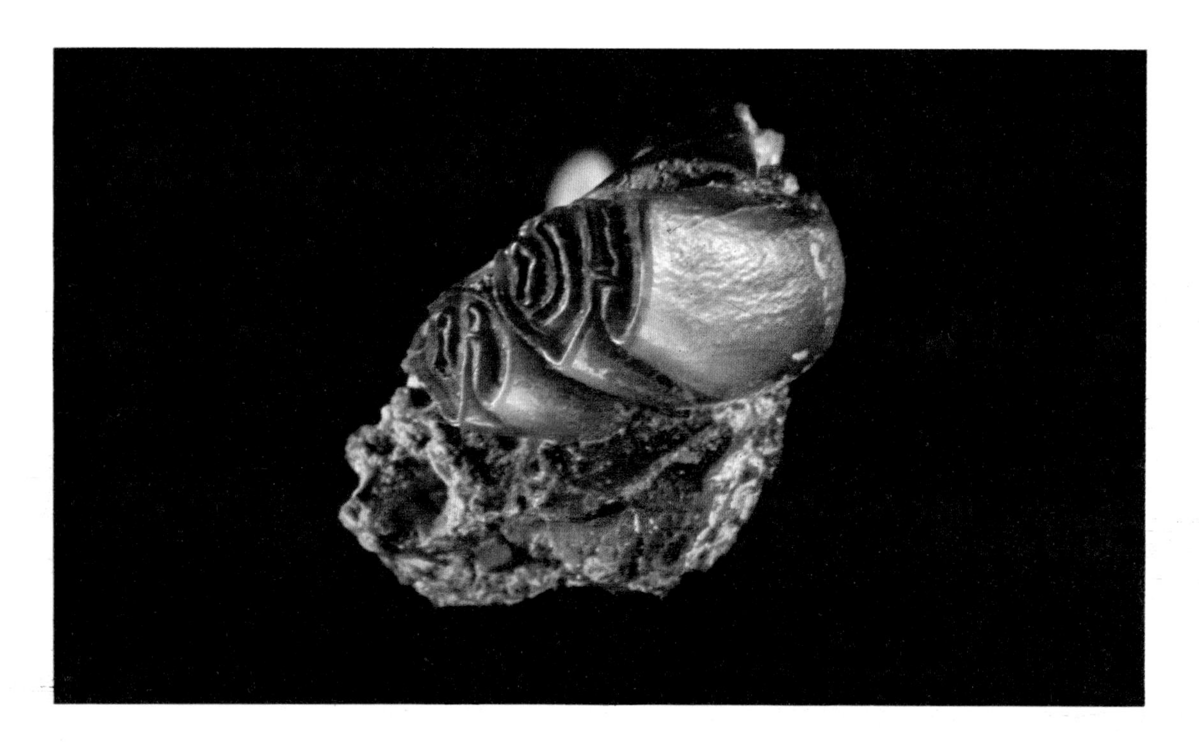

These beaver molars are from the famous Miocene Period formation of Sansan (in the French region of Gers), and show the phenomenon of hypsodonty: while the crown of the teeth was worn down, they kept growing throughout the life of the animal (height of each tooth: approx. 8 millimetres ($\frac{1}{3}$ inch)).

The capybara is the largest of all the living rodents, and certain specimens can weigh over 50 kilograms (110 pounds). It lives in packs in the swampy regions that border South American rivers.

Palaeoparadoxia of the Miocene Period is sometimes compared to a marine hippopotamus. As for *Desmostylus*, it looks more like a sealion with four tusks. These similarities of body shape were not because these animals were related; the aquatic environment exerts great stresses upon the anatomy of the animals which dwell there, and many groups evolve similar body designs in order to overcome the problems. It is for this reason that most birds have adopted similar body shapes in order to fly.

Rodents

Within the mammals, the order of rodents is of particular significance, since it alone represents half the total number of mammal species in the class. Also, it includes a huge number of individuals. Evolving very rapidly, new species of rodents followed each other at an enormous rate in the course of the Cainozoic Era. The history of these creatures therefore is one of intricate detail and complexity, which cannot be explained fully in this book. Suffice it to say that the ease with which we are able to collect their teeth by washing and sifting deposits, together with the small life-span of individuals of the species, makes them valuable indicators of the fossil record for palaeontologists.

The first rodents appeared during the Upper Palaeocene Period in North America. The members of the family to which *Paramys* belonged, and which were found at the beginnings of the Eocene Period, look like types of small squirrel. From then on, the rodents diversified, occupying ecological niches left by the disappearing Multituberculata (see page 239). The method of chewing used by different rodents enables us to divide them into four types: a primitive type, a 'squirrel' type, a 'mouse' type and a 'porcupine' type. *Paramys* belongs to the primitive type that has persisted into modern times, and is close to other curious genera such as *Epigaulus* of the Pliocene Period, a large burrowing animal with strong claws and horns on its head.

The second type includes the squirrels, marmots and beavers. Some of the latter were burrowers, like *Paleocastor* of the

In 1840 the British Palaeontologist Richard Owen named this little animal from the early Eocene Period *Hyracotherium*, believing it to be related to the modern *Hyrax*. Much later on, Marsh identified it as a primitive horse and called it more accurately *Eohippus* ('dawn-horse'). However, palaeontologists always use the earliest name and so the animal is still called *Hyracotherium*.

Miocene Period whose corkscrew-shaped galleries have been found in the ground. We should also mention the theridomorphs which, although now extinct, were once the dominant group in Europe during the Oligocene Period.

The 'mouse' type covers the dormice, the lerots, the hamsters, the jerboas and the rats, and is a group that began to expand from the Eocene Period onwards. Man has both intentionally and unintentionally allowed the rats and mice to multiply. Furthermore they seem to have an uncanny ability to adapt and survive greater than any other mammal.

The 'porcupine' type is interesting in that it appeared simultaneously in Africa and South America during the Oligocene Period. Later on, the porcupines invaded Eurasia. It is likely that these rodents are African in origin and that they moved to South America along with other small creatures, clinging on to natural rafts. They survived their enforced voyage because the South Atlantic was still fairly narrow at the end of the Eocene Period (see pages 232–233). Their adaptive success was considerable in this new world, which they colonized with many different species. One such was the giant among the order of rodents. This was *Phoberomys* of the Pliocene Period, which was the size of a cow. In modern times the record-holder is the capybara which is 1 metre ($3\frac{1}{4}$ feet) long from nose to tail. Other examples include the guineapigs, the agoutis, the chinchillas, the pacas and, of course, the porcupines.

Contrary to popular opinion, the rabbits and their relatives are not rodents, but belong to a different order, the lagomorphs, and this is a distinction which goes back as far as we can trace. Both the rodents and the lagomorphs gnaw with their incisors, which are in a constant state of growth and which are self-sharpening. The rodents have only one pair of upper incisors, but the lagomorphs have two. Both orders lack canine teeth, and like the rodents (see page 248) some lagomorphs possess 'hypsodont' molars. The most ancient lagomorphs date from the Palaeocene Period of Central Asia.

The history of the horse

In European and North American deposits of the early Eocene Period have been found the astragal bones of ungulates which do not correspond to the condylarth type (see page 242). In some, a part of the bone called the

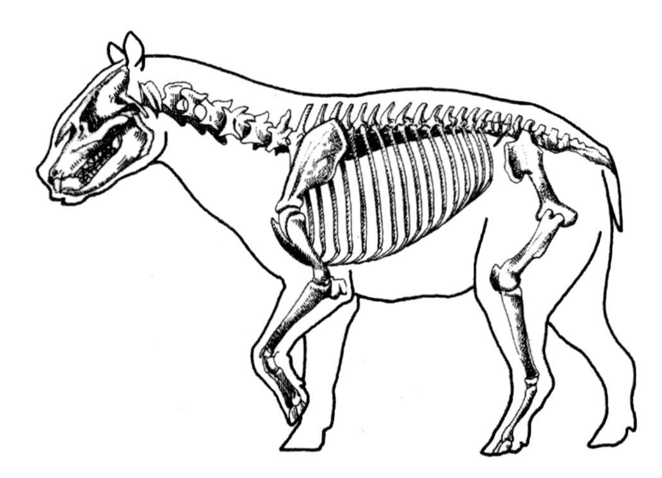

A descendant of forms close to *Hyracotherium*, *Paleotherium magnum* is one of the most famous fossils identified by Cuvier in the Upper Eocene Period around Paris. It looked rather like a tapir, and Cuvier even gave it a trunk, although it probably only had an overhanging upper lip. Unlike the true horse, the paleotheres had well developed canines.

lower condyle had been replaced by a flat surface which does not allow any movement, while in others it was replaced by a deeply hollow 'pulley'. In both cases – in the groups called the perissodactyls and the artiodactyls – the foot had become much more rigid. These animals no longer moved on the soles of their feet and hands, and were equipped with limbs that were longer and stronger than those of the condylarths: their adaptation to running was considerably enhanced as a result. This acquisition is most important when one considers that, when faced by a predator, the medium-sized herbivores' only chance of escape was in flight.

The perissodactyls are represented today by the rhinoceros, the tapir and above all by the horse and its cousins. They were all characterized – unlike the artiodactyls – by the predominance of a central digit which extended the axis of the limb. Perissodactyls originated from condylarth stock, and they can be traced back to the Asian Palaeocene Period. About 50 million years ago Europe and North America were the home of a small mammal some 25 centimetres (10 inches) high at the shoulder which is called *Hyracotherium*. This creature was a distant ancestor of the horse, as amazing as this may seem. It still had four digits on each foreleg and three on each hind leg, and the short teeth covered in tubercules indicated that it was herbivorous. In Europe it became the forerunner of a whole series of new genera which eventually died out. The full history of the horse was to be played out in North America, which finally split off from Europe during the Eocene Period. Up to the end of the Oligocene Period the successors of *Hyracotherium* regularly increased in size: *Mesohippus* for instance was more or less the size of a large dog. They acquired three digits on each limb, and changes to the premolars increased the chewing surface, and hence the efficiency, of the teeth. This increase in chewing area corresponded in ratio with the growth in the creature's body mass.

The Oligocene–Miocene Period boundary saw the separation of two distinct lines of evolution. One line was first represented by *Anchitherium*. *Anchitherium* retained its forest-dwelling habits, preferring humid locations where the vegetation was most tender. Crossing over from Alaska to Siberia, this animal eventually came to inhabit Asia and Europe, where it finally disappeared at the end of the Miocene Period. In North America it was replaced by other genera: *Hypohippus* and *Megahippus* with their large, heavy bodies, and *Archeohippus* which was somewhat smaller in stature. The second line proved to be more adventurous.

Hipparion gracile, an elegant little horse with three digits which lived in Europe at the beginning of the Upper Miocene Period.

The changes in climate during the Miocene Period were chiefly reflected in a general 'drying-out', in the course of which the forests gave way to vast, grassy expanses. This change in the landscape obviously had its effects on the mammals living at that time, a fact that we shall return to in due course. In the meantime the horses had to adapt to this new prairie environment and, more specifically, needed to solve the problem posed by the new, grass-like vegetation. This had much silica contained in the stalks, which wore away the teeth at a considerable rate. The molars of *Parahippus* increased in height and became more like millstones thanks to the appearance of 'cement' between the enamel crests of their teeth. Then, in *Merychippus*, the dental growth was prolonged and hypsodonty (an increase in length in the crown of the teeth) appeared for the first time. Some of these animals still retained their three digits, as in *Hipparion* (see above), which colonized Eurasia at the end of the Miocene Period and even crossed

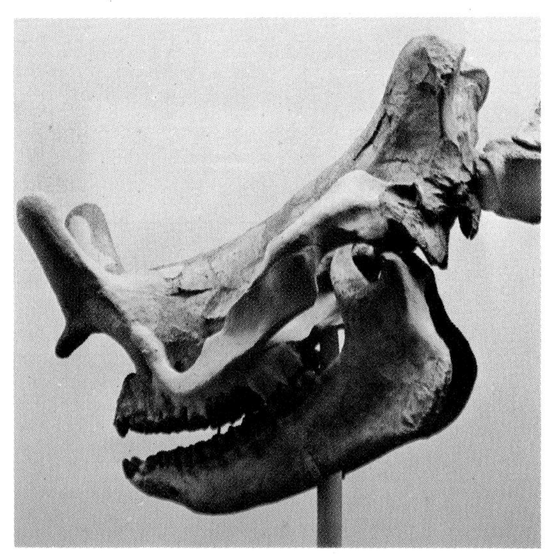

The cranium (above) and a reconstruction (above right) of *Brontotherium* (Lower Oligocene Period). This was the giant of the brontothere group whose very sudden extinction during the Oligocene Period has given rise to many possible explanations – epidemics, change in fauna or climate, etc. – none of which have yet been proved.

into Africa. Other genera went even further in the reduction of the number of digits, resulting finally in the modern horse: *Equus*. During the Pleistocene Period this new breed also colonized Eurasia while the horses curiously disappeared from the American continent that had been the scene of their evolution. The horse was reintroduced to America by Man comparatively recently. *Equus*, a highly efficient running animal, only places the nail of a single digit on the ground, although sometimes foals are born with well-developed lateral 'fingers', a feature which is a throw-back to ancestral forms, and reminds us of their past evolution.

Brontotheres and chalicotheres

The first horses developed alongside other perissodactyl groups, some of which have not completely died out.

The Brontotheres appeared at the beginning of the Eocene Period with small forms not very different from *Hyracotherium*. Their short teeth were adapted to a diet of soft vegetation and never altered throughout their history. They rapidly increased in size, and became as large as a rhinoceros by the Oligocene Period (see above). At the same time, the limbs which had four digits at the front and three at the rear, were transformed into thick 'pillars', able to support the huge weight of the body. While the muzzle of members of the horse family lengthened at the expense of the brain-pan during their evolution, the opposite was the case with the brontotheres. The most surprising feature of their skull, however, were the bony outgrowths, such as the nasal fork of *Brontotherium*. A remarkable fact is that these structures were well developed in the giant

This is how the chalicotheres are thought to have looked. This is *Moropus* from the Lower Miocene Period of North America. It was a perissodactyl which looked rather horse-like but whose limbs and feet – with their long claws – were not made for running. Its appearance was very different in fact, and today we know, among other things, that it rested on the back of the front digits.

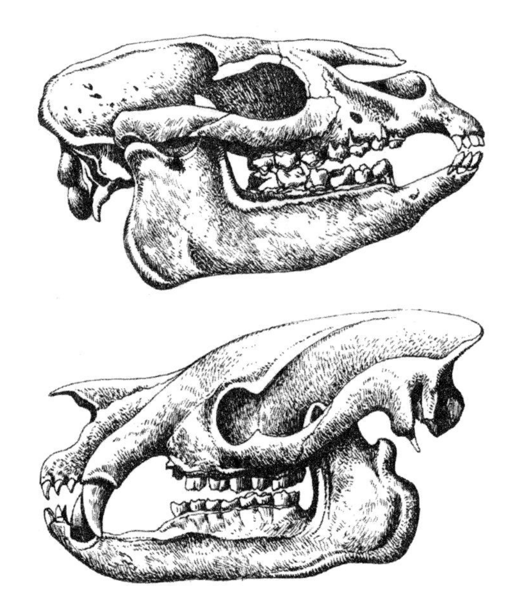

The nasal notch in the skull of this *Protapirus* from the Oligocene Period (top left) was positioned further forward than in a modern tapir, and because of this we can see (above) that it certainly had no nasal appendage like its present-day cousins.

Although *Protapirus* was a small animal, *Lophiodon* of the Eocene Period (left) sometimes grew to the size of a rhinoceros.

species only, while the skulls of the smallest and oldest among them have none at all. This is known as allometry: during the evolution and increase in size of a group of animals, one particular organ acquires a correspondingly increased importance. We believe that the chromosomes which govern mutations (see page 37), and which trigger the increase in the overall size of an animal at the same time, govern the appearance of such organs. A perfectly 'useless' anatomical structure may thus emerge, if the genetic instructions for its appearance are linked to that of a characteristic which is important and subject to strong selective pressure. Following a very rapid evolution in North America and Asia, the brontotheres died out somewhat mysteriously during the Middle Oligocene Period, at a time when they were still flourishing.

The chalicotheres might well have walked straight out of a palaeontologist's nightmare, yet they actually existed. As the illustration on page 252 shows, a horse-like head suggests the group to which they were related, and their teeth confirm this. Their molars were short, and the forward-most teeth were not well developed; in certain cases they were altogether absent. It is probable that the lips were prehensile, in other words they could move and grasp vegetation. As we stated earlier the head was attached to a rather unexpected body. The feet had three digits and were armed with enormous claws which were much longer at the front than at the rear, having once been the same size and clawless in the ancestral forms of the Eocene Period. Many reconstructions were attempted in the past, one of

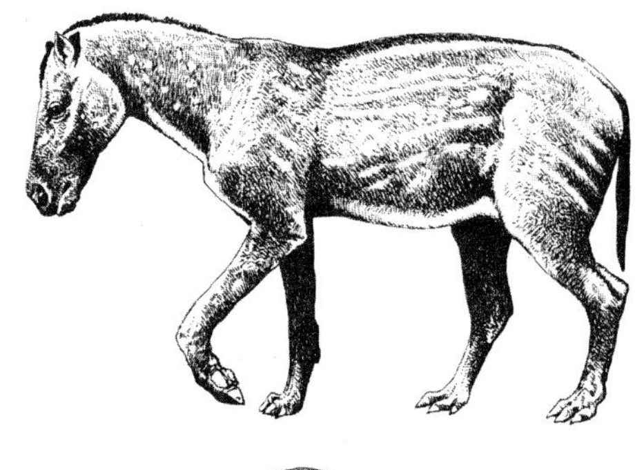

Hyrachyus and *Metamynodon* (from the Eocene and Oligocene Periods respectively). *Hyrachyus* was a similar shape to the primitive tapirs which lived at the same time. *Metamynodon* was more closely related to the rhinoceroses; it was a 'hippopotamoid' type without a horn and was very widespread over the globe during the first half of the Cainozoic Era.

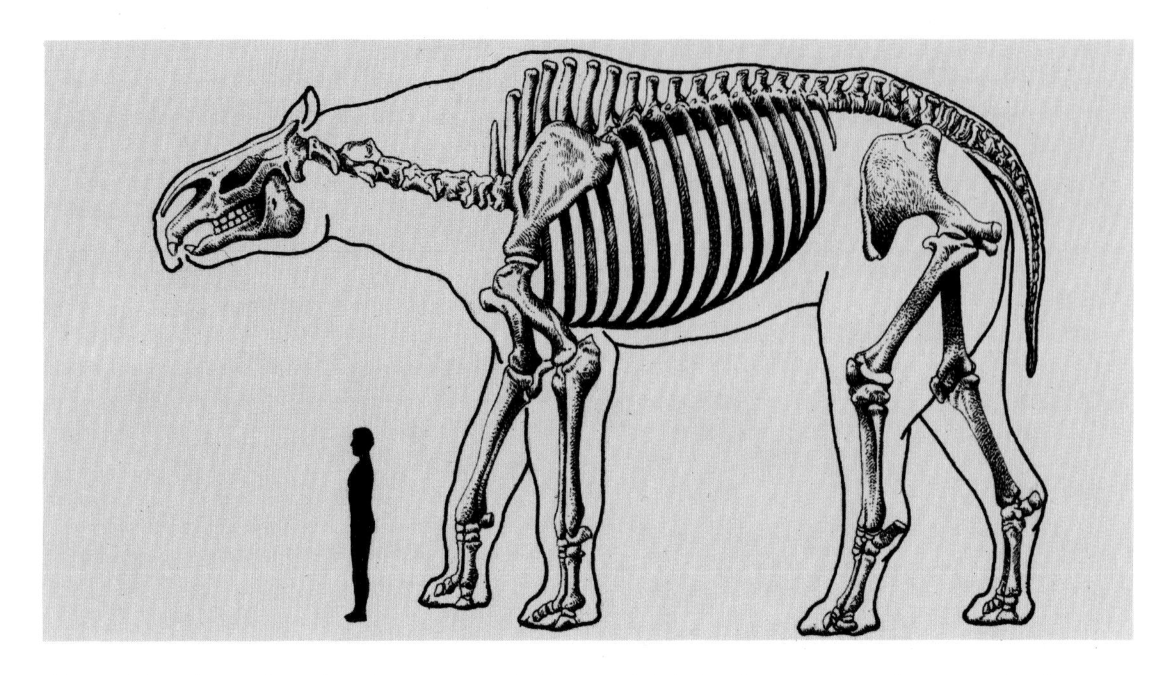

Above: *Indricotherium* from the Oligocene Period of Asia. This massive creature of course never saw Man, but a human silhouette gives some idea of its size. Its dimensions varied from one adult to another, but the biggest specimens were higher than 5 metres (16½ feet) at the shoulder. This genus holds the absolute weight record among the land mammals: its total weight has been estimated as 16 tonnes, or three times that of a large African elephant.

Below: The very short feet of *Teleoceras*, a Miocene–Pliocene Period rhinoceros from North America, gave it a hippopotamus-like appearance; at times people have thought that it lived in water. The appearance of a short-foot mutation can be seen in certain domestic mammals, especially in the dog and the sheep.

which is shown here, but certain features continued to defy the imagination. Today, *Chalicotherium* continues to confound and confuse: the specialists consider it to be a sort of adaptive hybrid between horse and great ape, a large creature capable of standing on its hind legs, supporting itself against trees, the foliage of which it grazed on. Apparently forest-dwellers only, the chalicotheres are rather rare in fossil deposits, although they lived a long time since they have also been found in sediments from the Quaternary Period of Africa.

Tapirs and rhinoceroses

The superfamily of tapirs is limited in modern times to the equatorial forests of South America and Indonesia, but in the Tertiary Period they inhabited North America and Europe. They are also descended from small animals of the Lower Eocene Period similar to *Hyracotherium* and, apart from their increase in size, proved to be rather unremarkable. Their relative importance among the mammals was once far greater than it is today, however. In particular, we should mention *Lophiodon* (see page 253), remains of which abound in certain Eocene Period layers and which possessed powerful canines in the form of 'tusks'. The true tapirs appeared in the Oligocene Period with *Protapirus* (see page 253). They were still rather small at this time, and although they became larger, they became increasingly less important. Their nasal bones became smaller and a short trunk developed. This regression was even further advanced in a giant tapir of the Pleistocene Period, *Megatapirus*, than in the modern genus, *Tapirus*.

The rhinoceroses, today on the verge of extinction, had their heyday in the Cainozoic Era. There were some very varied species, quite different from those living today and usually devoid of the famous 'horn'. Despite its name, this 'horn' is in fact a dense mass of long hairs. The origins of the rhinoceroses are close to those of the tapirs, and both groups are known by the term 'ceratomorphs'. The very earliest rhinoceros must have looked like a small running ungulate. Some, like *Hyrachyus* (see page 253) possessed four digits on each forelimb, while others had three only. This second, slender type survived to the Oligocene Period, and *Hyracodon* is a representative genus. It was a kind of small pony that occupied the place in nature which was eventually taken over by the horse family in the Miocene Period.

In the Upper Eocene Period the 'running rhinoceroses' were joined by heavier forms: the amynodonts, whose ecological niche is today occupied by the hippopotamuses. *Metamynodon* of the Oligocene Period, shown on page 253, was a large, low-slung beast that frequented fresh water, while within the group itself the resemblance was not restricted to body shapes and habits: the canines of *Metamynodon* had developed into tusks very like those of a hippopotamus, with a bevelled front edge. The amynodonts were essentially North American and Asiatic creatures, although one genus did cross from Asia into Europe in the Middle Oligocene Period.

In the midst of the variety of Oligocene rhinoceroses, there is one group that is especially astonishing: the baluchitheres, which were all Asiatic. The increased size of

Despite its appearance, *Arsinoitherium* from the lower Oligocene Period of Fayyum, Egypt, is not related to the rhinoceroses. Its strange anatomy distinguishes it from all the other placental mammals, and a separate order had even to be created to classify this unique genus: the embrithopods.

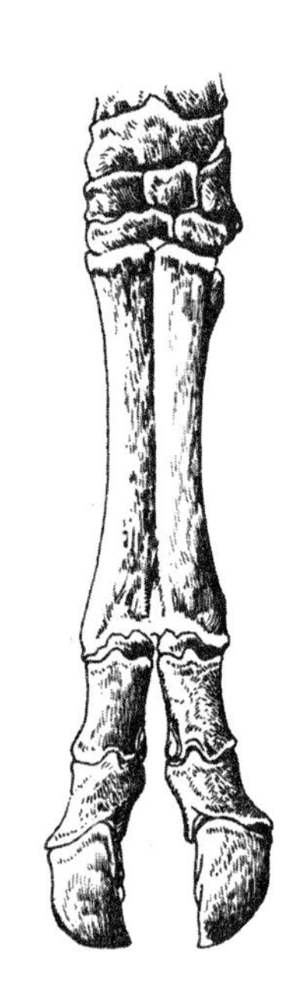
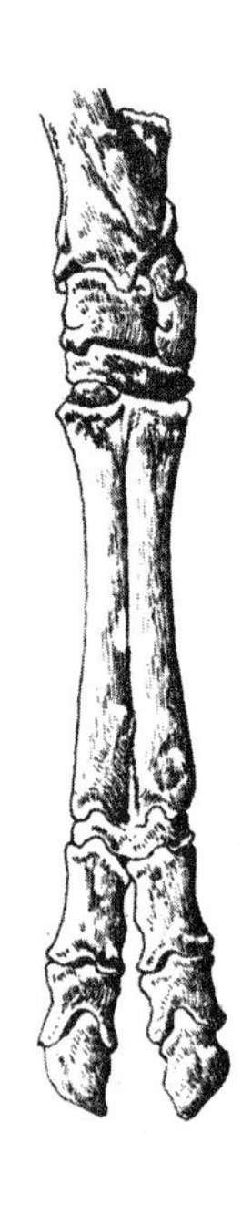
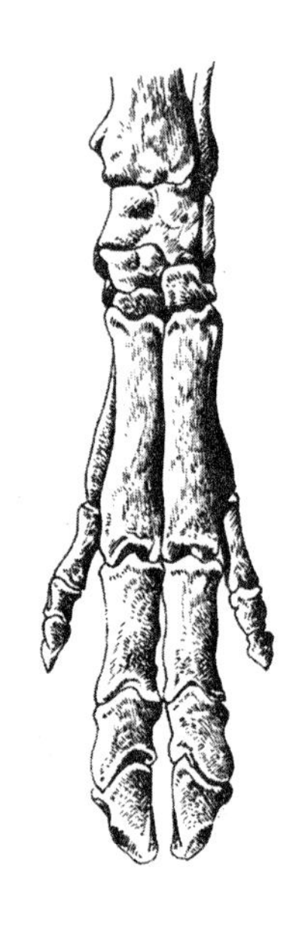

Above: On the left: fore and hind foot of a bovine. On the right: fore and hind foot of a pig.

Below: The Oligocene Period genus *Archaeotherium* belongs to the group of entelodonts. Its conical front teeth and laterally compressed premolars were those of an omnivore which probably ate flesh occasionally. The large entelodonts had a hump on their backs which made them look rather like a present-day bison. The genus *Dinohyus* from the end of the Oligocene and beginning of the Miocene Periods in North America measured as much as 2 metres ($6\frac{1}{2}$ feet) at the shoulder.

these creatures – which were also 'hornless' – was very rapid, and culminated in the genus *Indricotherium*, the largest terrestrial mammal that has ever existed (see page 254). With the aid of its long neck it was able to tear off the branches of trees at a height of 6 metres (20 feet) above ground, much like a giraffe does today. In this task it was assisted by its very powerful upper lip. *Indricotherium* obviously had very powerful legs, but despite the animal's incredible weight, these still retained certain features which were like the limbs of the running rhinoceros.

The family to which the modern rhinoceros belongs first appeared in the Upper Eocene Period, and by the Oligocene Period it was well represented by a variety of genera, many of which by now had acquired 'horns'. *Diceratherium* even had two horns side by side. This huge family of animals had a

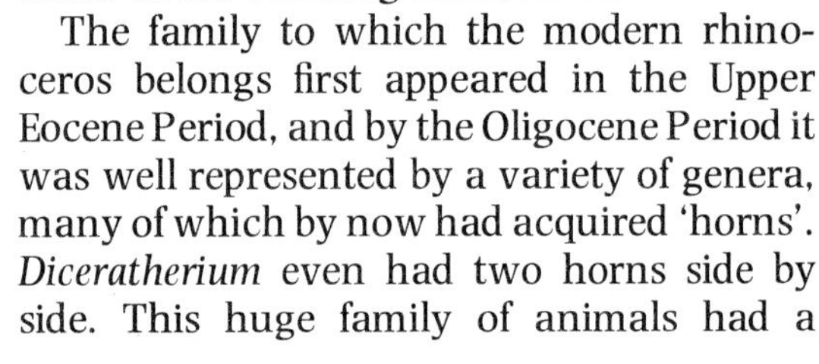

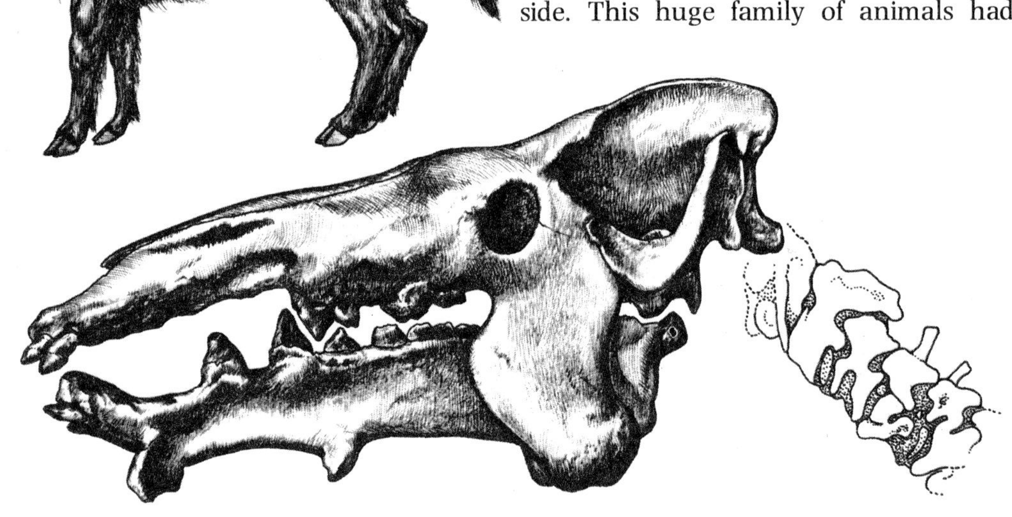

rather complex history, declining from the Pliocene Period onwards, a period which also saw the disappearance of the rhinoceros from North America. They were thus unable to colonize South America, a feat that was undertaken by the horses and tapirs when the Panama isthmus (the land bridge connecting North America and South America) was established.

Cloven-hoofed animals

The second important group of ungulates which flourished during the Tertiary Period were the artiodactyls which appeared a little later than the perissodactyls, and were distinguished from the latter by a number of anatomical differences, for instance, the very specialized shape of the astragalus. Still, the feature that immediately identifies an artiodactyl is the form of its feet. In these, the axis of symmetry runs between two main digits, numbers 3 and 4, and not through a main central digit as in the perissodactyls (see page 256). This forked arrangment is, of course, the one found in the cow, goat or pig, to mention but a few familiar domestic animals.

The history of the artiodactyls is rather complicated and their evolution and relationships are still far from clear. During the Eocene Period, a number of small forms arose, while some grew to quite large sizes. *Anoplotherium*, one of the mammals from the Montmartre Gypsum identified by Cuvier, looked like a short-legged donkey equipped with a long tail appendage. Towards the end of this period they became very important, with forms all over Europe and North America.

Those living in Europe (called cainotheres) were small, finely-contoured quadrupeds about 20 centimetres (8 inches) high. Their limbs ended in four elongated digits front and rear, and their teeth were adapted to a diet of tender vegetation. Their life-style has often been compared with that of modern rabbits, and it indeed seems likely that they were capable of sitting on their hind legs, using their front paws to carry food to the mouth. Their heyday came in the Oligocene Period, although they are found right into the Miocene Period.

The second group, the oreodonts of North America, also flourished, during the Oligocene Period in particular, where very varied species – some the size of a pig – inhabited the great plains (see below).

Suiforms Among the various artiodactyl lines of evolution we can distinguish two principal groups, the suiforms and the ruminants. The suiforms include both primitive forms and genera very reminiscent of the modern pigs and hippopotamuses. The ruminants consist of families that have ac-

Merycoidodon was a typical oreodont of the North American Oligocene Period. Its remains have been found in such vast quantities in the badlands of South Dakota that the layers have been nicknamed the Oreodon Beds, where they are more numerous than all the other fossil mammals put together. Most of the Oligocene–Miocene Period oreodonts looked rather like present-day pigs, although some species were well adapted to running and were rather gazelle-like. The latest genus of the group, *Brachycrus*, probably had a short trunk.

quired a highly developed form of digestion known as rumination or chewing the cud.

Of the suiforms which developed at the end of the Eocene Period we should mention the entelodonts (see page 256), whose skull was very elongated and bulky, as well as other well-known suiforms which included the anthracotheres, so named because the first of them was identified in a coal mine. They look rather like the entelodonts and, like them, were of Asiatic origin. *Anthracotherium* of the Oligocene–Miocene Period was similar in size, and probably lived in a similar way to a hippopotamus; its canines were transformed into very impressive hooks which crossed like those of a carnivore and not like the tusks of a wild boar.

The Oligocene Period of North America has revealed an especially curious genus – *Agriochoerus*. While the dentition of *Agriochoerus* was undoubtedly that of a herbivore, its very primitive short legs were armed with long claws reminiscent of a flesh-eater. Was this low-slung animal with its long tail capable of climbing trees? Or perhaps the claws reflect the habits of a burrower.

From the Oligocene Period onwards we detect ancestors of the Suidae (pig family) in Europe, represented today by the pig and wild boar. In North America we find ancestors of the Tayassuidae which were represented in the Old World by just two species and of these, only the peccaries are the modern survivors. The hippopotamuses on the other hand appeared rather suddenly at the end of the Miocene Period and were very similar to those living today.

In the Miocene Period a rich collection of ungulates lived on the North American prairies. It included several rhinoceroses and small horses, but it was the artiodactyls which gradually came to dominate, like the tragulins shown here in the foreground (*Synthetoceras* and *Syndyoceras*) or the camel-giraffes (*Alticamelus*) drinking in the background.

Ruminants We have seen that the evolution of the perissodactyls was accompanied by first a sudden increase in the number of species followed by an increasingly marked decline. This gradual regression was basically caused by the very keen competition from the artiodactyls, starting in the Oligocene Period and reaching a peak during the Miocene Period. Today, while the former are on the verge of extinction, the latter are flourishing, especially the ruminants. The perissodactyls are limited to a few large species, all the small and medium-sized representatives having died out. The suiforms themselves are restricted to very particular life-styles, and do not make up a large number of species.

The ability to ruminate has enormous advantages for a herbivore. Vegetable food is relatively low in energy content, and therefore large amounts of food must be eaten. However if the animal has to chew each mouthful, it will have to spend a lot of time in the open, and may fall victim to a carnivore

at any moment. A ruminant, on the other hand, very rapidly swallows a substantial volume of grass – the belly of a cow may contain up to 200 litres (44 gallons) – then, having taken refuge, it is able to regurgitate small amounts of food by means of contractions of its second stomach and chew them at its leisure before swallowing them directly into its *psalterium* or third stomach. In this way the fast-grazing ruminants with their more efficient digestion, were able to outcompete the non-ruminants while, at the same time, escaping more easily the unwelcome attentions of the giant predators. At the beginning of the Cainozoic Era the astragalus bone of these herbivores changed its shape, making them faster runners, and so the carnivores were forced to become increasingly efficient in order to capture them.

This beneficial adaptation is more perfected in the tylopods (camels, llamas and vicunas) than in the ruminants, and it is likely that it has appeared at various times during evolution. In the Eocene Period, when North America and Europe were still joined, the tylopods were probably present in both continents, but after the opening of the North Atlantic their evolution continued in North America only. *Poebrotherium* of the Lower Oligocene Period of the U.S.A. already looked like a miniature llama – 50 centimetres (20 inches) at the shoulder – its limbs ended in two spread digits only. The Oligocene and the Miocene Periods saw the emergence of strange types of 'camel-giraffes' (see page 259) and later 'camel-gazelles'. At the same time, the group leading to the modern tylopods continued its development, and in the Pliocene Period some of its representatives migrated towards Asia, in the region of the Bering Straits, while others moved towards South America via the now emerged Panama Isthmus. The Quaternary Era saw the extinction of the camels in their original regions while the emigrants continued to evolve in the deserts of the Old World and the mountains of South America.

The tragulids which today make up only two genera, the Asian *Tragulus* and *Hyaemoschus* of Africa were once rather more important. From the Oligocene to the Miocene Periods we find strange beasts in which the skulls of the males were decorated with long bony outgrowths. These creatures were particularly common in North America. The structures were found on the forehead and end of the muzzle, where they were forked (see page 259). In the males, the upper canine and first lower premolar formed tusks. These ruminants of the Tertiary Period must have looked far more spectacular than the small musk-deer of our own times, an American family which is sometimes thought to belong to the tylopods.

The group of pecorids, whose types

The reindeer (*Rangifer tarandus*) was frequently hunted by prehistoric man at the end of the Pleistocene Period. Both males and females had antlers, a unique feature among the Cervidae. The size and shape of the antlers can vary widely within the species from one herd to another.

showed an almost universal tendency to develop all kinds of cranial ornamentation, includes most of the modern ruminants: giraffes, deer, bovines and antilocaprids. The first of these groups, which is not widespread today, has small frontal horns covered in skin in the giraffe proper, and the outline of a horny appendage in the okapi (a creature that was not discovered until the early 20th century – rather amazing for an animal the size of a horse). These cranial outgrowths, however, were much more imposing in the fossil species. *Sivatherium* especially, from the end of the Cainozoic Era, possessed huge slightly twisted horns located behind a pair of conical protuberances. This animal was more thick-set than the modern giraffe and had a rather shorter neck. It was contemporary at least with the first hominids in Africa.

In the Cervidae (the group which includes the deer) the very elongated top canines and the antlers played the part of defence mechanisms. The canines were highly developed in the antler-less primitive genera, but became smaller as the antlers increased in size and finally disappeared in nearly all the modern Cervidae. The first antlered cervid appeared in the Lower Miocene Period with the genus *Procervulus*, although it did not shed its antlers like modern antelopes. A second stage was marked by genera like *Dicrocerus*: only the upper part of the antler fell, albeit somewhat irregularly, as in the Asiatic muntjac. Finally, the loss of the antlers became annual, and increasingly complex replacements grew each year.

The horns of the Bovidae on the other hand are always perennial, with a solid case of bone surrounding a bony core. The little Miocene Period *Eotragus* possessed only a modest pair of horns but his successors showed many different forms. The curved and twisted horns of the bovines, antelopes and wild goats are extremely varied. This family has increased ever since it first appeared, and is extraordinarily widespread, representing along with the Cervidae the overwhelming majority of modern ungulates.

Finally, the uniquely North American antilocaprids presented an original combination: a horny case decorated their frontal appendages in the manner of the Bovidae, but this was lost each year like Cervidae antlers. Only the 'antelope-goat' (*Antilocapra*) has survived to modern times, but from the Miocene to the Quaternary Periods the American antilocaprids were highly diversified, with horns that were sometimes twisted or branched, and on occasion even resembled deer antlers.

Mastodonts, elephants and dinotheres

These are all attributed to the order of the proboscidians. The proboscis, or trunk, is of course the elephant's most distinctive feature, but the most primitive proboscidian, *Moeritherium* of the Upper Eocene Period, was certainly trunk-less. Its remains have been found in Egypt in the extraordinary deposits of Fayyum, a region in which it led a semi-aquatic existence. Still modest in pro-

Right to left: *Moeritherium* (Upper Eocene Period) and *Phiomia* (Lower Oligocene Period); 60 centimetres (24 inches) and 1.2 metres (4 feet) long respectively.

Left to right: *Platybelodon* (Upper Miocene Period) and *Amebelodon* (Pliocene Period); 1.3 metres ($4\frac{1}{2}$ feet) and 2 metres ($6\frac{1}{2}$ feet) high respectively at the shoulder.

portions (about the size of a tapir) this creature's body was low-slung and, if some reconstructions are to be believed, its body was very long – even longer than our illustration on page 261. This creature was still a long way in time from the huge profiles of the African elephant! But the differences did not just concern the overall form: the skull was low, the dentition almost complete and all the teeth were functional at the same time, a point which we shall see had its importance. The canines had disappeared, and the second incisors were strong and prefigured the four tusks, two upper and two lower, observed in other Fayyum genera such as *Palaeomastodon* and *Phiomia* of the Lower Oligocene Period (see page 261). The cranium was higher and the overall size, especially of *Palaeomastodon*, was much bigger than that of *Moeritherium*. The nasal bones and shape of the nasal passages already indicate the beginnings of a trunk. These were the first representatives of a group that was to prosper throughout the whole of the second half of the Tertiary Period. At the beginning of the Miocene Period the northward shift of the African continental plate enabled the proboscidians to migrate into the rest of the World.

The term 'mastodont' derived from the very special shape of these animals' molars. They are found in Europe from the Lower Miocene Period onward, and persisted in North America for long after the appearance of Man. The molars in fact formed a succession of transversely arranged nodules that tended to increase in number during the course of evolution. The mode by which the molars were replaced is curious, since they now grew one after another rather than at the same time. The tusks were powerful, and still number four in the Miocene Period gomphotheres, a group that later led to the modern elephants. Some mastodonts featured a stange adaptation: the lower tusks were modified to form 'buckets' that were very wide in *Platybelodon* of the Upper Miocene Period and very long in *Amebelodon* of the Pliocene Period (see page 262). This phenomenon was in fact nothing more than an exaggerated form of a cranial characteristic possessed also by *Phiomia*, and was connected with the digestive system.

The size of the mastodonts was very comparable with that of our own elephants, which also increased in size in the course of their history. Initially, the trunk was short and the animal's forelegs were rather longer than their hind legs, while an almost non-existent neck supported a short but high skull. In the primitive elephant genus *Stegodon* the bottom tusks had disappeared while the upper had assumed majestic proportions. On the molars, the tubercules joined together to form crests, and this tendency continued, culminating in the structure of a typical elephant molar (see page 263, top). The crests have multiplied in number and have hollowed to form a series of enamel blades separated by the cement that was already present in the grooves separating the molar crests of *Stegodon*. The utilization of these grinding teeth in succession has been pushed to its limit, and in a modern elephant there are never any more than two functional semi-molars in each jaw half. The

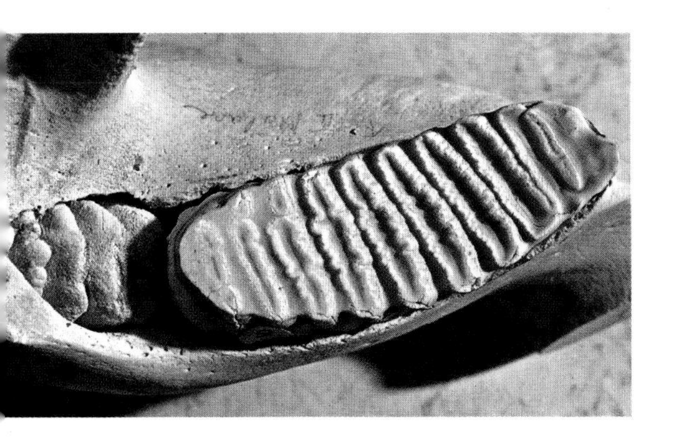

Left: The upper half-jaw of an Indian elephant: only a single molar is in use at any one time. We can clearly see the next one coming along, still hidden in the jaw-bone. As the first tooth gradually wears down the next one pushes it towards the front and replaces it when it falls out.

third molar replaces the second after about 30 years, and continues to function alone until completely worn down. Its loss then inevitably brings about the death of its owner which is no longer able to feed itself.

Parallel to this flourishing array of proboscidians another group, less rich in terms of numbers of species, was also invading Eurasia during the Miocene Period. It never succeeded in getting as far as the New World, however. Apart from variations in size, the appearance of the dinotheres remained stable up to the Pleistocene Period, when the last species died out in the African lands that had seen them develop in the first place. *Deinotherium* can be thought of as an elephant whose tusks, carried by the mandible, pointed downward. The purpose of these devices is still debatable, while the animal's molars were curiously like a tapir's.

Marsupials

We must thank a German meteorologist called Alfred Wegener for some of the most original and at the same time most fruitful

Elephas meridionalis, *Elephas antiquus* and *Elephas primigenius* were the most widely spread elephants during the European Pleistocene Period. The latter, the famous mammoth, died out only recently and its anatomy is perfectly known thanks to the cave drawings of Upper Palaeolithic man and to the discovery of frozen specimens in Siberia.
The modern Indian elephant is its nearest living relative.

Variation in size among the dinotheres: *Deinotherium bavaricum* from the Middle and Upper Miocene Period (2.5 metres (8 feet) at the shoulder) and *Deinotherium gigantissimum* of the Upper Pliocene Period (4 metres (13 feet) at the shoulder).

ideas about the history of earth sciences. He did much work on the study of the shifting continents, of which we have made repeated mention in this book. Using mainly geological and palaeontological arguments, Wegener proposed that the modern continents had once been united to form a 'supercontinent' called Pangaea. Pangaea began to split up in the Palaeozoic Era (see pages 232–233). This phenomenon was for a long time rejected by geophysicists who today, however, are enthusiastic supporters of the theory. Like those of many of his predecessors, Wegener's ideas were still regarded as crazy by many scientists when he died tragically in 1930 during an expedition to Greenland. Since then, palaeontology, and especially that of the vertebrates, has provided a constant stream of fresh evidence to prove the opinions of Wegener. One result of this is that experts have now ceased to explain many of the animal migrations which took place on the belief that strips of land emerged from the seas at certain times to provide 'bridges'.

The geographical upheavals had, as we have seen, a great effect on the future lives of many of the mammals – a fact that becomes clear when we consider the groups which inhabit South America and Australia. For many years, zoologists noticed how different both the animals and plants in these regions were. However, they also saw certain similarities which seemed impossible to explain in view of the huge distances separating the continents. Apart from the North American opossum, all modern marsupial mammals are concentrated in the regions of South America and Australia.

The answer to this puzzle is to be found in the changes which occurred in the continental masses and the distribution of animals at each of these times. In the Cretaceous Period (see pages 232–233) America, Antarctica and Australia formed a more or less continuous land mass separate from Africa. While the placental mammals were evolving in Eurasia, it was on the great America–Antarctica–Australia land mass that the marsupials appeared, and the subsequent break up of this land mass isolated the marsupials. In North America, the two groups (marsupials and placentals) lived together peacefully, but in the early Tertiary Period, after some placentals had made the passage to South America, the bridge between these two continents was cut for a long time and the placentals dominated their rivals. Before the north Atlantic Ocean was present, some marsupials spread to Europe. Some species of marsupials may even have managed to reach Asia in the Upper Cretaceous Period.

After the splitting up of the great land mass of the southern hemisphere, Australia became almost completely isolated. The marsupials there prospered in the absence of any competition from the more adaptable and successful placentals, occupying ecological niches filled in other areas by the placentals. The result was a series of spectacular convergences like the one which produced the thylacine (Tasmanian wolf), a

We need to be able to reconstruct the changes in the geography of our World in order to understand the distribution of its animals. Of course, the study of modern and fossil distributions can also help to clarify the evolution of the great land masses in the course of geological time. The distribution of the marsupials in the Australo-Papuan region as well as South America suggests that these two regions were much closer in the remote past. The opossum (below) is one of the few marsupials that had some success outside these regions.

marsupial whose appearance greatly resembles that of the wild dogs and placental wolves. 'Marsupial cats', 'marsupial moles' and 'marsupial mice' are also known. One large herbivorous marsupial the size of a rhinoceros, *Diprotodon* (see page 267), even survived into a recent period of the Australian Quaternary Era.

In the course of its 'voyage', the Australian continent moved northward, approaching Indonesian regions. Some Asiatic placentals such as bats and rodents managed to penetrate this hitherto solely marsupial terrain by travelling from one island to another until they reached Australia. The phalangers managed to invade new lands by going in the opposite direction (see page 228). The most spectacular upheaval, however, was caused by Man, who started by introducing dogs and rats into Australia, and then other creatures such as the rabbit. The rabbit ravaged the countryside, while the foxes later imported to deal with this catastrophe ignored their intended prey and instead contributed to the extermination of the smaller marsupials. The larger species also suffered, this time from the growth of stockbreeding.

South America – a divided continent

The history of South America is made even more complicated by the extent to which placentals and marsupials were present before its isolation, a timespan that covers most of the Tertiary Period. Still, this isolation did not stop the arrival of certain immigrant animals by sea, and eventually ended by a re-establishment of communications with the northern part of the continent.

Among all the placental mammal population there was not a single carnivore, and so there was a sort of division of labour, with the placentals remaining as herbivores while the marsupials diversified, especially as omnivores or predators. The major group of South American marsupials is represented by the didelphids, which includes the opossum. Their diet is highly irregular, although they are basically carnivores of small prey. Nevertheless, their ranks also included omnivorous forms, which occupied the place of the primates at the start of the Tertiary Period before being eliminated by the arrival of their placental competitors. The same kind of phenomenon occurred in another marsupial family which played the part of rodents before the immigration of the first real rodents. These creatures are part of a second

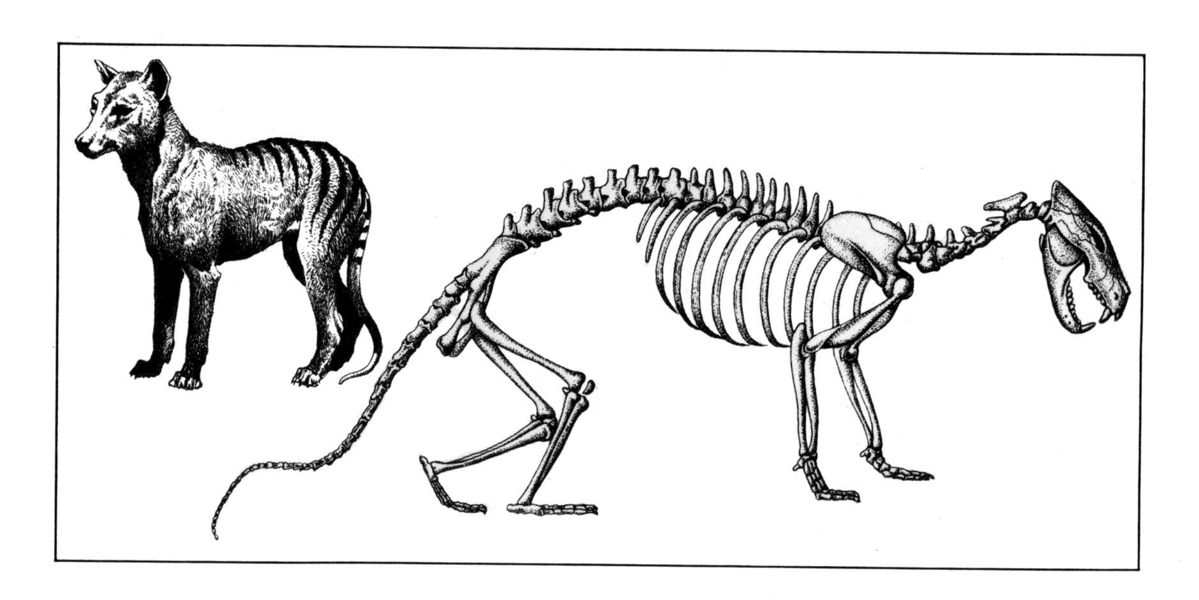

Two marsupial 'false dogs': the Australian thylacine (left), probably extinct today, and *Prothylacinus* from the Lower Miocene Period of South America. Despite their resemblance to each other they belong to two different super-families.

important super-family, the coenolestids whose 'marsupial shrew' is a modern member. Of the fossil genera, *Abderites* provides us with another example of convergence, this time with the Multituberculata.

Last but by no means least there were the borhyaenids, which competed with the great flightless birds for the position of major carnivore. Some of these were small, like our own weasels and foxes, but *Prothylacinus* of the Lower Miocene Period looked like a thylacine, as its name suggests. *Borhyaena* itself, appearing in the Upper Oligocene Period, was even larger and doubtless capable of taking on very sizeable prey. But the most extraordinary of these carnivorous marsupials was without doubt *Thylacosmilus* of the Pliocene Period which looked almost exactly like a sabre-tooth feline. All these marsupials had low-slung bodies, and the enamel of their teeth was quite thin compared with that of their placental equivalents – it wore down much more quickly as a result, and this is surely one of the reasons why the marsupial sabre-tooths soon became extinct when the two groups confronted each other.

The placentals were no less astonishing than the marsupials. The condylarths are the only ones we have met so far, and are well represented in the early Tertiary Period, persisting into the Miocene Period, in other words, much later than in the rest of the

Above: *Thylacosmilus*, a borhyaenid of the South American Pliocene Period, not only had sabre-tooth canine teeth like those of certain cats, but also was able to open its mouth very wide to strike at its victims with its deadly teeth.

Below: The largest of all the marsupials – living and extinct – is an Australian phalangeroid: *Diprotodon*. This heavy quadruped was some 4 metres (13 feet) long and weighed about the same as a rhinoceros. It was a herbivore with very modified teeth and powerful upper middle incisors which grew continuously and had just a single strip of enamel in front. Complete skeletons of this animal have been excavated from the beds of the great salt lakes in which they had been drowned.

Left: *Macrauchenia*, a litoptern of the Pleistocene Period, was found all over South America. It seems that this genus, which grew to 1.8 metres (6 feet) at the shoulder, was adapted to life in a steppe environment.
Below: *Propachyrucos* (Upper Eocene–Lower Oligocene Period) was one of the 'lagomorph' type of South American notungulates, and was about the size of a tapir.
Bottom: *Pyrotherium*, a pyrothere of the Lower Oligocene Period of Patagonia, had many features in common with the proboscidians, in particular with *Palaemastodon*.

World. Convergence with families of animals in Europe were frequent among the placentals, too. The litopterns, which evolved from the condylarths, conformed to two types. A 'horse' type led to forms well adapted to running, and their digits gradually fell in number until the genus *Thoatherium* evolved, which had just one. The single digit nature of the small creatures (around 60 centimetres/2 feet at the shoulder) was not only perfected before that of the Equidae but was also more advanced, since the 'stylets' (remains of the lateral digits 2 and 4), disappeared almost completely. Another type was clearly reminiscent of the Camelidae, of which *Macrauchenia* shown at the top of this page was one of the best examples. This was a three-fingered ungulate whose nasal bones were very much reduced, a characteristic that has been variously interpreted. Some see it as the beginnings of a trunk (as in the reconstruction shown), while others regard it as the sign of an aquatic life, but it is likely that the nostrils were surrounded by lips which could close, thus protecting them from the dust borne on the winds of the steppes.

The notungulates, the richest of the groups of South American ungulates, also provided horse-like types; these were three-fingered animals whose teeth were hypsodont from the Upper Oligocene Period onward. Other, heavier types evolved the type of body shape which approached that of the rhinoceros. In the Miocene and Pliocene Periods, some even bore a large frontal horn. Like the chalicotheres, some of the larger notungulates had their hoofs replaced by claws which they used to dig or to seize branches. Finally, a whole series of genera arose, which bore an astonishing resemblance to the rabbits and hares (see above). The notungulates were not exclusively South American; they appeared in China during the Upper Palaeocene Period and in North America in the Lower Eocene Period, but were unable to survive for long in those areas.

From the Palaeocene to the Upper Miocene Periods, the South American continent was also inhabited by the strange group of

astrapotheres. *Astrapotherium*, whose largest specimen approached 1.4 metres (4½ feet) at the shoulder, was remarkable in many ways. Its four-tusked head undoubtedly featured a small trunk; it frequented swampy areas and may even have been amphibious, as is suggested by several of its anatomical characteristics such as the enlarged feet and hands. Other large creatures, the xenungulates of the Upper Palaeocene Period, looked something like the dinoceratans, with whom they may well have shared common origins. The pyrotheres, however, from the Eocene to Lower Oligocene Periods were in no way related to the proboscidians despite their appearance (see page 268, bottom).

While all these groups are extinct today, the toothless xenarths are still well represented by the armadillos, sloths and anteaters. Only the anteaters are in fact completely toothless, but in all these groups the dentition is very reduced and has lost its enamel. Still, the teeth of the armadillos may be very great in number. These three modern groups were also present in ancient times, although their fossils are far more numerous and varied than the few surviving species. Of those surviving groups, the armadillos are the most successful. Some 20 more or less omnivorous species still exist. Of their extinct cousins, some were evidently carnivores, while the giant forms were decidedly herbivorous in diet. This was also the case with the glyptodonts; the carapace of these animals, unlike that of the armadillos from which they evolved, was not hinged but formed a rigid dome. Their claws were transformed into hoofs and their hypsodont teeth enabled them to browse on the vegetation that formed their food. In the absence of dental enamel, folds of dentine of varying toughness constituted the principal structure of the teeth of this genus. The glyptodonts had their heyday during and after the Miocene Period, and some became gigantic: in the Pliocene Period, for instance, the shell alone of the large *Glyptodon* measured 2.5 metres (8 feet) in length.

Another group of xenarths included the sloths. These spectacular creatures were mostly devoid of hard shells. While modern sloths are tree-dwelling, their fossil relatives were terrestrial and also possessed huge claws. Their characteristically twisted feet rested on the outer surface. *Megatherium*, whose skeleton is shown above, was the giant of the group. A remarkable fact is that we even have the fur of some of these creatures, as they died out comparatively recently and some specimens were able to mummify under conditions of extreme drought. In certain caves, therefore, fragments of skin and fur were intermingled with the bones of these animals.

Megatherium, one of those South American creatures which used the Panama Isthmus to invade the northern part of the continent in the early Quaternary Era. It was a kind of gigantic sloth covered with long hair, and was able to rear to a height of 6 metres (20 feet) to graze on its diet of foliage.

The end of isolation

Except for the arrival in the middle of the Tertiary Period of the rodents and African primates which we shall discuss in Chapter 9, the animals of South America evolved in isolation up to the end of the Pliocene Period. In the Upper Miocene Period at the very earliest, some procyonids were able to bridge the marine gap that separated the two Americas, but at the Tertiary–Quaternary boundary the re-establishment of communications between these two land masses brought about a total reversal of the situation. South America was gradually invaded by successive waves of placentals. We have already listed the fissiped families which thus penetrated the lands of the south, and to these were added the tayassuids, the horses, the mastodonts, the tapirs, the Camelidae, the Cervidae, the lagomorphs and even more rodents. The intertropical zone, however, functioned as a kind of ecological barrier preventing certain families from undertaking this conquest. The Bovidae and elephants, for instance, never got further than Central America, while the beavers were never to leave North America.

This invasion involved a number of extinctions, especially among the ranks of the marsupials: all the borhyaenids disappeared before the onslaught of apparently more competitive fissipeds. Nevertheless, the movement of animals was not all in one direction – since the xenarths spread northward – and although the glyptodonts and ancient sloths died out in the north as in the south towards the end of the Pleistocene Period, the nine-banded armadillo penetrated as far as central U.S.A. South American rodents followed the same route, but on the whole, the south–north migrations were fewer in number: only eight southern families moved north, while 16 North American families went south.

Ice Ages

One of the chief factors which impeded the northern movement of the South American mammals, most of which were tropical, was the climate. In the course of the Pleistocene Period, the currently temperate zones of Eurasia and North America underwent a series of glaciations which seriously affected the overall climate. We have thus been able to divide the Quaternary Era into a number of glacial and interglacial periods, the exact number of which has yet to be finally established. There were at least five major cold periods to which the names of central European rivers have been given: Biber, Günz, Mindel, Riss and Würm. The actual details are far more complex than we could hope to set forth in this relatively brief account. Suffice it to say that each of these periods was affected by changes in the climate with the result that in a 'cold' period the climate may have been simply cool or even temperate, with a freezing spell now and again. Similarly, the 'hot' periods were also very irregular. The cause of these fluctuations is still not known for sure, but current theories suggest cosmic causes. The glaciations do not appear to be specifically recent occurrences either, since glacial traces have also been found in far more distant epochs in the Earth's history. But those of the Quaternary Era have, of course, left the 'freshest'

The mammoth and the woolly rhinoceros were both perfectly adapted to the cold glacial climate of Europe during the Würm. The stomach contents of frozen specimens found in Siberia have included the ingredients of their last meal, which was based on pine needles and the shoots of shrubs. No doubt, using its anterior horn, the rhinoceros was able to dig up roots to add to its diet. The mammoth showed several adaptations to a cold climate: as well as its coat it had tiny ears and a muscular flap at the base of its tail which covered its anus and surrounding parts.

The cave bear (*Ursus spelaeus*) was well over a third larger than even the biggest of the modern brown bears. Furthermore its front feet were much longer and its forehead was very much more domed. Its rounded teeth also indicate basically vegetarian feeding habits. Many of its remains have been extracted from caves, especially in the Alps and Pyrenees of Europe, although others have been found in open terrain.

marks, and yield most information.

It is not hard to imagine the impact that these upheavals had upon the animal life. The animals living in the warm conditions at the end of the Tertiary Period had to adapt or migrate to the south during the initial glaciations, to return later on when conditions improved. In Europe, however, the possibilities of migration were restricted by the Mediterranean, and so the disappearance of sub-tropical forms was made even more rapid. Nevertheless, certain mammals were able to adapt extensively, and families which are familiar to the layman as being quite tropical or equatorial were capable of surviving alongside arctic foxes, musk oxen and reindeer which inhabited Europe right down to its southernmost points. One such example was the woolly rhinoceros, so abundantly portrayed by men of the Upper Palaeolithic. As we have managed to ascertain from their works of art and from frozen specimens, this animal was protected by a thick black fleece covered in long reddish hairs. Among the elephants, the mammoth is one which is particularly well known. Contrary to popular opinion, it was not all that much larger than a large modern elephant, but its tusks were indeed gigantic structures (see below).

Giants and dwarfs

Unlike the mammoth, numerous Pleistocene Period mammals did exceed the size of their living cousins. Among the ranks of the carnivores, for instance, the cave bear (see above) and cave hyaena were clearly larger than their modern relatives – a fact verified within several groups. This is an observation that can be set beside another in living nature: within a genus or species of mammal, the weight as well as the physical

proportions vary with latitude, and populations established in the coldest environments seem to produce the most massive individuals. This is a phenomenon known since the 19th century as Bergmann's Law. Thus it is that the largest tigers are those that inhabit Siberia, while the southernmost varieties, those of Indonesia, are smaller than the average type as represented by the Indian tiger.

In addition to this, we observe that in various groups the length of the extremities (tail, ears, limbs) diminishes according to how close the species is to the poles. This is Allen's Law and, like Bergmann's, it explains the degree of adaptation of a mammal to a hot or cold environment (by varying the ratio of body volume to body surface area). This may explain why the fennec fox of the Sahara has huge ears compared with those of the Arctic fox, which are tiny.

The case of the tiger is interesting for another reason, since the smallest specimens come from the Sunda Islands. Now the dwarfism of island species of mammals is one of the more spectacular generalized phenomena which Pleistocene Period palaeontology has revealed. For over a century the remains of minute elephants have been identified in the larger islands of the Mediterranean, first on Sicily and Malta, then on Cyprus and Crete. These amazing creatures, some of whom did not exceed 80 centimetres ($31\frac{1}{2}$ inches) in height, are the descendants of *Elephas antiquus* (see page 263), a continental genus that exceeded 4 metres (13 feet). They lived alongside other dwarf animals, and in particular a minute cousin of *Megaceros* (see page 273). This phenomenon was not restricted to the Mediterranean, since the same period saw a *Megatherium* the size of a poodle living on the Antilles! And that is not all: as if better to emphasize these extravagances, the rats, dormice and field-mice assumed enormous proportions in these islands.

All of these animals were no less 'normal' than their more conventional relations: they

Above: The cave lion (*Felis spelaea*) is so called because its remains have been found in caves, but it is in fact unlikely that it actually lived in them. It probably frequented the pine forests and steppes that covered Europe some 10,000 years ago.

Right: The tarpan, the wild horse of Eurasia, is now extinct. The last animals were killed in the middle of the last century, 5,000 years after they had provided the strain that led to the first domesticated horses. There is another horse that is truly wild – never having been domesticated by Man, unlike the American mustangs – this is Przewalski's horse which was identified in 1879 by a Russian explorer travelling in Mongolia. A number of zoos now have specimens, but there is little doubt that the last wild herds have probably interbred with domestic horses.

had simply adapted to very special conditions. The elephants, for example, in the absence of large predatory carnivores, were able to adapt better to a rocky terrain and sparse vegetation by becoming smaller, while the rodents were able to escape their old enemies, the birds of prey, by increasing in size. The final glaciation, and the most extreme, caused the retention of huge volumes of water in the form of ice in the polar caps, lowering the level of the seas considerably. The islands, now linked to the main continents, saw their highly unusual animal groups wiped out by carnivores or larger, more successful members of the same genera which invaded from the surrounding lands.

Megaceros, a magnificent animal of the Pleistocene Period, was a giant deer the size of a horse. Its antlers attained a majestic span of up to 3.5 metres ($11\frac{1}{2}$ feet). Complete skeletons have been found in the peat-bogs of Ireland. These perfectly preserved specimens truly deserve the name of 'peat stag'.

Chapter Nine

How Man Evolved

How man evolved

The world of the primates

The origins of the human race go back to the primates, whose own history has been traced back as far as the end of the Mesozoic Era. The fossil remains of these creatures are rather rare, but common enough to retrace the major features of their evolution. In addition, the anatomy and physiology of living primates provide information which, as we shall see, allows us to understand the connections between the various groups.

One of the factors that contributed to the relative rarity of fossil primates was no doubt their basic adaptation to a tree-dwelling existence, implying a forest habitat hardly favourable to fossil preservation. Today, a number of species of primate spend most of their time on the ground; on two feet in the case of humans, on four feet in most other species. Nevertheless even these forms have retained features that give away their initial life-style. The limbs are very mobile. The clavicle (collar bone) that has disappeared in other mammals persists in the primates. The hands and feet have kept the primitive five-digit arrangement, while the thumb can be closed on to the other fingers, making for a firm grip on branches. The claws have generally been replaced by nails. As regards the senses, sight is all-important and the large eyes tend to be set in the front of the skull, resulting in stereoscopic vision and the consequential ability to judge distance, a useful skill when it comes to leaping across the void from tree to tree! The brain also increases in volume, and its visual areas are very extensive unlike the olfactory.

The first fossil that is attributed to the order of primates comes from the Upper Cretaceous Period of Montana, and has been called *Purgatorius*, a reference to the Purgatory Hill formation where American palaeontologists experienced much trouble digging out these very ancient remains. This genus is encountered again at the beginning of the Tertiary Period, and is in no way a direct ancestor of modern primates but certainly was not very far from that ancestor by its appearance, which was that of a sort of tree-dwelling shrew. It formed part of the now completely extinct group of plesiadapiforms that are found again during the Eocene Period of Europe and North America; it is possible that they inhabited Africa as well. The plesiadapiforms which inhabited Europe are very well known and are still very primitive: their thumb was not opposable, they had claws instead of nails and their eyes were set arranged to give stereoscopic vision. Their powerful incisors were doubtless used to crush hard shells in much the same manner as is done by the squirrels of our own forests (see page 278).

A question of noses

Modern primates are divided into two major groups, the strepsirhinians and the haplorhinians. These complicated terms refer to the different ways in which the noses are designed. The former, of which the lemur and the loris are principal representatives, possess a nose split in the manner of a dog's, while the tarsiers and the monkeys, which belong to the second group, have a more 'highly evolved' nose without a split. The two groups are further separated by other physical and physiological differences, and this basic division into two groups must have taken place by the dawn of the Cainozoic Era, since the two branches were already totally distinct in the Eocene Period.

The sense of sight is highly developed among the primates and some, like this nocturnal African galago (a strepsirrhine) have very sophisticated eyesight indeed. At the back of the eye the golden layer formed by the layer of cells called the tapetum reflects the incoming light, passing it through the retina twice and so doubling its sensitivity.

The place from which the strepsirhinians first evolved is usually taken to be Africa. One group within them, the adapiforms, penetrated Laurasia (a land mass once comprising North America and Eurasia) in the early Eocene Period where they existed for over 30 million years before dying out completely (see left). In the meantime, the lemuriforms continued their African evolution, soon inhabiting Madagascar where they continue to flourish to this day. Elsewhere, and subjected to competition from the other primates, the lemuriforms have regressed and are today represented only by the few pottos and galagos of the African forests and by the Asiatic loris (see the illustration on page 277). Sheltered as they were on Madagascar, the lemuriforms became widely diversified. This initially nocturnal group produced some secondarily diurnal species (diurnal means active by day), some of which were reminiscent of the absent monkeys. The current fauna of Madagascar can only give us a sketchy picture of its former variety, a factor surely influenced by the relatively recent arrival of Man. Still, remains have been recovered from lake deposits and certain caves, revealing the most extraordinary creatures as well as more familiar lemurs, indris and aye-ayes. In particular we have found an enormous animal, *Megaladapis*, whose cranium might have been 30 centimetres (1 foot) in length and which seems to have been to the lemurs what the gorilla is to the apes.

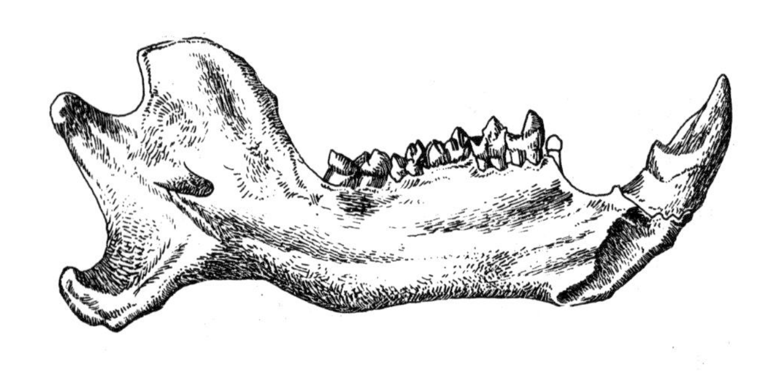

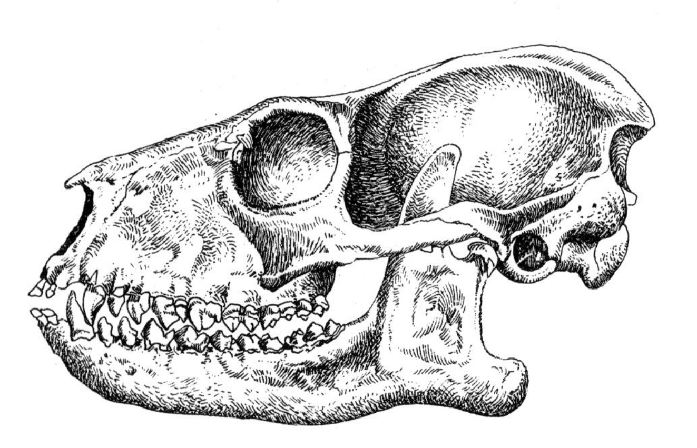

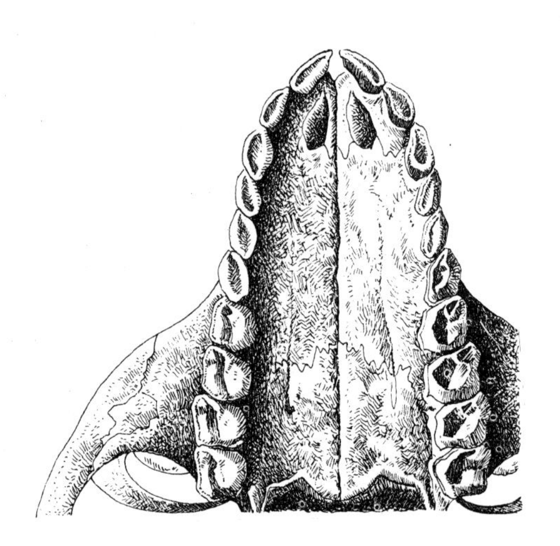

From top to bottom: Jawbone of *Plesiadapis* from the Upper Palaeocene Period of North America (× 1.8). The plesiadapiforms were an early group of primates that evolved towards a type of rodent.

Skull of *Notharctus* from the Eocene Period of North America (× 1 approx.). We have a very detailed knowledge of the skeleton of this adapiform.

The upper teeth of *Adapis* (× 1.3). Cuvier described this European genus on the basis of a portion of a skull from the Montmartre gypsum. The teeth were very primitive, the only teeth missing from those of basic placental mammals were a pair of incisors.

Although the tarsiiforms once spread into the whole of Laurasia (North America and Eurasia), today they include just one genus found in the Sunda Islands and the Philippines. The tarsiers are small nocturnal or twilight animals with extremely long tails. They move among the trees with amazing agility, hunting insects and lizards for food. The tarsiers have the extraordinary ability to turn their heads completely to the rear without moving their bodies. Unlike the strepsirrhines, however, their huge round eyes do not have the tapetum one might expect them to possess.

Tarsiers and monkeys

The haplorhinians, on the other hand, were certainly diurnal from the outset, and it is even possible that the segregation between the two major groups of primates may have been based on this ecological difference. As is more generally the case, however, the split between the tarsiiforms (tarsiers) and the simiiforms within the haplorhinians was caused by geography. The former developed in Laurasia while the latter evolved first of all in Africa. Tarsiers have been found from the Lower Eocene Period in Europe and North America, but their history is patchy during the rest of the Tertiary Period. It is probable that their regression was the result of the Miocene Period invasion of African apes and lorises. Today the sole genus *Tarsius* (see page 279) is restricted to the islands of South-east Asia. It has acquired a nocturnal life-style, but its outsize eyes have retained the characteristics of organs adapted to diurnal vision.

The very first simiiforms are still unknown, although they must have originated in the forests of Africa and go back at least to the Lower Eocene Period, since at this time their sister group, the tarsiers, had already produced individual forms in the northern hemisphere. Nevertheless, the oldest fossil traces date from the Oligocene Period only, appearing simultaneously in Africa at Fayyum, and in South America. As has been suggested by R. Hoffstetter, it is likely that the ancestors of the Bolivian *Branisella* crossed the South Atlantic – far narrower then than now – on natural rafts formed by intertwined tree-trunks and branches which the great equatorial rivers tore from their banks and carried hundreds of kilometres to the coasts, where favourable currents and the tradewinds did the rest. The simultaneous invasion of South America by the African rodents, which must surely have used the same means of transport, is another argument in support of this theory.

From very remote times, therefore, the apes were divided into two major groups which evolved independently, one in the New World, one in the Old World. These two branches bear the names platyrrhines and catarrhines. In America, the apes had laterally positioned nostrils separated by a very thick cartilaginous wall, while in all the other simiiforms the nostrils are much closer together and point either to the front or downward. The platyrrhines prospered in the immense South American forests but were less diversified than their cousins which had remained in Africa. They remained solely tree-dwelling and small in size: the largest among them, the howler monkey, weighs around 8 kilograms ($17\frac{1}{2}$ pounds), far from the massive 200 kilograms (440 pounds) of an adult male gorrilla.

Of all the peculiarities of the American monkeys, the prehensile tail of the spider monkeys, woolly monkeys, howler monkeys and, to a lesser extent, of the capuchins, is the most unusual. This is an amazing organ which enables the animal to suspend itself in mid-air and sometimes even to seize an object – a feat no ape of the Old World is able to perform. Another characteristic shared by all the platyrrhines is the three premolars per half-jaw that separate the canine from the first molar. This is a primitive characteristic inherited from the African 'pioneers', but in that continent it was very quickly lost, and *Oligopithecus* from the Fayyum Oligocene Period already had a dental formula identical to our own – 32 teeth. It is little wonder, therefore, that the South American palaeontologist Ameghino took, at the end of the last century, a fossil from Patagonia as our direct ancestor. He called it *Homonculus* ('little man') in his enthusiasm, but sadly he was incorrect in assuming that this was an ancestor of Man. We must move to Africa in order to take up the thread of events and species which actually led to the appearance of Man at the end of the Cainozoic Era.

The spread of the catarrhines

The Fayyum deposits have revealed a whole series of fossil apes. Of these, some are related to the ancestors of the platyrrhines, while others such as *Oligopithecus* represent the oldest known catarrhines. In this second group the experts have attempted with varying degrees of success to locate the roots of modern species. *Aeolopithecus*, for instance, is reminiscent of the gibbons in certain ways, while other species have been put forward as the origins of the present-day apes and even of Man himself. But these connections are highly disputed, especially in the latter instance, and particularly in the light of the

Like all South American monkeys, this pinche is the distant descendant of the early 'pioneers' from Africa. This species, which lives in the high equatorial forests of Colombia, is slightly larger than the other silky marmosets. Its body is just 20 centimetres (8 inches) long, but its tail is almost twice that length.

sparseness of the remains concerned.

It is not until the Miocene Period that we are able to discern first the hominids, as represented by the gibbon, the siamang, the orang-utan, the gorilla, the chimpanzee and Man, and second the cercopithecoids which included all the other apes of the Old World (macaques, baboons, cercopithecus, proboscis monkeys, colobus and langurs). The latter are smaller creatures than the former, and all have a tail. Their palaeontological history poses many fewer problems than that of the Miocene Period hominids, which are a puzzle that is all the more irritating since it concerns our own direct ancestors, and has been investigated by so many experts.

As a result of the continental shift, Africa at the end of the Lower Miocene Period was close enough to Eurasia to bring about substantial changes to the fauna, as we have seen. The hominids soon assumed a vast geographic spread, a fact that only complicates the tangle we have to unravel. *Dendropithecus* from the Lower Miocene Period of Africa and *Pliopithecus* which did not appear in Europe until the Middle Miocene Period are often suggested as belonging to the line which led to the modern gibbons, but this concept is already the object of much criticism. And when we come to the quartet orang-utan, gorilla, chimpanzee, Man, the row really begins in earnest! Until quite recently, the science of classification put the first three among the family of Pongidae, and the fourth was entitled to a family all of its own: Hominidae. What a relief to be able to answer in this way the question asked repeatedly ever since Darwin: 'Are we really descended from apes?' Or to search with an easy mind among the Miocene Period fossil hominids for the trace of both these lines. *Proconsul* and *Limnopithecus* of East Africa have been joined by the later European and Asiatic *Dryopithecus* to form the main members of the Pongidae whose history would thus go back more than 20 million years. At the same time, another group was being organized around *Ramapithecus*, claimed to be the oldest hominid.

Various biochemical blood-tests give the same 'dendrogramme' among the different groups of simiiforms. This shows that they are all quite closely related. Left to right: marmoset (platyrrhine), macaque (cercopithecoid), gibbon, orang-utan, chimpanzee and a representative of the genus *Homo*.

Man: a great ape?

By analyzing the similarities in certain large molecules from one species to the next, especially the molecules contained in the chromosomes, we are able to trace a tree indicating the degree of 'chemical' kinship between the different groups, and can even date the moment of their crystallization with some degree of certainty. Although much criticized, and confronted by the comparative anatomies of fossil and living forms, this is a device that no one wishing to retrace the history of the living world can afford to ignore. These techniques in fact confirm the interlinking of the various phases in the evolution of the primates which we have seen so far. But as soon as Man and the great apes become the object of our studies, certain difficulties arise, to say the least. What are the biochemical facts? Let us start with something we can all easily accept: the family of the gibbons first separated off from the rest of the hominids, even though the age of that separation (around 13 million years ago) would appear to be rather late. The gorilla and the chimpanzee on the other hand are very close to each other and we tend to unite them in a single genus (*Pan*). Agreement stops there, however, for the result of all these events is as follows: *Pan* is a much closer relative of Man than the orang-utan (see page 282). This conclusion obliges us to separate the Pongidae family whose sole member is now the orang-utan, and to put the gorilla and chimpanzee in with the hominids. And, what is more, the latter are claimed to have a common ancestor, with man only 5 million years old!

What can we conclude from all this? It may be that the biochemical methods are not very accurate and that we have to revise the dates of the various splits. Basically, however, studies into the behaviour of the primates end up by backing the idea of a close proximity between Man and the chimpanzee, and by putting the orang-utan at a greater distance from ourselves. Perhaps these often passionate controversies surrounding matters of classification also reflect a certain intellectual attitude which can affect the mind of those who ponder Man's past, pushing them unconsciously to brush aside inconvenient 'simian' ancestors in favour of more noble ancestry.

Oreopithecus bambolii found in the lignite mine of Monte Bamboli (Tuscany) is one of the more mysterious primates of the Miocene Period. This creature, little more than 1 metre ($3\frac{1}{4}$ feet) in height, lived in the European forests some 8 million years ago, swinging through the trees much like a present-day gibbon. It was first thought to be a cercopithecoid, a type of hominid, but this classification is now debatable.

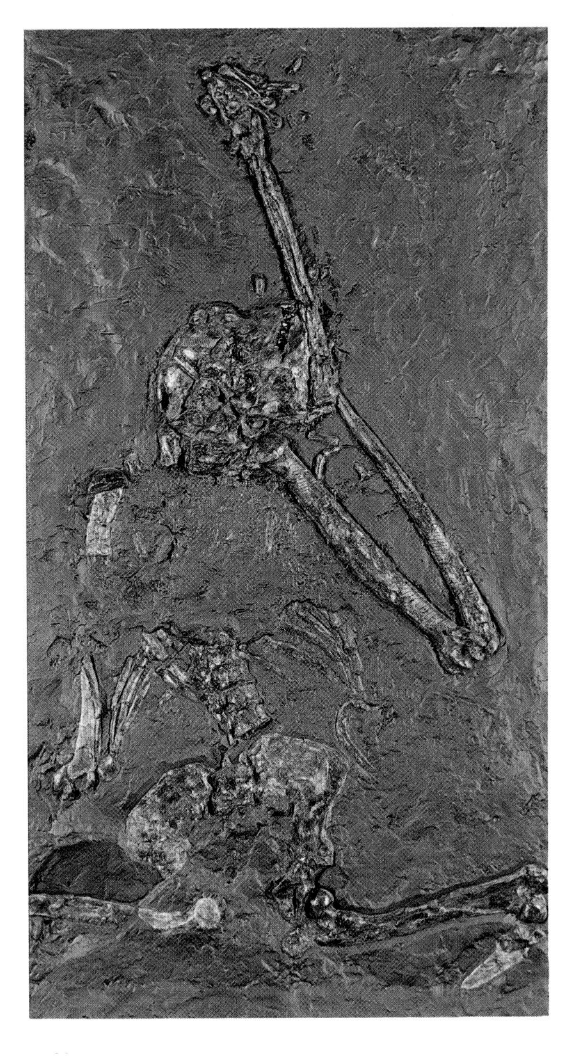

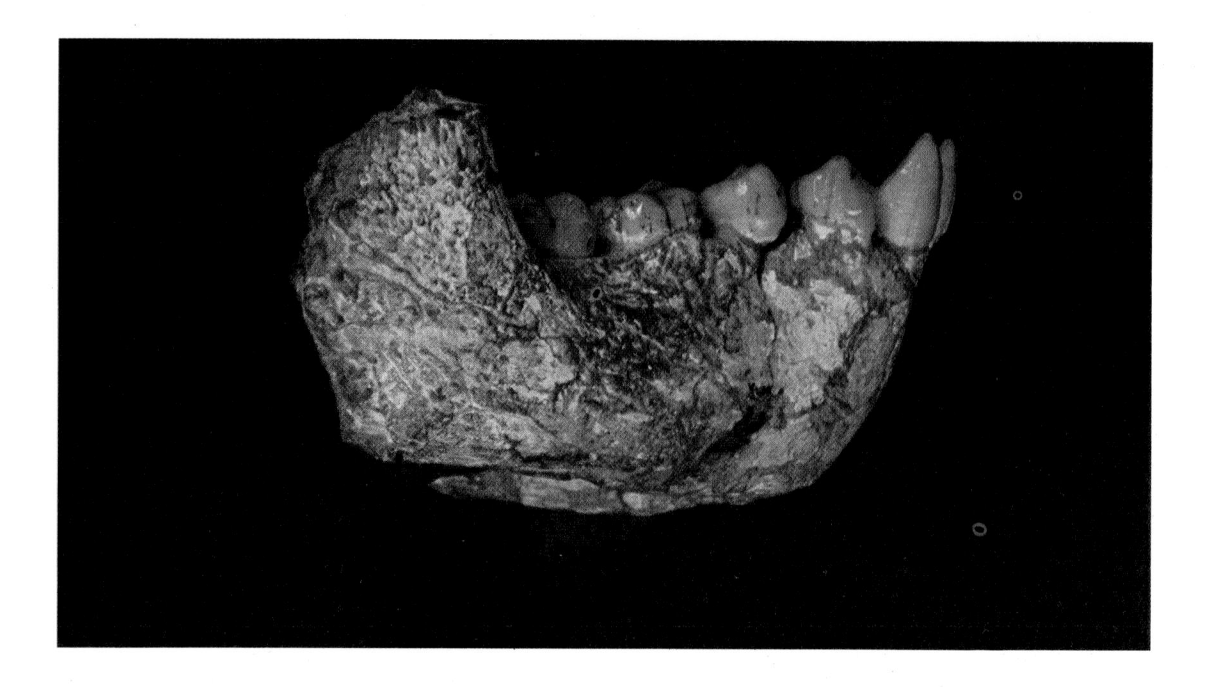

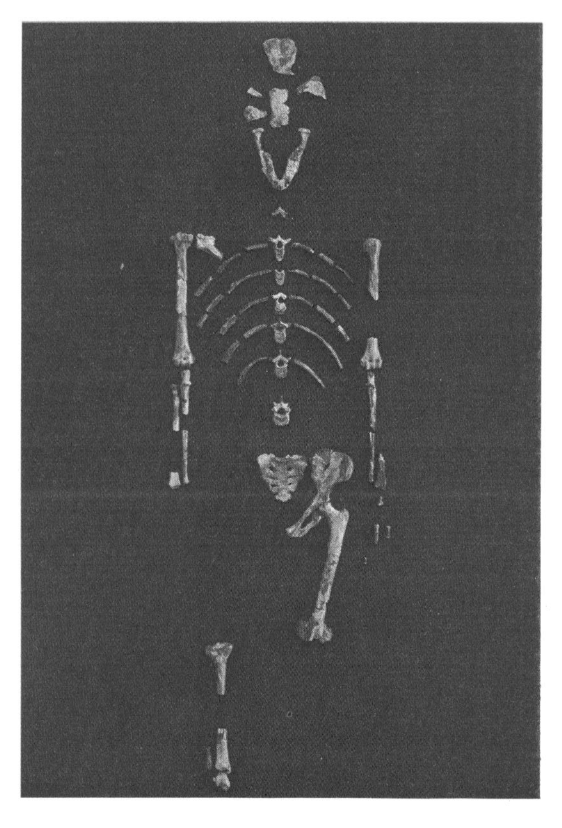

Above: This jaw is one of the fossils of the genus *Ouranopithecus* found in the Macedonian beds known as the 'ravine of rain'. In the Upper Miocene Period this primitive hominid lived in a savannah region along with giraffes, antelopes, *Hipparion*, hyaena and mastodonts. This specimen, about 8 cm (3 inches) long, belonged to a female.

About 3.5 million years old, this specimen is AI-288-1 from the beds of Hadar (Ethiopia), better known simply as 'Lucy'. This exceptional specimen, probably a young female, was found in 1974 by a team from France and America. It is the most complete australopithecine skeleton that has ever been excavated.

From forest to savannah

The Miocene Period underwent a climatic modification that caused the disappearance in Europe of vast forests leaving behind savannahs and prairies in their place, an event which, as we have seen with the Equidae, had a profound effect on the animal life. Among the hominids this drying process worked in favour of a lineage adapted to this new environment, and *Ramapithecus* is known from the Middle of the Miocene Period onwards in China, Pakistan, eastern Africa and Europe. Their teeth were in many ways like those those of Man: the canines were small, the premolars were used for crushing and grinding, while the molars were more square rather than rectangular and they seem to have emerged from the jaws over a prolonged period of growth. The enamel on these molars was thick, making the teeth even stronger, illustrating the adaptation to a diet based on roots, seeds and other harder materials than that offered by wet forests.

Whether or not *Ramapithecus* is our direct ancestor, the fact remains that this adaptive mechanism was certainly the one which guided the first tentative steps down the path of human lineage. Studies of the great apes show that the more open environment of the savannahs is far more likely to cause change than is the more protective world of the forests. Several attempts were made by the hominids parallel to those of the *Ramapithecus*. *Ouranopithecus* of the Upper Miocene Period of Greece (see page 284) is a more robust form and prefigures an extraordinary genus, *Gigantopithecus*, the first of whose identified teeth were purchased in 1935 from Chinese pharmacies in Hong Kong and Canton. Fossils found in the Kuansi were sold for therapeutic purposes; after the discovery of similar specimens researchers were able to locate a cave in the Liu-Cheng district which within its Middle Pleistocene Period deposits concealed much more complete examples, especially several mandibles. Since then, other specimens of *Gigantopithecus* have been found in both China and the Upper Miocene Period of India. As its name suggests, it was a huge creature, the largest known primate with a height often put at 3 metres (10 feet). This makes it larger than the biggest ape living today, the gorilla.

Australopithecines

It would appear that the Pliocene Period episode of the human race was played out in Africa only. Certain finds in China and Java seem to contradict this statement, but they are not yet sufficiently well established, although an Asiatic connection must not be ruled out. It may just be that this African supremacy was only due to exceptional geological conditions and to extraordinary prospecting and digging effort. Southern Africa, where the first *Australopithecus* was exhumed in 1924, and more recently East Africa, have in fact seen a succession of armies of palaeontologists all bent upon the discovery of Plio–Pleistocene hominids.

Among these fossils two groups were quickly recognized which were given the rank of species by some and genera by others. Their most essential common feature was a marked degree of bipedal motion, although their walking mechanism had not yet attained the degree of sophistication of modern Man. Primary importance was attached to this erect posture, since it freed the hands from the requirements of locomotion and allowed them to concentrate on using tools in particular.

Australopithecus (Paranthropus) robustus was about 1.5 metres (5 feet) tall and weighed around 60 kilograms (132 pounds), although certain individuals surpassed this average. Two varieties are known, one South African, and one East African which is often raised to the rank of species: *Australopithecus boisei*. This creature's skull shows a lack of proportion between a small brain-pan

Fossil beds in the valley of the Omo (Ethiopia). Lake deposits and volcanic activity have together laid down huge layers of sediment during the African Pliocene Period and Quaternary Era. Movements of the Earth's crust have since thrown up these layers which have been revealed by erosion. These geological archives have recorded the evolution of Man since his first appearance, as well as the changes that have taken place in his environment.

Left: These two australopithecine jaw-bones are from the Shungura formation in the valley of the Omo in Ethiopia. The more slender specimen is thought to be a form close to *Australopithecus africanus* and is about 2.5 million years old, while the more sturdy example (*Australopithecus boisei*) is more recent, dating from around 1.9 million years ago.

About 2 million years ago two very different types of hominids lived in Africa. While the heavier australopithecine were mainly vegetarian, our distant ancestors hunted and made tools out of stone.

encasing a brain of approximately 500 c.c. volume (one third that of modern Man) on the one hand, and extremely powerful jaws on the other. Enormous chewing muscles enveloped a cranium equipped with solid bony crests right to its summit, while the molar and premolar teeth formed a grinding system capable of crushing the stubbornest food. In East Africa this group, first identified 2.3 million years ago, evolved little, constituting a stable type that was adapted to an almost solely vegetable diet and capable of living in very dry savannah environments. The last of these 'robust' *Australopithecus* died out less than 1 million years ago and left no descendants behind them.

Australopithecus africanus was a smaller, less specialized creature which seems to have opted for a rather wet environment consisting of high-grass savannah and the forests bordering rivers and lakes. Its skull is more rounded but still of small capacity (450–550 c.c.), and the face, now surmounted by the beginnings of a forehead, was not as impressive. The front teeth (incisors and canines) were now clearly more important compared with the molar and premolar teeth. This may be seen as the sign of a more omnivorous diet in which meat was also eaten from time to time. Accentuating a tendency expressed in modern chimpanzees which hunt smaller animals and are occasionally carrion eaters, *Australopithecus africanus* added another human feature to bipedal motion: a pronounced taste for flesh. This *Australopithecus* is more ancient than the robust forms and

may even have given rise to them before developing on its own in parallel. Its origins were clarified by the discovery in 1978 of a new species: *Australopithecus afarensis* which plunges us more than 3 million years into the past. This creature is known from a complete skeleton (see page 284) whose pelvic bones and lower limbs show all the signs of bipedality. The dental characteristics of this small being are very primitive, and it is interpreted as being a landmark between *Ramapithecus* and the more recent *Australopithecus*.

First tool, first man Long bones, jaws still carrying their teeth, tusks, horns, clubs and even stones were certainly part of the equipment used by the first hunters. But these items are not always preserved, and since most of them were never fashioned they are hard to identify as having been tools. The mark of human ingenuity can be discerned, however, the moment when two hands strike two stones together, to break and shape them, an event that occurred in East Africa some $2\frac{1}{2}$ million years ago. Stones chipped to form a crooked cutting edge, and little quartz flakes with sharp corners – this is the sum total of the first attempt at stone-masonry (see page 296).

But who was the craftsman? This distinc-

tion is usually given to *Australopithecus africanus*, but it is time to introduce a new species: *Homo habilis*, the first true representative of the human race. It is a subject on which at least two schools of thought dispute with each other. There are those who regard *Homo habilis* as only a more evolved type of *Australopithecus africanus*. He is larger, his brain has an increased capacity and ability, while the teeth show the omnivorous adaptation of an increasingly efficient hunter. His age – around 2 to 1.8 million years according to the more complete remains – would fit quite well into this theory.

For others, however, *Homo habilis* is far older, perhaps even as much as 4 million years old. He is claimed to have developed alongside *Australopithecus afarensis* and his descendants, or alongside the latter only, and the theory is that he, and he alone, was the founder of the very first 'industries'.

It is a puzzle that defies solution, for the moment at least. The geo-chemical datings

Right: *Homo erectus* did not hesitate to attack large game but probably with rather less bravery than was imagined in this illustration from the last century. Young or weak animals were often attacked, especially elephants or rhinoceroses, and fire and natural traps such as swamps were used to catch them.

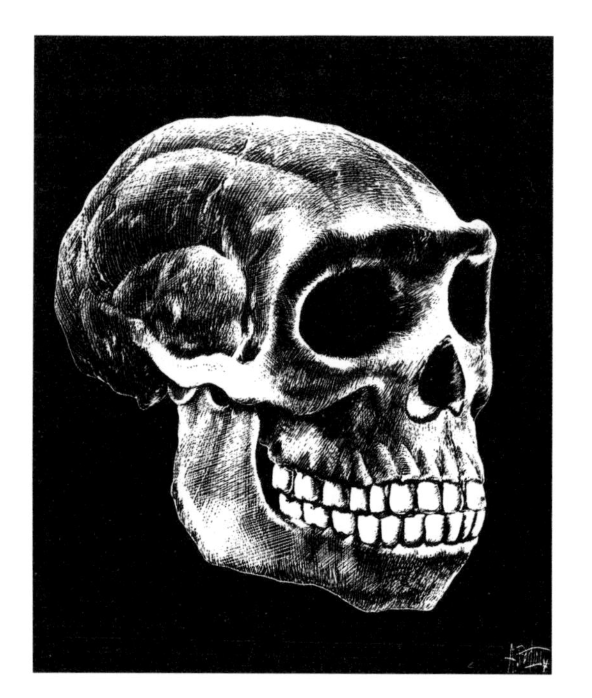

Left: Spread over most of the Old World, the various tribes of *Homo erectus* quickly acquired regional characteristics. The 'pithecanthropan' was an Indonesian variety and 'sinanthropus' (shown here) was a Chinese variety which was first discovered in a traditional Chinese pharmacy. Digs carried out from 1921 in and around Choukoutien, 50 kilometres (30 miles) from Peking, have revealed a whole series of remains, mainly brain-pans and mandibles some 400,000 years old. Unfortunately this marvellous collection disappeared in 1941 when the Japanese invaded China and has never been recovered. Fortunately fresh 'sinanthropans' have since been found both at Choukoutien and at Lan-Tian where a skull dated about 700,000 years old was excavated in 1964.

on which these contradictory arguments are based are the object of frequent and drastic revisions, and it must be admitted that the signs pointing to *Homo habilis* in very ancient formations are pretty slender. In addition, the obvious sensationalism that sometimes surrounds this type of discovery, and a natural urge that often pushes palaeontology towards the concept of an 'Adam' as ideal as he is mysterious and remote, do nothing to help clarify the problem in an objective way.

Homo erectus

Much earlier we gave an account of the fortunes of Dubois and his *Pithecanthropus* of Java (see page 30). This human type is now known as *Homo erectus*. The second term means 'upright', and was coined by Dubois himself who thought he had found the very first biped, a belief that was shattered with the finding of *Australopithecus*. *Homo erectus*, the result of the evolution of *Homo habilis*, was the first man of whom undeniable traces were found outside Africa. In Java, a very young skull is dated as being some 1.9 million years old, but this is most uncertain. In the French Massif Central a few tools resembling the old African stone implements may be as much as 1.8 million years old, but we have no idea of what their maker looked like. Nevertheless, the traces of *Homo erectus* multiply the closer we come to modern times – he is found in East, North and South Africa, in Europe and China (see page 288). Only America, Australia, the isolated islands and the coldest regions of Eurasia were not yet peopled by Man.

So far as we can tell, the body and limbs of *Homo erectus* were very much like our own (see above). His bones were extremely robust but their size and shape were much like those

African bifaces and axes. The biface, was invented by *Homo erectus* but was never widely used. Certain populations, especially in Asia, did not use such tools but in Europe many such splinter tools are being found which date from the Lower Palaeolithic Period. These were neglected by the early explorers because they were not spectacular.

of modern skeletons. His skull and jawbone on the other hand were rather special – once the erect posture had been established and perfected, it was the turn of the head to undergo continued remodelling.

Pursuing the path trodden by his ancestors, *Homo erectus* slowly acquired an increasingly large skull, while his face and chewing apparatus tended to diminish in size. His own descendants continued this process, the intellectual functions developing more and more while the teeth and jaws receded as many of their former tasks were done by tools instead.

For the moment we are still dealing with a creature which has a massive cranium with thick walls and featuring a heavy brow above deep-set orbits. The forehead was receding, the cranial dome low and narrow. The volume of the brain varied between 750 and 1250 c.c., almost attaining that of modern man. But volume is not everything: the structure also became increasingly complex. The face was powerful and the lower jaw had a chin. The proportions of the dental arch were more or less identical to those of modern dentition but all the dimensions were larger, although the molars were clearly smaller than those of *Australopithecus.*

The mastery of stone All down the huge expanses of time, cultural and technical progress was at first practically nil, and two fashioned pebbles separated by hundreds of thousands of years are hardly different from each other. Nevertheless, from the middle of the Pleistocene Period things developed with increasing rapidity, and it is to *Homo erectus,* and more specifically the last of them, that important technological innovations were due.

Tools became very varied, and the fashioned pebbles were joined by more elaborate objects. The expanding art of chipping and dressing the stone gave rise to a new type of tool, the biface (see pages 290 and 296), a sort of almond made of flint, quartz, obsidian or any other hard rock. In the course of time variations on this basic form led to many types each of which responded to specific functions. These artifacts were apparently never joined to a handle or haft but were hand held, sometimes perhaps by a leather strap. Modern experiments have shown that, provided with this kind of accessory, it is quite easy to cut up the carcase of even a large animal in a relatively short space of time, and many other uses can also be imagined. The flakes and splinters chipped from a block of raw material can also go to form lighter tools. Once dressed, they provide scrapers, more or less rough spikes, 'combs' and other notched devices used to work leather, bone and wood. We must remember that the abundance of stone tools by which the distribution of *Homo erectus* is chiefly known to us is above all a reflection of the semi-permanence of this material; in addition to the flints, however, which may not perhaps have been all that numerous, the implements of the hunter and his family must surely have included utensils fabricated from perishable materials, especially wood. Some outstanding finds like broken spears, miraculously preserved by a peaty environment at Clacton-on-Sea, Britain, or by carbonization such as at Torralba, Spain are reminders of this.

The constant improvements achieved in the working of stone gradually led to the adoption of new techniques. The stone hammer which, used with sufficient skill, can create rock objects as beautiful as they were functional, was replaced for 'finishing' by a long bone, a deer antler or even a piece of hardwood. The resulting splinters were thinner and longer, while the biface became more graceful and its cutting edge straighter.

Hearth and home

The traces of what seem to have been actual dwellings have been found at Olduvai in Tanzania and at Melka Kunturi in Ethiopia, in layers respectively dated as being 1.8 and 1.6 million years old. These are circular assemblies of stones, some of which form a low wall while others probably formed the floor of branch huts around 3.5 metres ($11\frac{1}{2}$ feet) in diameter. *Homo habilis* may therefore have been the architect of very basic buildings.

For a long time *Homo erectus* preferred to dwell in the open air, on the shores of rivers, lakes and seas close to sources of the raw materials from which he fashioned his imple-

Like other pre-Würmian fossils of western Europe, the man of 'Caune d'Arago' from the eastern Pyrenees of France preceded typical neanderthalians. His primitive characteristics were amplified by the 'swelling' of the maxillary bones that typified neanderthal man.

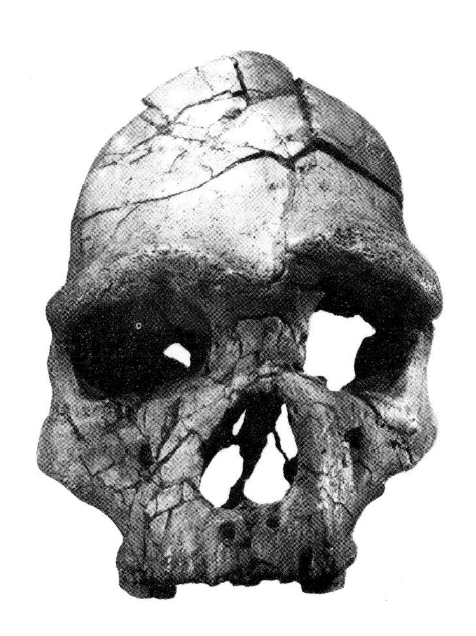

ments. But when the climate was particularly hostile he, too, found refuge among the rocky escarpments in caves or even at the foot of cliffs which offered a natural shelter. He attempted to add comfort to these dwellings by means of increasingly elaborate fixtures. Near Nice on the Mediterranean coast of France, two examples of well-organized dwellings have been found. Around 380,000 years ago the beach at Terra Amata must have provided a site for branch huts which bear witness to the brief passage of Acheulian hunters. These structures were rather vast and were supported on posts, and had an oval layout that attained as much as 15 metres (50 feet) in length. The remains of a tent with an area of some 35 square metres (376 square feet), probably made of skins and erected against the back wall of the Lazaret cave, are very much later.

It was around 400,000 years ago, or perhaps a little earlier, that a fresh page in the adventures of the human race was turned – the conquest of fire. The oldest evidence of this innovation has been found in France (Lunel–Viel, Terra Amata), Italy (Torre in Pietra), Spain (Torralba and Ambrona), Hungary (Vertelsszöllös) and China (Choukoutien), while in Africa similar traces only go back 100,000 years, perhaps owing to a friendlier climate or soil conditions unfavourable to their preservation. Once mastered, Man had fire to give him light, warmth, cook his food and temper the points of his spears. Did he actually light the first flame, or merely gather it and keep it burning following some natural conflagration? In any event, it is likely that our stone-mason did not take long to realize that a glowing splinter produced by the impact of two flints, or better still of a flint and a piece of pyrites, could set fire to twigs, brushwood or dried moss. Once under control and kept going by means of a cunningly devised air draught, the hearth became an integral part of Man's dwelling which he now organized around it.

A man called wise

Our own species is called *Homo sapiens* ('wise man'), and succeeded *Homo erectus* in such a continuous manner that it becomes increasingly difficult as new finds come to light to identify in fact a boundary between the two, let alone date it. The suggestion has even been made that we abandon purely and simply the distinction between them, and include *Homo erectus* in with *Homo sapiens*. (The latter being a *Homo erectus* that has progressed the evolutionary tendencies of the type, for instance greater cerebral development and reduction of chewing apparatus. He has also generally lightened his skeleton.) These aspects can be observed everywhere where *Homo erectus* lived, but with different regional rhythms and patterns. Europe is a very interesting and significant case in point. Thanks to a long palaeontological tradition, we possess a series of fossil documents that are relatively well dated and more plentiful in this region than anywhere else in the World. An additional attraction of this group is that it led to a very singular type of *Homo sapiens*: Neanderthal Man.

The 'classic' *Homo sapiens neanderthalensis* lived in Europe between 80,000 and 35,000 years ago at least. When we examine his European predecessors, however, we find that they possess features typical of neanderthalians. These are little reflected on the most primitive fossils, but were nonetheless present from the Mindel–Riss interglacial onwards. We must therefore regard these forms as ancient specimens of *Homo sapiens neanderthalensis*.

Neanderthal

Neanderthal man is without doubt the best-known of all primitive *Homo sapiens*. We have complete skeletons of him, and are thus able to form a fairly accurate image of his appearance (see page 294). Short in stature, he was solidly built and his muscles were powerful. His skull (see page 293) was large, wide and elongated to the rear as well; among other primitive attributes it retained its rather thick walls and a rounded brow forming a sort of 'visor' above the eyes. The chin was not yet distinct on the jaw. The cranial capacity on the other hand was compatible with that of modern Man; other specific features included the lack of cheekbones and the very marked projection of the central portion of the face towards the front. A broad, prominent nose contributed to give the neanderthalians their familiar appearance.

The extensive knowledge we have of these people is based on a major cultural factor. The human remains we have so far discussed were fossilized only as a result of some fortuitous chance, but the result was usually the dislocation of the skeleton, and Lucy (see page 284) is a case unique to this day. Neanderthal man was to be the first to bury his dead, a fact that explains the frequency of almost complete specimens. This new custom reflected metaphysical and perhaps even religious preoccupations which bring these men very close to ourselves. The neanderthalian burials, most of which were regrettably uncovered at a time when the techniques of archaeological excavation were not as sophisticated as they are today, seem to have been accompanied by complex rituals and varieties of sacrifice. The skull discovered at Monte Circeo in Italy is perhaps yet another manifestation of belief in a

On the 3rd August 1908 near the village of Chapelle-aux-Saints in the French Region of Correze, in Mousterian layers situated inside a small cave, an almost complete neanderthalian skeleton was discovered. These remains helped palaeontologists to make the first precise anatomical study of this human type. Abbots Bardon and Bouyssonie collected the remains from a trench 1 metre by 1.45 metres ($3\frac{1}{4}$ feet by $4\frac{1}{2}$ feet), dug to a depth of 30 centimetres (1 foot) in the substrate: obviously this was a deliberate human burial. The man from Chapelle-aux-Saints was 'old': over forty, he had lost many of his very worn teeth as a result of inflammation of the gums. A detailed examination of the remains also showed a deformed left hip, arthritis of the neck vertebrae, one crushed toe and a broken rib. This last injury, sustained shortly before he died, was perhaps the cause of a fatal wound to the chest.

The neanderthalians adapted well to the cooling of the climate, using the skins of reindeer which had been driven into southern Europe by the weather as very basic clothing. This at least is what can be deduced from several flint punches and bone awls found in the Mousterian layers.

supernatural order: it had been placed like some cult artifact or trophy in the middle of a stone circle, its base broken no doubt to extract the brain in the same way as, until recently, the Melanesians did.

Nor are we without signs of the high cultural level which neanderthal man achieved. His dwellings have supplied us with quantities of dyes, especially ochre, sometimes in the form of small pencils. He must have used a lot of these but no trace of the use to which they were put has come down to us: they may well have been used to decorate perishable artifacts, wooden objects, leather clothing and perhaps even the skin of the men themselves.

The origins of modern man

The special characteristics of the neanderthalians which consolidated in the course of their evolution and put them further and further away from the morphology of modern man (*Homo sapiens sapiens*) have led a good many palaeontologists to discount them altogether as the latter's ancestors. For once at least, truth and fiction would seem to coincide; still, the line leading to modern humanity remains hypothetical. Ever tempted to believe that their origins could have been nowhere else but in Europe, men have used the most fragmentary and poorly dated specimens to support the idea of European 'pre-sapiens'. At least one of these finds, however, was relatively complete – fragments of a very ancient skull and jawbone extracted from the Piltdown gravelpits in Britain. After 40 years of speculation, the awakening was indeed rude. In 1953 an injection of fluorine into the remains revealed all: Piltdown man had been a fake, made up of a recent human skull and the jawbone of an orang-utan into which teeth had been glued! Today the European 'pre-sapiens' have returned to obscurity, and it is to other horizons that we must now turn.

Middle East confrontations

In the caves of Amud and Tabun in Israel, Shanidar in Iraq and further east at Teshik-Tash in Uzbekistan, archaeologists have found remains of neanderthalians which, while different in detail from those that populated Europe, still possess their essential characteristics. For a period of time that is difficult to estimate, they inhabited these regions alongside quite different populations, a situation that may not have lasted long, and it is likely that one or other of these groups was not of local origin.

The fact remains that men of a new type have been discovered at Skhul and Djebel Qafzeh (Israel) who produced the same type of artifacts as the neanderthalians, using ochre and burying their dead. At Qafzeh in particular (a formation extraordinarily rich in the number of specimens it has yielded up and recently explored using modern methods) it has been possible to study these men in detail. Their burials (see page 295) indeed contain offerings: a child's hands hold the antlers of a deer to accompany it into the beyond. Their anatomy was almost identical with that of modern man. Their bones were more robust and their teeth still very strong;

The Mousterian strata of the Qafzeh deposits near Nazareth in Israel have so far revealed seven adult and eight infant skeletons. One of the most remarkable finds was that of a burial containing the remains of a young woman and a child. The hands of another child (above) held on to the antlers of a deer in some ritual we cannot understand.

some of the characteristics were primitive, but their skull was rounded, their forehead high, their cheekbones prominent and their chin quite distinct. These eastern contemporaries of neanderthal man evidently spread westwards and eventually superseded him in Western Europe.

Confirming certain suspicions, the discovery which came in 1979 of Saint Cezaire Man (Charente, France) demonstrates that the first artifacts of the Upper Palaeolithic Period, during the Würm II/III interglacial, were still the work of neanderthalians who nevertheless disappeared, possibly overtaken by the new arrivals. This is not impossible given the doubtless modest populations of the tribes, and it is even possible to estimate that the inhabitants of a country such as France during the Palaeolithic Period numbered no more than a few thousand individuals.

The hunters of the Upper Palaeolithic Period

The second half of the Würm saw the expansion of a man who is our equal. This unique hunter and gleaner spread across our planet and colonized lands still unknown to mankind at that time. In the north he inhabited the coldest regions of Siberia, while towards the east he crossed the Bering Strait and set foot in the American continent. At Lewisville in Texas, his traces have been dated as 40,000 years old, and increasingly

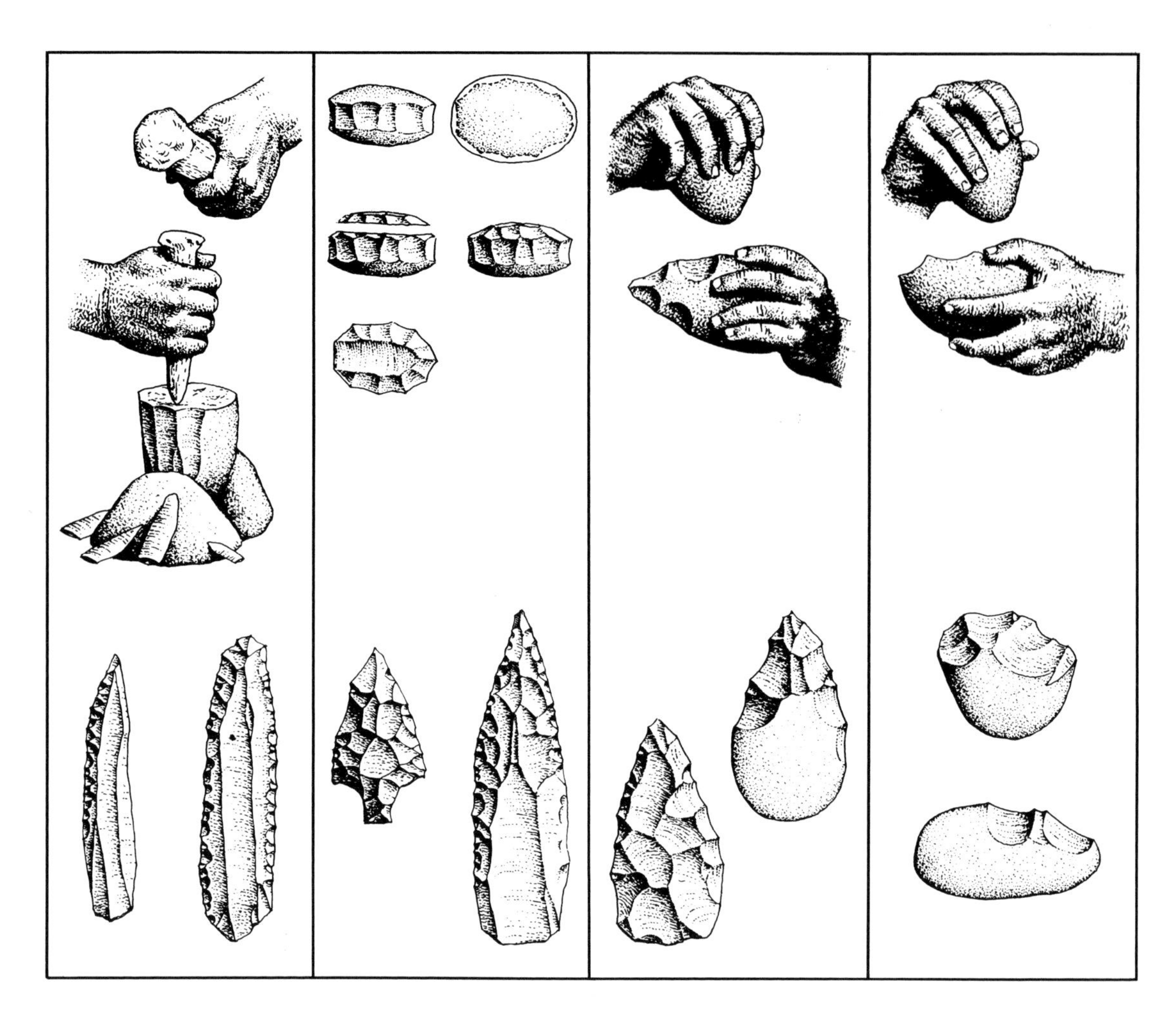

Left to right: The chipped pebble is the simplest possible stone tool, and a rough cutting edge can be achieved just by shearing off a few flakes.

More extensive chipping on both sides gives a biface which, when chipped all over, produces a more deadly implement, such as an arrowhead.

The preparation of what is known as a levallois core is more complex. The block is rough-hewn until it is the shape of a tortoise-shell. Further chipping operations on the upper face make it possible to pre-select the shape of a large splinter which will either be used as such or dressed by the craftsman to his taste.

To produce blades, the best method is to carry out preliminary preparation leading to the appearance of an initial vertical edge on the core, then finishing by means of a bone 'strop'. This technique makes it possible to select the exact location of each point of impact, thus obtaining highly regular blades.

recent sites mark his descent towards Tierra del Fuego which he reached 10,000 years ago. On the other side of the Pacific, in Australia, his presence is attested by remains from Lake Mungo that are at least 30,000 years old.

On the technological side, stone-working took on fresh dimensions. While the Middle Palaeolithic Period saw the development of flake and splinter artifacts, the Upper Palaeolithic Period is typified by the more generalized use of long blades (see above). This increasingly economical utilization of the raw materials made it possible to obtain a much greater effective cutting length for the same weight, hence a much lighter tool. A further step in this direction was taken when these were attached to bone or wooden supports to make the first composite utensils.

The 'bone industry' flourished in spectacular fashion: spearheads, needles with eyes, barbed harpoon tips and a host of other instruments, as well as pendants, pierced teeth and sometimes real works of art. And here is the great innovation: the neanderthalians were already picking up odd-looking rocks and spiral fossils and have even left a few scrawls on bones and stones. But the first truly non-utilitarian objects burst upon the cultural scene with the coming of cro-magnon man.

Geometric symbols and animal outlines were cut into familiar objects, while rows of notches were fashioned along a bone: perhaps a primitive abacus? Or calendar? Or even a 'rosary' on which the fingers recounted the episodes of some mythical tale? There is no lack of interpretations. As for the engraved slabs and sculptures in the round, they were certainly not intended just to please the aesthetic sense, and like the reindeer, horses and bison painted on the walls of caves, they bear witness to highly elaborate mythological and religious concepts. Man himself is rarely depicted, a fact which indicates that his art was not gratuitous and that he was following a now-forgotten code of belief. Apart from a few caricatures, human portrayals are mainly limited to sculpted or modelled genitalia and the statuettes of females. These are usually devoid of facial features and their arms and legs are barely discernible, while the generous proportions of their breasts, bellies and buttocks are very exaggerated. These representations are without doubt references to the mysteries of reproduction and fertility, but are surely linked with cave art and formed part of a cultural whole whose degree of complexity we can barely guess at. It was not to decorate his home that Man reproduced the image of the beasts that surrounded him, sometimes portraying them in the very depths of almost inaccesible caverns. Intermingled and accompanied by mysterious symbols and hands with muti-

In the Upper Palaeolithic Period of Europe over a hundred burials have been found. In these cases the burial rites are much more complex than for neanderthal man. In many cases these corpses lie upon a bed of red ochre and the deceased was sometimes covered in ornaments, perhaps the ones worn during life. This cro-magnon man from the Grimaldi cave in France took a head-dress of tiny gastropods to the grave.

lated or crooked fingers, these images were not arranged haphazardly but in accordance with a pre-established protocol that made these caves the sanctuaries of a religion whose deeper meaning we may never fathom.

The neolithic revolution

Towards the end of the last glaciation the Middle East was the scene of the last great change to precede the beginning of recorded history – this was the emergence of agriculture. This development was relatively gradual and the various species of plants and animals were domesticated in successive stages.

From the Upper Palaeolithic Period onward, a wolfcub occasionally brought back to camp by the hunters may have been

Above: A painting of a black ibex from a cave in Niaux (France). This cave, which includes fine examples of Palaeolithic art, is one of the best-known and most visited. Most cave paintings are found in the south-west of France and northern Spain. The men of the time used torches and tallow lamps to light their caves and several of these have been found. At Niaux the clay has preserved the imprints of the bare feet of these cave dwellers. Beneath a low vault we can still observe the footprints of children that splashed in a puddle of water some 14,000 years ago.

Right: The 'Venus of Lespugue' from the Upper Garonne in France is a statuette 14.7 centimetres (nearly 6 inches) high made from the ivory of a mammoth. These prehistoric 'Venuses' tell us little about how Palaeolithic women must have looked in real life. Some of the most recent are particularly stylized; they have become a swollen object merely bearing the engraved triangle of a pubis.

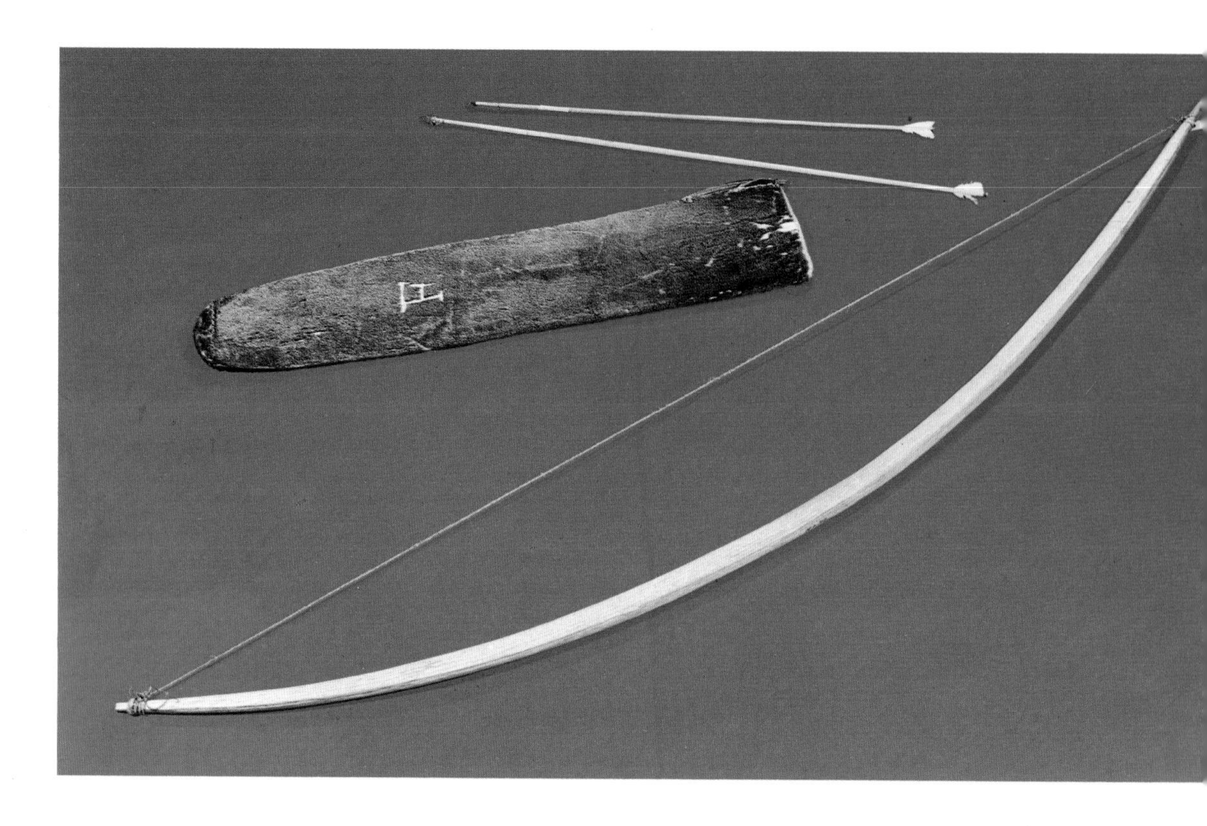

tamed, leading to the domestic dog which made its appearance a long time ago. The dog, however, is a special case: as a hunting companion and guardian of the dwelling, it never represented a source of food. In the same way a young deer may have been kept alive for some time – a kind of walking larder – hobbled or tethered to a large pierced stone like the one found in the Aurignacian of La Quina (Charente). While this had little in common with modern stockbreeding, it was the kind of experiment that much later on formed the basis for the first attempts at domestication.

Ten thousand years ago the Middle East presented a landscape in which wheat and barley grew in the wild state. The hunters, who were also gleaners, may well have exploited this food source in their wanderings, making primitive millstones and sickles for the purpose. Attending to this vegetation offered them by nature, they progressed slowly from mere picking to growing, while they ceased their wanderings and established permanent villages.

The sheep, the first to be domesticated, were joined by goats, pigs, cattle and other animals. A study of the skeletons of these animals shows from the distribution of their

At the end of the Upper Palaeolithic the perfection of the bow was an extremely important development, and slings such as the bola gradually disappeared in favour of this formidable new weapon. F. Bordes has even linked the changes in the population that mark this age with the appearance of this invention. Neolithic sites provide us with many arrowheads, which were a weapon of war as well as of the chase.

ages that they were indeed the product of deliberate breeding and not of hunting. Other revolutions followed on from the first: with a sedentary life came pottery that was too fragile and cumbersome for the nomad; with the necessity of cultivating came the polished axe which was added to the existing array of implements. This neolithic culture spread outwards from centres of ingenuity such as the Middle East (there were others, in particular in America and perhaps in the Far East as well). It was followed by another wave of creativity in which metal took its turn as the raw material for tools and artifacts. Commerce, wars, fortified towns and an increasingly hierarchical social structure are but the consequence of the abandonment of an economy of gathering for one of production. Finally, the art of writing marked the start of recorded history some 5000 years ago.

Man has ceased to live in nature in order to live from nature. Throughout his history he has gradually replaced a purely biological adaptation by technological adaptation, an achievement which, to a certain extent at least, liberates him from the laws that govern the rest of the living World.

The pestle and mortar symbolize the gradual adoption of a productive life-style organized around the cultivation of cereals – the neolithic revolution. However, despite the breeding of domesticated animals for food, hunting remained an important source of food for a long time.

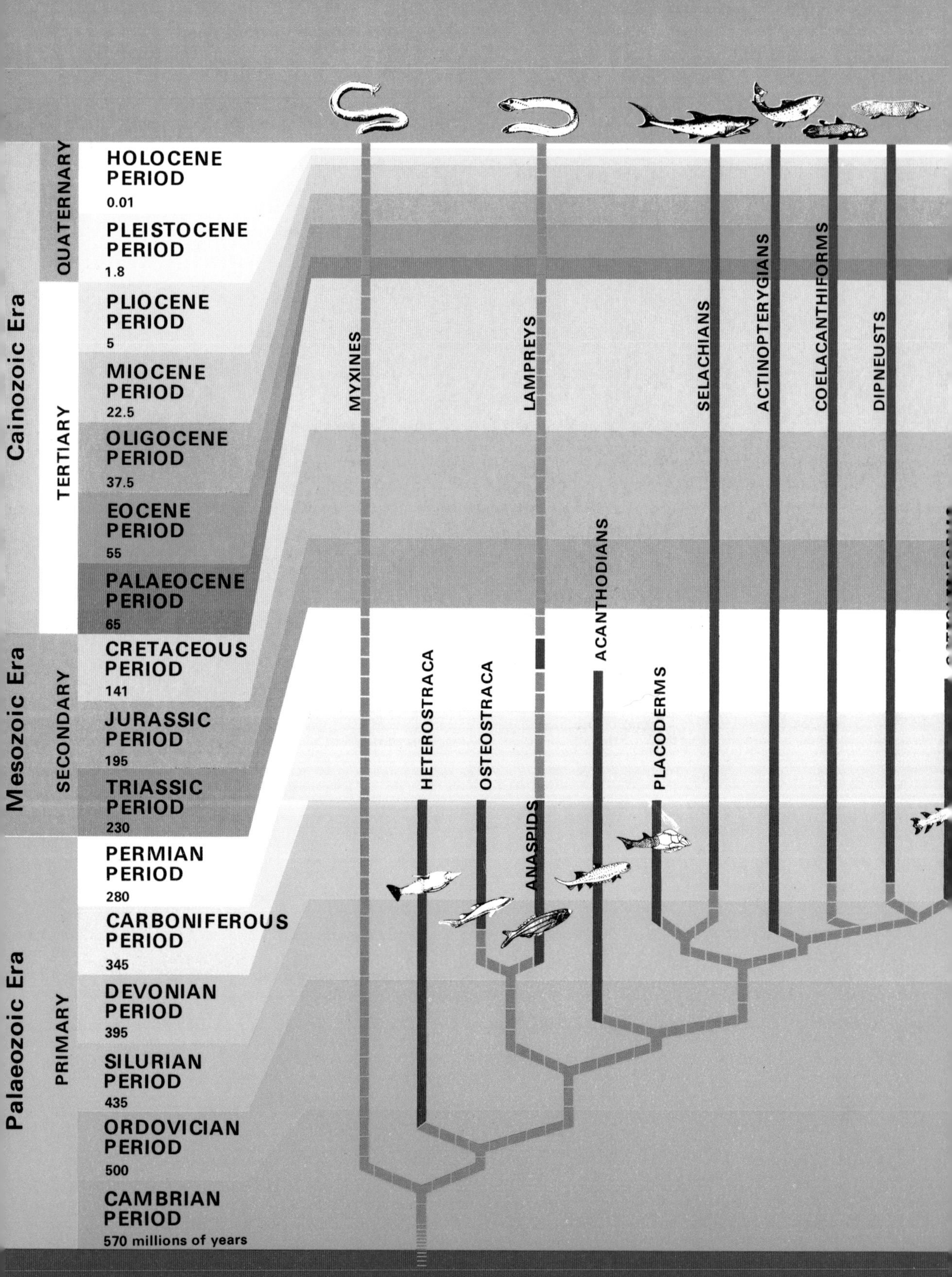

Cainozoic Era
QUATERNARY
HOLOCENE PERIOD
0.01
PLEISTOCENE PERIOD
1.8
TERTIARY
PLIOCENE PERIOD
5
MIOCENE PERIOD
22.5
OLIGOCENE PERIOD
37.5
EOCENE PERIOD
55
PALAEOCENE PERIOD
65
Mesozoic Era
SECONDARY
CRETACEOUS PERIOD
141
JURASSIC PERIOD
195
TRIASSIC PERIOD
230
Palaeozoic Era
PRIMARY
PERMIAN PERIOD
280
CARBONIFEROUS PERIOD
345
DEVONIAN PERIOD
395
SILURIAN PERIOD
435
ORDOVICIAN PERIOD
500
CAMBRIAN PERIOD
570 millions of years
PRECAMBRIAN
MYXINES
LAMPREYS
SELACHIANS
ACTINOPTERYGIANS
COELACANTHIFORMS
DIPNEUSTS
ACANTHODIANS
HETEROSTRACA
OSTEOSTRACA
ANASPIDS
PLACODERMS

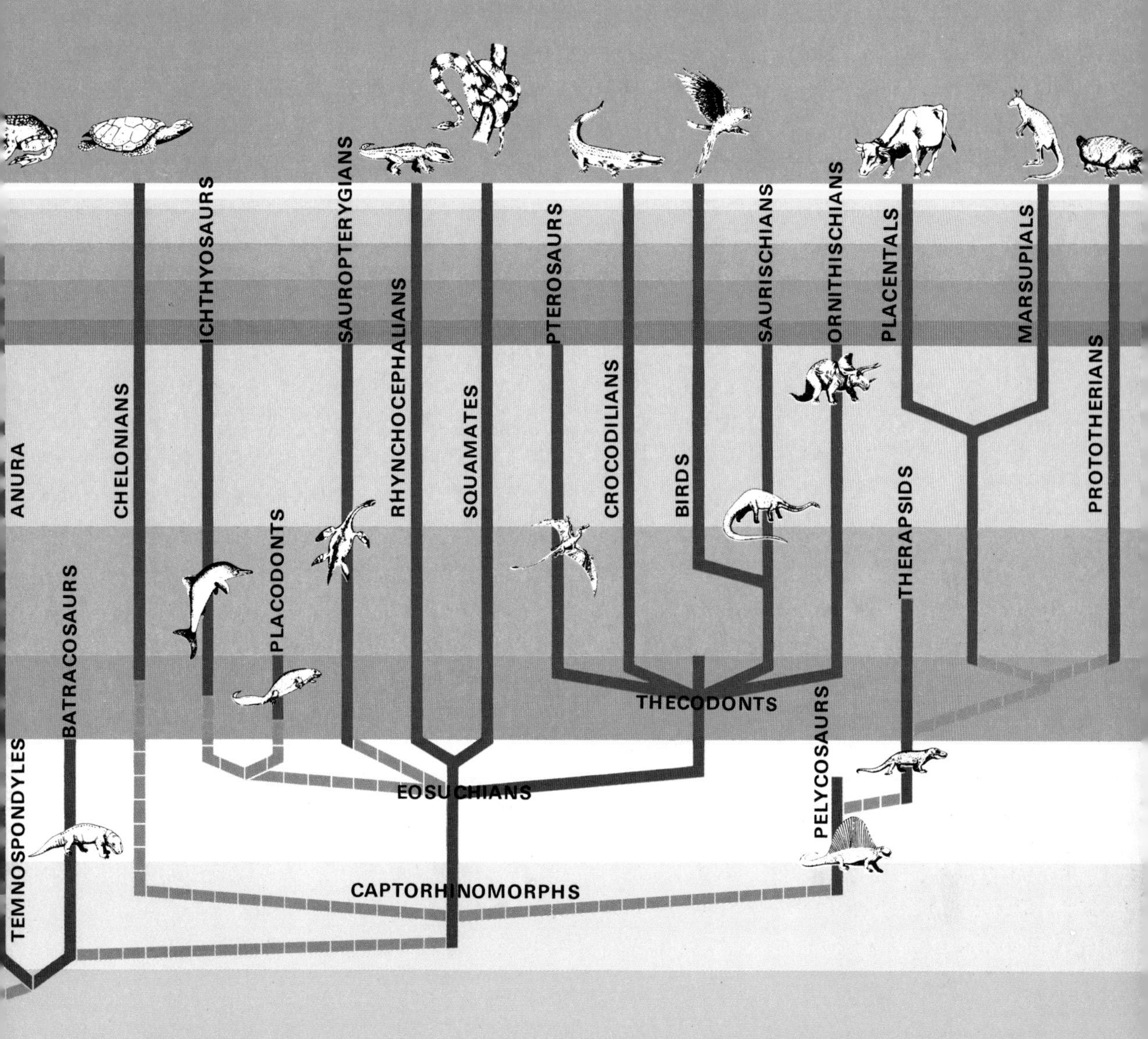

The evolution of the vertebrates

On the left of this chart are shown the various geological periods which sub-divide the eras, as well as their approximate dates. Not all the periods are listed, for they are too numerous. For example, the Lower Jurassic Period, or Liassic, is just included in the Jurassic Period. The figures underneath each period indicate the number of years, in millions, from the present day.

Our phylogenetic tree shows the links which we believe exist between the groups; the broken lines indicate that there is a lack of palaeontological evidence. Where three, four or more lines converge upon a 'stock group' (for instance the thecodonts) this means that here there are phases in the history of the vertebrates where the relationships between various lineages are not yet fully understood.

Glossary

Acanthodian – type of ancient fish that was shark-like in appearance. A typical acanthodian was *Climatius*. It lived during the early Devonian Period and each of its many fins, except the tail, was stiffened by a strong spine.

Actinopterygian – type of fish characterized by having angular fins in which the bones fan outwards from the girdle. The actinopterygians were also called the ray-finned fishes.

Allantoic sac – surrounds the developing embryo in the amniotic egg. It contains the waste products that would otherwise poison the embryo.

Ammonoid – term usually given to the type of suture found in the ammonites. The suture marks the position where the chamber wall meets the outer shell. In ammonites the suture is much folded into complex lobes and saddles.

Amniota – animals that have an amniotic egg belong to this group. They include the reptiles and birds.

Amniotic egg – egg that is enclosed by a number of protective membranes and a shell. The developing young (embryo) is enclosed in a liquid, and has its own food source called the yolk.

Anapsid – type of reptile that has a solid skull roof. The group that includes the turtles.

Archosaur – group of reptiles that includes the crocodiles, dinosaurs, thecodontians and pterosaurs.

Artiodactyl – one of two types of hoofed mammal. They are also called the even-toed ungulates, and include the pigs, camels and deer.

Astragalus – name given to the ankle bone that articulates with the tibia in mammals.

Avian – bird-like; belonging to birds in general.

Batrachosaur – extinct form of amphibian known from the Eocene and Miocene Periods, approximately 50 to 15 million years ago.

Belemnite – extinct animal related to the squids and cuttlefishes. Known from the presence of their bullet-like guards in Mesozoic Era rocks.

Bilateral symmetry – is when an imaginary line can divide an animal or object into two equal halves.

Bipedal (locomotion) – term used to describe animals that walk on their hind legs, for instance man or various dinosaurs.

Branchial – gill regions in vertebrates; usually stiffened by a series of cartilaginous or bony bars. These support muscles that are used in the opening and closing of the gill slits.

Captorhinomorph – primitive type of reptile that first appeared in the Carboniferous Period.

Carapace – outer shell of a crab or turtle.

Catarrhine – term used to describe the living monkeys of Asia and Africa; the apes and man.

Cephalothorax – name given to the combined head and body region found in certain animals.

Chondrichthyan – term used in the description of sharks and rays; fishes that have a cartilaginous skeleton.

Class – grouping of animals or plants based on certain common characteristics. All mammals are included in the class Mammalia, for instance.

Condyle – name given to the rounded articulation region at the end of long bones or at the base of the skull.

Convergent evolution – occurs when two groups of animals or plants adopt the same overall shape or appearance.

Cranium – is simply another name for the skull which houses the brain.

Creodont – primitive type of meat-eating mammal that dominated the Eocene Period.

Cynodont – a type of mammal-like reptile that ate meat, and had numerous small 'nipping' teeth and 'dog-like' canines.

Dentary – bone found in the lower jaw; in mammals the dentary forms the whole jaw.

Deuterostome – animals such as the echinoderms possess an anus whose position in the juvenile is marked by an opening called the blastopore. This condition is termed *deuterostomatous*, and the animal is called a deuterostome.

Diapsid – type of skull found in crocodiles and dinosaurs; it has two openings behind the eye or orbital opening.

Didelphid – term used in the description of the opossum (*Didelphis*) and its close relatives.

Digit – in four-legged animals the hands and feet possess a number of fingers or toes – these are called digits.

Diurnal – term used to describe animals that are active by day.

Dorsal – refers to the top or upper region of an animal.

Dysodont – term used to describe the tooth arrangement found in certain bivalve molluscs.

Entelodont – extinct form of even-toed mammal that resembled a wild boar.

Eoscuchian – was a primitive type of reptile related to the snakes and lizards. Eosuchians are known from the Permian Period of Africa and Europe.

Epicercal (tail) – is one of several terms used to describe the shape of the tail in fishes. In the epicercal tail the upper lobe is larger than the lower lobe.

Eucaryote – an organism in which the cells have a nucleus surrounded by cytoplasm.

Euryapsid – type of skull found in plesiosaurs and placodonts. Only one opening is present behind the eye, and this is placed high on the skull roof.

Exoskeleton – protective outer layer, often mineralized, that occurs in crabs, bivalve molluscs and many other animals without backbones.

Fauna – term used to describe the animals that live in an area at a particular time.

Foraminiferan – microscopic animal that has an external, mineralized skeleton. Important in the study of rocks.

Gamete – reproductive or sex cell. Two gametes fuse during reproduction to form an egg.

Genus – taxonomic group that is made up of a number of related species. (Plural is genera.)

Guard – hard, bullet-shaped structure that formed part of the internal skeleton of a belemnite.

Hermaphrodite – organism possessing both male and female organs, and producing both egg and sperm.

Heterocercal (tail) – type of tail found in sharks, in which the backbone is tipped up posteriorly.

Heterodont – means 'different-tooth', and is used to describe those animals that have a variety of different types of teeth in their jaws or, in the case of bivalve molluscs, on their shells.

Iguanodontid – term applied to the various features typical of *Iguanodon* and related dinosaurs.

Jugal – one of several bones that surround the eye socket in vertebrate skulls. It forms part of the lower margin of the orbital opening.

Lateral – refers to the side of an object.

Lepidosaur – reptiles such as the snakes and lizards that have a double-arched type of skull. They are also referred to as the 'scaly' reptiles.

Labyrinthodont – amphibians that were common during the late Palaeozoic and early Mesozoic Eras.

Mandible – another name for the lower jaw.

Marsupial – mammal that gives birth to very small young that often mature inside the mother's pouch.

Matrix – term applied to the sediment that surrounds a fossil.

Maxillary – bones forming a portion of the front

end of the skull in all vertebrates.

Mesonychid – mammals that possessed triangular upper cheek teeth with three blunt cusps. They were flesh eaters, and lived during the Palaeocene–Oligocene Period.

Mesosuchian – crocodiles that were common during the Jurassic and Cretaceous Periods. They possessed a number of primitive features not found in living forms.

Miacid – mammals that lived during the Palaeocene Period. They were among the first flesh-eating mammals, and are considered ancestral to most living carnivores.

Mosasaur – one of a group of large, flesh-eating sea lizards from the Upper Cretaceous Period, possessing a large head and a long, serpent-like body.

Multiturberculata – mammals that existed for 100 million years from the late Jurassic Period to the Eocene Period. They were probably plant eaters.

Nothosaur – primitive aquatic reptile that lived during the Triassic Period.

Notochord – thin rod of tissue that provides support in animals such as *Amphioxus*. It runs along the top or back of the organism, and is surrounded by a tough sheath.

Olfactory – term used in the description of the nasal region in animals with backbones.

Opposable – toes in the primates that allow these animals to take a firm grip on branches. The length of one toe is reduced so that the foot forms a pincer to grasp or hold.

Order – division in the classification of animals or plants; several families may be present within an order.

Ornithischian – dinosaurs that were characterized by the 'bird-hipped' type of pelvic girdle. The group includes *Iguanodon* and *Parasaurolophus*.

Ornithopod – dinosaurs that had both a 'bird-hipped' pelvic girdle and a bird-like foot. The *iguanodon* was an ornithopod.

Osteolepiform – describes the condition or shape characteristic of primitive lobe-finned fishes such as *Osteolepis*.

Oviparous – term applied to egg-laying animals such as the reptiles and birds.

Palaeonisciform – fishes that had a bony skeleton and were ray-finned. They were of modest size with thick, shiny scales and a heterocercal tail.

Palaeontologist – geologist who studies the fossil remains of animals and plants.

Palate – name given to the roof of the mouth. A secondary palate may occur in many reptiles and in the mammals.

Pangaea – name given to the late Palaeozoic Era 'supercontinent' formed by the collision of all the major land masses.

Pantothere – type of mammal found in the Upper Jurassic Period. The group is thought to be ancestral to living marsupials and placentals.

Pelagic – term applied to open sea environments and to creatures found swimming in them. The rocks deposited in these areas often contain planktonic organisms.

Pelycosaur – reptiles such as *Dimetrodon* and *Edaphosaurus* which thrived during the Permian Period. They were among the earliest of synapsid reptiles.

Perissodactyl – hoofed animal with an odd number of toes.

Phylum – major division within the plant or animal kingdoms. Insects, trilobites, crabs and lobsters all belong to the phylum Arthropoda, for instance.

Phytosaur – type of thecodontian reptile that resembled the living crocodiles. Phytosaurs lived during the Triassic Period.

Placental – mammals that give birth to live young and which were, as embryos, attached to the mother by the placenta.

Planktonic – animals and plants that float near the surface of the sea. The majority are microscopic in size.

Platyrrhine – group name given to the New World monkeys.

Plesiadapiform – term applied to the rodent-like early primates including *Plesiadapis*.

Porolepiform – term applied to a type of lobe-finned or crossopterygian fish found in the Lower Devonian Period.

Proboscidean – term used to describe a member of the order Proboscidea; the elephants and extinct mastodonts.

Procolophonid – name given to a type of extinct reptile thought by some palaeontologists to be linked to the turtles.

Protosuchian – well-armoured variety of early crocodile found in the Late Triassic Period of North America.

Prototherian – primitive mammal. The subclass Prototheria include the duck-billed platypus and spiny anteater.

Pterobranch – worm that possesses a small notochord, and are distant cousins of the graptolites and the vertebrates.

Quadrate – name given to one of the bones that forms the posterior region of the skull. The quadrate borders the optic region.

Quadrupedal – term applied to the type of locomotion in all four-footed animals.

Quaternary Period – covers the last 1.5 million years of geological time.

Radial symmetry – Condition found in echinoids and certain other invertebrates, in which the body can be divided radially into equal halves.

Rudista – name given to a coral-like group of bivalves that lived during the Jurassic and Cretaceous Periods.

Sarcopterygian – group of bony fishes characterized by a fleshy or lobed fin. They include the lung-fishes and coelacanths.

Saurian – general term used to describe lizards and animals with lizard-like features.

Sauropod – dinosaurs that were characterized by a 'lizard-hipped' pelvis and a 'beast-like' foot. Sauropod means 'beast-footed'.

Sauropterygian – reptiles that flourished during the Mesozoic Era. They include the plesiosaurs, pliosaurs and nothosaurs.

Selection pressure – relates to the evolution of a species, and refers to the environmental or competitive pressures that affect its survival.

Seymouriamorph – amphibians with a resemblance to the reptiles and which were considered by some palaeontologists as their possible ancestors. The seymouriamorphs showed several adaptations to a life on land.

Simian – term applied to the man-like apes including the gibbon, orang utan, chimpanzee and gorilla.

Species – groups of animals and plants that are distinct from others. Members of a species can breed with each other.

Spicule – name given to small, rod-like or branched skeletal elements found in sponges and some corals. They can be made of calcite or silica.

Squamate – reptiles that include the lizard and snakes. They are the most successful order of living reptiles.

Stock group – term used to describe a group of animals that may be considered ancestral to several more advanced families.

Stratigraphic column – sequence of geological periods characterized by different layers of fossil-bearing rocks.

Stromatolith – colony of blue-green algal material encrusted with sediment. Stromatolites first appeared 3,500 million years ago.

Synapsid – reptiles that are characterized by the presence of a single temporal opening in the skull. They include the mammal-like reptiles or therapsids.

Taxodont – term applied to the arrangement of teeth found in the shells of certain bivalves.

Taxonomy – the science of placing or classifying organisms into different groups or taxa.

Temnospondyl – type of amphibian that flourished during the Permian–Triassic Period.

Temporal – refers to the region of the skull above and behind the eye.

Tertiary Period – includes the Palaeocene, Eocene, Oligocene, Miocene and Pliocene. It began 65 million years ago and ended approximately 1.5 million years ago.

Test (of echinoderm) – name given to the skeleton of a sea urchin. It consists of many plates of calcite.

Tetrapod – describes a creature that has four feet; amphibians, reptiles and mammals can be described as tetrapods.

Therapsid – major group of synapsid reptile with mammal-like characteristics. They thrived during the late Permian–early Triassic Period.

Triconodont – mammals that were characterized by cheek teeth that bore 3 cusps, the largest of which occurred in the centre.

Tubercule – nob-like swelling found on the shells of several invertebrate animals.

Urodele – amphibians such as the newts and salamanders. They first appeared in the Cretaceous Period.

Valve – term applied to one of two halves of a bivalve or brachiopod shell.

Ventral – term applied to the lower surface of an organism.

Viviparous – describes the process of live births which occur in mammals and occurred in the extinct ichthyosaurs.

Index

Numbers in *italics* refer to pages on which illustrations appear.

A

B

C

D

E

F

G

H

I

J

K

L

M

N

O

P

Q

R

S

T

Photographic Acknowledgements
Archivio Fabbri, from Aquarium des Tierparks-Hellebrunn, Monaco p.133; Documents Donald Baird p.231; Bourreau p.50; Bucciarelli p.153; Yves Coppens pp.31, 33, 284 (bottom), 285, 286; P. Cross p.54, 55; Musée de l'Homme pp.274-5, 291 (J. Oster), 293, 297; Pio Mariani Desio pp.75, 82-3, 94-5, 96; De Bonis p.284 (top); E.P.S. p.265; Geological Museum, London p.247; Gorsen pp.196, 257; Dott. Leonard, Dott. Pinna p.152; Jean-Jacques Hublin pp.70, 71, 135, 245, 248; R. Maltini, P. Solaini pp.102-103; Margiocco, Genoa p.277; Museo Civico, Desio pp.89, 117; Museo Civico Archeologico Ligure p.301; Museo Geologico dell' Università di Pavia p.98; Museo di Storia Naturale, Milan pp.20, 79, 80, 84, 85, 86, 106, 124, 126-7, 145, 158, 161, 172-73, 227, 263, 283; Museum National d'Histoire Naturelle, Paris pp.251, 267, 269, 272, 273; Museo di Storia Naturale, Verona pp.126-7; Natural History Museum, London pp.16, 36, 40, 41, 128-9, 154-5, 164-5; NASA pp.60-1; Marie Pierre Aubry p.64; Willis Peterson p.260; Paul Popper p.229; Roebild-Müller p.279; Roebild-Emu p.281; A.P. Rossi, Milan p.249; Royal Institut des Sciences Naturelles de Belgique, Brussels pp.23, 175; Jacques Six pp.206-207; Yves Tanguy-Rolland, Explorer pp.16-17; Guy Teste p.92; Bernard Vandermeersch p.295.

Illustrations by Fedini, Diagram Visual Information Ltd, Sergio.

Date Due

ILL
FEB 16 2006

PRINTED IN U.S.A. CAT. NO. 24 161 BRODART